Die Grundlehren der mathematischen Wissenschaften

in Einzeldarstellungen
mit besonderer Berücksichtigung
der Anwendungsgebiete

Band 208

H. Elton Lacey

The Isometric Theory of Classical Banach Spaces

Springer-Verlag
New York Heidelberg Berlin 1974

H. Elton Lacey
University of Texas at Austin, Department of Mathematics
Austin, TX 78712/USA

AMS Subject Classification (1970)
Primary 46E05, 46E15, 46E25, 46E30
Secondary 46B05, 46J10, 47B05, 47B55

ISBN 0-387-06562-8 Springer-Verlag New York Heidelberg Berlin
ISBN 3-540-06562-8 Springer-Verlag Berlin Heidelberg New York

 Library of Congress Catalog Card Number 74-394.
Printed in Germany. Typesetting and printing: Zechnersche Buchdruckerei, Speyer.
Bookbinding: Konrad Triltsch, Würzburg.

To Bonnie

Preface

The purpose of this book is to present the main structure theorems in the isometric theory of classical Banach spaces. Elements of general topology, measure theory, and Banach spaces are assumed to be familiar to the reader.

A *classical Banach space* is a Banach space X whose dual space is linearly isometric to $L_p(\mu, \mathbb{R})$ (or $L_p(\mu, \mathbb{C})$ in the complex case) for some measure μ and some $1 \leqslant p \leqslant \infty$. If $1 < p < \infty$, then it is well known that $X = L_q(\mu, \mathbb{R})$ where $1/p + 1/q = 1$ and if $p = \infty$, then $X = L_1(\nu, \mathbb{R})$ for some measure ν. Thus, the only case where a space is obtained which is not truly classical is when $p = 1$. This class of spaces is known as L_1-*predual spaces* since their duals are L_1 type. It includes some well known subclasses such as spaces of the type $C(T, \mathbb{R})$ for T a compact Hausdorff space and abstract M spaces.

The structure theorems concern necessary and sufficient conditions that a general Banach space is linearly isometric to a classical Banach space. They are framed in terms of conditions on the norm of the space X, conditions on the dual space X^*, and on (finite dimensional) subspaces of X. Since most of these spaces are Banach lattices and Banach algebras, characterizations among theses classes are also given.

Both the real and complex cases are treated in general (however, some of the geometric results are for the real case only).

There is, of course, a corresponding isomorphic theory of classical Banach spaces which is under intensive investigation. This theory studies properties of these spaces invariant under linear isomorphisms and, in particular, isomorphic embeddings and complementation problems. The techniques and results of this theory are, in general, vastly different than those of the isometric theory. For a survey of the major theorems and themes in the isomorphic theory see the recent lecture notes of Lindenstrauss and Tzafriri [297].

Since most of the structure theorems for Hilbert spaces are well known and readily accessible in various texts, they are not included herein.

The excercises are, for the most part, directly related to the section or chapter in which they appear. They represent alternate approaches or extensions of the development, or in some cases they consist of filling in the details of the proof of a theorem in the text. As a general rule, theorems whose proofs are left to the reader in the text are not cited as exercises. Various open problems are also stated, but no attempt at a collection of open problems is made.

Several authors graciously provided me with pre-prints of their works. Among them are: J. Baker, J. Bednar, E. G. Effros, H. Fakhoury, J. Hagler, A. J. Lazar, D. R. Lewis, J. Lindenstrauss, A. Pełczyński, H. P. Rosenthal, C. Stegall, and M. Zippin.

I am grateful to Galen Seever for his interest, lively discussions, and various proofs. M. Zippin graciously explained to me some aspects of his work on L_p spaces. My special thanks goes to S. J. Bernau who was instrumental in the formulation of the theory of contractive projections on L_p as presented in chapters 5 and 6 and who helped me understand abstract L_p spaces.

During the academic year 1972—73 I was priviledged to receive support from the National Academy of Sciences and the Polish Academy of Sciences while in residence at the Institute of Mathematics in Warsaw, during which the final version of this text was written.

I am deeply indebted to the University of Texas at Austin for its unfailing support during the years of preparation of this work and I am especially grateful to Judy Bowman for coordinating the typing and to Linda White, Joyce Bell, and Nita Goldrick for their patient conversion of my scratches to typescript.

Above all my appreciation goes to Bonnie Lacey who has given me encouragement and moral support throughout our years together, especially during the preparation of this work.

Finally, the author is solely responsible for any errors, mathematical or otherwise, in the text.

Warsaw, Poland
March, 1973

H. Elton Lacey

Table of Contents

Chapter 1

Partially Ordered Banach Spaces

The theory of classical Banach spaces rests solidly on a foundation of partially ordered Banach spaces. In particular, Banach lattices play a key role in its development. In this chapter we present the results from partially ordered Banach spaces which are relevant to this development. (See [43], [197] and [256] for a more general treatment.)

In section 1 we recall the basic properties of vector lattices. In particular, we study the general properties of bands, band projections, order complete vector lattices, atoms, and finite dimensional Archimedean vector lattices.

Section 2 is devoted to the study of partially ordered normed linear spaces (for technical reasons we do not always assume the spaces are complete). The results used in the study of classical Banach spaces have to do with properties on the norm and order and the corresponding dual properties on the conjugate space.

Finally, in section 3 we apply the results of sections one and two to normed linear lattices to obtain some general structure theorems which will be used throughout the text.

§ 1. Vector Lattices

A convex set W in a real vector space X is said to be a *wedge* if $W+W\subset W$ and $aW\subset W(a\geqslant 0)$. A *positive cone* is a wedge C such that $C\cap(-C)=\{0\}$. Positive cones induce a natural partial ordering on X given by $x\leqslant y$ if and only if $(y-x)\in C$. This partial ordering clearly has these two additional properties: 1. $x,y\geqslant 0$ implies $x+y\geqslant 0$; 2. $x\geqslant 0, a\geqslant 0$ implies $ax\geqslant 0$. Of course, any partial ordering $\leqslant$ on X which satisfies 1 and 2 above comes from the positive cone $C=\{x:x\geqslant 0\}$ in the described manner. The pair (X,C) (or $(X,\leqslant)$) is called a *partially ordered vector space*. In practice, if the cone C (or the ordering $\leqslant$) is understood, we just say X is a partially ordered vector space.

Suppose (X,C) is a partially ordered vector space. A nonempty set $D\subset X$ is said to be *bounded from above (from below)* if there is an $x\in X$

with $x \geqslant y$ whenever $y \in D$ ($x \leqslant y$ whenever $y \in D$). By the *least upper bound of D (the greatest lower bound of D)* we mean an upper bound x (lower bound x) such that $x \leqslant z$ for all other upper bounds z ($x \geqslant z$ for all other lower bounds z). Of course, such may not exist for a given D, however, if it does we denote it by $\sup D (\inf D)$. Clearly $\sup D$ exists if and only if $\inf(-D)$ exists and in such a case $\sup D = -\inf(-D)$. If $D = \{x_1, \ldots, x_n\}$ and $\sup D$ exists, we also use $x_1 \vee \cdots \vee x_n$ for $\sup D$ ($x_1 \wedge \cdots \wedge x_n$ for $\inf D$ if it exists).

We shall, at times, use the duality between sup and inf without explicit mention or we may say that a formula follows by duality. For example, it is easy to see that if x, y are in X and for some $z_0 \in X$ $(x+z_0) \vee (y+z_0)$ exists, then for all $z \in X$, $(x+z) \vee (y+z)$ exists and is equal to $x \vee y + z$. The dual formula is, of course, $(x+z) \wedge (y+z) = x \wedge y + z$.

Definition 1. A *vector lattice* is a partially ordered vector space (X, C) such that $x \vee y$ exists for all $x, y \in X$.

Clearly $x \wedge y$ exists for all x, y in a vector lattice and, by induction, $x_1 \vee \cdots \vee x_n$ and $x_1 \wedge \cdots \wedge x_n$ exists for all $x_1, \ldots, x_n$. For $x \in X$, we put $x^+ = x \vee 0$ and $x^- = (-x) \vee 0$. Then x^+ is called the *positive part* of x and x^- is called the *negative part* of x. The *absolute value* of x is defined to be $|x| = x^+ + x^-$. We shall shortly see that $x = x^+ - x^-$ and $|x| = x \vee (-x)$. If $|x| \wedge |y| = 0$, we say x and y are *disjoint*.

Theorem 1. *Let (X, C) be a vector lattice and x, y, z, w be elements of X and $a \in \mathbb{R}$. Then the following formulas are valid.*

(1) $x - y \wedge z + w = (x - y + w) \vee (x - z + w)$; $x - y \vee z + w = (x - y + w) \wedge (x - z + w)$.
(2) $x + y = x \vee y + x \wedge y$.
(3) $x = x^+ - x^-$ *and this is the unique representation of x as the difference of positive disjoint elements.*
(4) $(ax) \vee (ay) = a(x \vee y)$ *and* $(ax) \wedge (ay) = a(x \wedge y)$ $(a \geqslant 0)$.
(5) $|ax| = |a|\,|x|$.
(6) $|x + y| \leqslant |x| + |y|$.
(7) $\big||x| - |y|\big| \leqslant |x - y|$.
(8) $|x| \wedge |y| = 0$ *if and only if* $|x| \vee |y| = |x| + |y|$.
(9) $x \leqslant y$ *if and only if* $x^+ \leqslant y^+$ *and* $x^- \geqslant y^-$.
(10) $x \vee y = (x - y)^+ + y$ *and* $x \wedge y = y - (x - y)^-$.
(11) $|x - y| = |x \vee z - y \vee z| + |x \wedge z - y \wedge z|$.
(12) *If* $x, y, z \geqslant 0$, *then* $(x + y) \vee z \leqslant x \vee z + y \vee z$ *and* $(x + y) \wedge z \leqslant x \wedge z + y \wedge z$.
(13) $(x - z) \wedge (y - z) = 0$ *if and only if* $z = x \wedge y$.

Let $\{x_i\}_{i \in I}$ *and* $\{y_j\}_{j \in J}$ *be two indexed families of elements in X such that* $\sup x_i$ *and* $\sup y_j$ *exists (dually,* $\inf x_i$ *and* $\inf y_j$ *exists).*

(14) $\sup x_i + \sup y_j = \sup(x_i + y_j)$ $(\inf x_i + \inf y_j = \inf(x_i + y_j))$.
(15) $x + \sup x_i = \sup(x + x_i)$ $(x + \inf x_i = \inf(x + x_i))$.
(16) If $a \geq 0$, $\sup a x_i = a \sup x_i$ $(a \inf x_i = \inf a x_i)$.
(17) $x \wedge \sup x_i = \sup x \wedge x_i$.

Proof. (1) From the formula $(u_1 + v) \vee (u_2 + v) = u_1 \vee u_2 + v$ we get $(x - y + w) \vee (x - z + w) = (-y) \vee (-z) + x + w = x - y \wedge z + w$.

(2) By (1), $(x - x + y) \wedge (x - y + y) = -x \vee y + x + y$. Thus $x \vee y + x \wedge y = x + y$.

(3) By (2), $x = x + 0 = x \vee 0 + x \wedge 0 = x^+ - x^-$. To see that $x^+ \wedge x^- = 0$ we note that from (1), $x - x^+ \wedge x^- + x^- = (x - x^+ + x^-) \vee (x - x^- + x^-) = x^+$. Hence $x^+ \wedge x^- = 0$. Suppose $x = y - z$ with $y, z \geq 0$ and $y \wedge z = 0$. Then clearly $y \geq x^+$ and $z \geq x^-$. Moreover, $(y - x^+) \wedge (z - x^-) \leq y \wedge z = 0$. Since $y - x^+ = z - x^-$, it follows that $y = x^+$ and $z = x^-$.

The verifications of (4)—(9) and (12)—(17) are routine.

(10) Now $x \vee y - y = (x - y) \vee (y - y) = (x - y) \vee 0 = (x - y)^+$.

(11) First note that $|x - y| = (x - y) \vee 0 - (x - y) \wedge 0$ $= (x \vee y - y) - (x \wedge y - y) = x \vee y - x \wedge y$. Thus $|x \vee z - y \vee z| + |x \wedge z - y \wedge z|$ $= (x \vee z) \vee (y \vee z) - (x \vee z) \wedge (y \vee z) + (x \wedge z) \vee (y \vee z) - (x \wedge z) \wedge (y \wedge z)$ $= (x \vee y) \vee z + (x \vee y) \wedge z - (x \wedge y) \vee z - (x \wedge y) \wedge z = (x \vee y) + z - (x \wedge y + z)$ $= x \vee y - x \wedge y = |x - y|$. ∎

The following theorem characterizes when a partially ordered vector space is a vector lattice. Its proof is routine.

Theorem 2. *Let (X, C) be a partially ordered vector space. Then the following are equivalent.*

(a) *(X, C) is a vector lattice.*

(b) *For each $x, y \in X$, $(x + C) \cap (y + C) = z + C$ for some $z \in X$.*

(c) *For each $x \in C$, $(x + C) \cap C = y + C$ for some $y \in X$.*

For example, (c) implies (a) goes as follows. For $x \in X$, $(x + C) \cap C = y + C$ for some $y \in X$. Note that $y \in C$ and $y \geq x$. If $z \geq x, 0$, then clearly $z \geq y$ since $z = x + (z - x) \in (x + C) \cap C = y + C$. Thus $y = x \vee 0$. From this one can easily see that $x \vee y = (x - y) \vee 0 + y$. ∎

We shall often need the following technical result known as the *decomposition lemma for vector lattices.*

Lemma 1. *Let (X, C) be a vector lattice and suppose x_i and y_j are positive elements of X for $i = 1, \ldots, n$ and $j = 1, \ldots, m$. If $\sum x_i = \sum y_j$, then there are positive elements z_{ij} in X such that $x_i = \sum_j z_{ij}$ and $y_j = \sum_i z_{ij}$ for all i and j.*

Proof. We shall do the case $n = m = 2$. The rest follows by induction on n and m. For this, let $z_{11} = x_1 \wedge y_1$, $z_{12} = x_1 - z_{11}$, $z_{21} = y_1 - z_{11}$ and

$z_{22}=x_2-z_{21}$. Clearly z_{11}, z_{12} and z_{21} are positive and $x_1=z_{11}+z_{12}$, $x_2=z_{21}+z_{22}$, $y_1=z_{11}+z_{21}$, and $y_2=z_{12}+z_{22}$. Hence it only remains to show that $z_{22}\geqslant 0$. Now $z_{22}=y_2-z_{12}$ and $z_{21}\leqslant z_{21}+y_2=z_{12}+x_2$ and $z_{21}=z_{21}\wedge(z_{12}+x_2)\leqslant(z_{21}\wedge z_{12})+(z_{21}\wedge x_2)=z_{12}\wedge x_2$, since $z_{12}\wedge z_{21}=0$. Thus $z_{21}\leqslant x_2$ and it follows that $z_{22}\geqslant 0$. ∎

A partially ordered vector space (X,C) is said to have the *decomposition property* if whenever x,y_1,y_2 are in C and $x\leqslant y_1+y_2$, we have that $x=x_1+x_2$ where $x_1,x_2\in C$ and $x_1\leqslant y_1$, $x_2\leqslant y_2$. It is easy to see that if (X,C) is a vector lattice, then it has the decomposition property. For, let $x_1=x\wedge y_1$ and $x_2=x-x_1$. Clearly $x_1,x_2\in C$ and $x_1\leqslant y_1$. Now $x\leqslant y_1+y_2$ implies that $x\leqslant x\wedge(y_1+y_2)\leqslant x\wedge y_1+x\wedge y_2\leqslant x\wedge y_1+y_2$. Thus $x_2=x-x\wedge y_1\leqslant y_2$. (It is, in general, not true that if (X,C) has the decomposition property then it is a vector lattice (see exercise 9).)

A *sublattice* of a vector lattice (X,C) is a linear subspace Y such that if $x,y\in Y$, then so is $x\vee y$ (and $x\wedge y$). An *ideal* in X is a linear subspace Y of X such that if $y\in Y$, $x\in X$ and $|x|\leqslant|y|$, then $x\in Y$. Clearly an ideal is a sublattice. A *band* in X is an ideal Y such that if D is a nonempty set in Y and $\sup D$ exists in X, then $\sup D$ is in Y. Since arbitrary intersections of sublattices (resp. ideals, bands) is again a sublattice (resp. ideal, band), we have that for any nonempty set A in X there is a smallest sublattice (resp. ideal, band) containing A (we call it the *sublattice* (resp. *ideal, band*) *generated by* A). For any set $A\subset X$, $A^+=C\cap A$. Clearly if Y is a sublattice of X, then $Y^+=\{y^+:y\in Y\}$. Note further that an ideal Y in X is a band if and only if for each nonempty set $D\subset Y^+$ such that $\sup D$ exists, $\sup D$ is in Y. The necessity is clear. Suppose that Y has the stated property and D is a nonempty subset of Y such that $x=\sup D$ exists. Then $x^+=x\vee 0=\sup\{y^+:y\in D\}$ and by hypothesis, $x^+\in Y$. Similarly we get that $x^-\in Y$.

The ideal generated by a set A has an easy description. It is, in fact, $\{y\in X:|y|\leqslant|a_1x|+\cdots+|a_nx_n|$ for some $a_1,\ldots,a_n$ in $\mathbb{R}$ and $x_1,\ldots,x_n$ in $A\}$. In particular, if $A=\{x\}$, then it is $Y=\{y\in X:|y|\leqslant|ax|$ for some $a\in\mathbb{R}\}$. In such a case we call Y the *principle ideal* generated by x.

The following theorem gives a description of the band $B(Y)$ generated by an ideal Y.

Theorem 3. *Let (X,C) be a vector lattice and Y be an ideal in X. Then for any $x\in B(Y)\cap C$, $x=\sup\{y\in Y:0\leqslant y\leqslant x\}$.*

Proof. Let $B=\{x\in X:x=\sup D$ for some $D\subset Y^+\}$. Clearly $B\subset B(Y)$ and it suffices to show that $Z=B-B$ is a band (so that $Z=B(Y)$) and $Z^+=B$. Now Z is a linear subspace of X since B is a cone. Suppose $x\in B$, $y\in X$, and $0\leqslant y\leqslant x$. Since $x=\sup D$ for some $D\subset Y^+$ and $y=y\wedge x=\sup\{y\wedge z:z\in D\}$, we get that $y\in B$. Moreover, if $x\in Z$, then

$x = \sup D - \sup E$ for some $D \cup E \subset Y^+$ and, thus, $x^+ \leqslant \sup D$, $x^- \leqslant \sup E$ imply that x^+ and x^- are in B. Hence $Z^+ = B$. Suppose $x \in Z$ and $y \in X$ with $|y| \leqslant |x|$. Then $|x| \in B$ and $y^+ \leqslant |x|$, $y^- \leqslant |x|$ imply that $y \in Z$. To see that Z is a band it suffices to show that $x = \sup D (D \subset B)$ is in B. But, for each $y \in D$, $y = \sup D_y$ for some $D_y \subset Y^+$. Thus $x = \sup(\bigcup D_y)$ is in B. ▮

Let X and Y be partially ordered vector spaces. A linear operator $T: X \to Y$ is said to be *positive* if $Tx \geqslant 0$ whenever $x \geqslant 0$. It is said to be *order preserving* if $Tx \geqslant 0$ if and only if $x \geqslant 0$. If X and Y are vector lattices, T is said to *preserves the lattice structure* if $T(x \vee y) = Tx \vee Ty$ for all $x, y \in X$ (clearly if T is one to one, onto, and order preserving, then it preserves the lattice structure).

Lemma 2. *Let X and Y be vector lattices and $T: X \to Y$ be a linear operator. Then T preserves the lattice structure if and only if $T(x \vee y) = Tx \vee Ty$ whenever $x \wedge y = 0$.*

Proof. The necessity is clear. For the sufficiency note that for any $x, y \in X$, $(x - x \wedge y) \wedge (y - x \wedge y) = 0$. Thus $T(x + y - 2x \wedge y) = (Tx - T(x \wedge y)) \vee (Ty - T(x \wedge y))$ which implies that $(Tx - T(x \wedge y)) \wedge (Ty - T(x \wedge y)) = 0$ (theorem 1(13)) and, hence, $T(x \wedge y) = Tx \wedge Ty$. ▮

Positive projections P $(P^2 = P)$ on X will play a large role in our development. Let us first note that if a vector lattice X is the direct sum of ideals Y and Z, then the associated projections are positive. For, let $x \geqslant 0$ be in X. Then $x = y + z$ with $y \in Y$ and $z \in Z$. Hence $x \leqslant |y| + |z|$ and by the decomposition property, $x = x_1 + x_2$ with $0 \leqslant x_1 \leqslant |y|$ and $0 \leqslant x_2 \leqslant |z|$. Thus $x_1 \in Y$ and $x_2 \in Z$ and since the decomposition $x = y + z$ is unique, we get that $y = x_1$ and $z = x_2$. Projections arise in a natural fashion as follows. For a nonempty set $D \subset X$ let $D^\perp$ (called the *polar* of D) be equal to $\{x \in X: |x| \wedge |y| = 0 \text{ for all } y \in D\}$. It is easy to see that $D^\perp$ is an ideal in X. Moreover, if $x = \sup E$ with $E \subset (D^\perp)^+$, then for any $y \in D$, $|y| \wedge x = \sup\{|y| \wedge z: z \in E\} = 0$ so that $x \in D^\perp$ and it follows that $D^\perp$ is a band. For an ideal Y in X it is often true that $Y \oplus Y^\perp = X$. In such a case the natural projections of X onto Y and $Y^\perp$ are positive. We call the projection P of X onto Y with kernel $Y^\perp$ a *band projection* and call Y a *projection band* (note that in such a case $Y = Y^{\perp\perp}$ so that it is a band).

The following theorem characterizes projection bands and gives a formula for the band projection.

Theorem 4. *Let X be a vector lattice and Y be a band in X. Then Y is a projection band if and only if for each $x \in X^+$, $x_1 = \sup\{y \in Y: 0 \leqslant y \leqslant x\}$ exists in X. In such a case, $x = x_1 + x_2$ where $x_2 = \sup\{y \in Y^\perp: 0 \leqslant y \leqslant x\}$ and $x_1 \in Y$, $x_2 \in Y^\perp$.*

Proof. Suppose Y is a projection band. Then for each $x \in X^+$, $x = x_1 + x_2$ with $x_1 \in Y^+$ and $x_2 \in (Y^\perp)^+$. If $y \in Y$ and $0 \leqslant y \leqslant x$, then $x - y = (x_1 - y) + x_2$ and $x - y \geqslant 0$ imply that $x_1 - y \geqslant 0$. Hence $x_1 = \sup\{y \in Y : 0 \leqslant y \leqslant x\}$. The proof that $x_2 = \sup\{y \in Y^\perp : 0 \leqslant y \leqslant x\}$ is similar.

Suppose that for each $x \in X^+$, $\sup\{y \in Y : 0 \leqslant y \leqslant x\} = x_1$ exists in X. Since Y is a band, $x_1 \in Y$ for all $x \in X^+$. Let $x_2 = x - x_1$. Clearly we need only show that $x_2 \in Y^\perp$. For any $y \in Y^+$, $y \wedge x_2 \in Y$ and $0 \leqslant y \wedge x_2 \leqslant x_2 = x - x_1$. If $z \in Y$ with $0 \leqslant z \leqslant x$, then $z \leqslant x_1$ and $y \wedge x_2 + z \leqslant y \wedge x_2 + x_1 \leqslant x$. Therefore $y \wedge x_2 + x_1 = \sup\{y \wedge x_2 + z : 0 \leqslant z \leqslant x, z \in Y\} \leqslant x_1$ and it follows that $y \wedge x_2 = 0$. ∎

An important special case is when Y is the band $B(u)$ generated by a single element $u \in X$.

Corollary. *Let X be a vector lattice and $u \in X$. Then $B(u)$ is a projection band if and only if for each $x \in X^+$, $\sup\{x \wedge n|u| : n = 1, 2, \ldots\}$ exists in X. In such a case, the component x_1 of x in $B(u)$ is equal to this supremum.*

Proof. Recall that the ideal Y generated by u is given by $Y = \{y \in X : |y| \leqslant n|u|$ for some $n = 1, 2, \ldots\}$. In particular, $B(u)$ is the band generated by Y. By the theorem, $B(u)$ is a projection band if and only if for each $x \in X^+$, $x_1 = \sup\{y \in B(u) : 0 \leqslant y \leqslant x\}$ exists in X. But, for each $y \in (B(u))^+$, $y = \sup\{w \in Y : 0 \leqslant w \leqslant y\} = \sup\{y \wedge n|u| : n = 1, 2, \ldots\}$. From this it is easy to see that $x_1 = \sup\{x \wedge n|u| : n = 1, 2, \ldots\}$ whenever either one of them exists. ∎

A vector lattice X is said to be *order complete* if for each nonempty set $D \subset X$ which has an upper bound in X, $\sup D$ exists in X. Clearly in such a case, each band is a projection band.

Theorem 5 (F. Riesz). *Let X be an order complete vector lattice. Then for any nonempty set D in X, $D^\perp$ is a projection band and $D^{\perp\perp}$ is the band generated by D.* ∎

For $u \in X$, we let $[u]$ denote the band projection of X onto $u^{\perp\perp} = \{u\}^{\perp\perp}$.

Corollary. *Let X be an order complete vector lattice. Then for any $u \in X$ and $x \in X^+$, $[u](x) = \sup\{x \wedge n|u| : n = 1, 2, \ldots\}$.* ∎

We shall need the following two theorems on linear operators with order complete range. The first one is a modification of the Hahn-Banach theorem and its proof is omitted.

Theorem 6. *Let X, Y be vector lattices with Y order complete. Suppose $p : X \to Y$ is a function such that $p(x + y) \leqslant p(x) + p(y)$ and $p(ax) = ap(x)$ for all $x, y \in X$ and $a \in \mathbb{R}^+$. If Z is a linear subspace of X and $T_0 : Z \to Y$ is a linear operator such that $T_0(x) \leqslant p(x)$ for all $x \in Z$, then there is a* *linear extension T of T_0* to X (i.e. $T|Z=T_0$) *such that* $T(x)\leqslant p(x)$ *for all* $x\in X$. ∎

Theorem 7. *Let X and Y be vector lattices with Y order complete. Suppose $T:X\to Y$ is a positive linear mapping and $x\geqslant 0$. There is a linear mapping $S:X\to Y$ such that $Sx=Tx$, $0\leqslant Sy\leqslant Ty$ $(y\in X^+)$, and $Sy=0$ if $x\wedge y=0$.*

Proof. Let $S(y)=\sup\{T(y\wedge nx): n=1,2,\ldots\}$ if $y\in X^+$. Clearly $S(ay)=aS(y)$ for $a\in\mathbb{R}^+$ and $y\in X^+$. If $y_1,y_2\in X^+$, then $S(y_1+y_2)$ $=\sup\{T[(y_1+y_2)\wedge nx)]: n=1,2,\ldots\}\leqslant\sup\{T(y_1\wedge nx): n=1,2,\ldots\}$ $+\sup\{T(y_2\wedge nx): n=1,2,\ldots\}=S(y_1)+S(y_2)$. On the other hand, $T(y_1\wedge nx)+T(y_2\wedge mx)\leqslant T[(y_1+y_2)\wedge(n+m)x]\leqslant S(y_1+y_2)$. Thus S can clearly be extended to a linear operator with the required properties. ∎

We need the following approximation theorem.

Theorem 8. *Let X be an order complete vector lattice and $0\leqslant x\leqslant u$. Then for each $\varepsilon>0$ there are $u_1,\ldots,u_n$ in X and $a_1,\ldots,a_n$ in $[0,1]$ such that*

(1) $u_i\wedge(u-u_i)=0$ *for* $i=1,\ldots,n$,

(2) $u_i\wedge u_j=0$ *for* $i\neq j$,

(3) $\sum_{i=1}^{n} a_i u_i\leqslant x\leqslant\sum_{i=1}^{n} a_i u_i+\varepsilon u$.

Proof. For each $0\leqslant r\leqslant 1$ let $P_r=[(x-ru)^+]$. Note that $P_0=[x]$ and $P_1=0$ and for $0\leqslant r\leqslant s$, $0\leqslant P_s\leqslant P_r\leqslant I$ (the identity). Now $(P_r-P_s)(x)-r(P_r-P_s)u=(I-P_s)(P_r(x-ru))=(I-P_s)(x-ru)^+\geqslant 0$. Moreover, $(P_r-P_s)(x)-s(P_r-P_s)(u)=P_r((x-su)-P_s(x-su))=P_r((x-su)$ $-(x-su)^+)\leqslant 0$. Hence $r(P_r-P_s)(u)\leqslant(P_r-P_s)(x)\leqslant s(P_r-P_s)(u)$.

Let $\varepsilon>0$ be given and choose n so that $1/n<\varepsilon$. Let $u_i=\left(P_{\frac{i-1}{n}}-P_{\frac{i}{n}}\right)(u)$ and $a_i=\dfrac{i-1}{n}$ for $i=1,\ldots,n$. Then $a_iu_i\leqslant\left(P_{\frac{i-1}{n}}-P_{\frac{i}{n}}\right)(x)\leqslant a_iu_i+\dfrac{1}{n}u_i$ and, hence, $\sum_{i=1}^{n}a_iu_i\leqslant\sum_{i=1}^{n}\left(P_{\frac{i-1}{n}}-P_{\frac{i}{n}}\right)(x)=(P_0-P_1)(x)=x\leqslant\sum_{i=1}^{n}a_iu_i$ $+\dfrac{1}{n}\sum_{i=1}^{n}u_i=\sum_{i=1}^{n}a_iu_i+\dfrac{1}{n}(u)\leqslant\sum_{i=1}^{n}a_iu_i+\varepsilon u$. ∎

We now turn our attention to atoms in vector lattices and their relationship to finite dimensional vector lattices. For this we will need the following additional axiom called the *Archimedean property*: For each $x\in X^+$, $\inf\dfrac{x}{n}=0$. All normed linear lattices have the Archimedean property. We note that if X has the Archimedean property, then for any $x\in X^+$ and any sequence $\{a_n\}$ of positive real numbers converging to 0, $\inf a_n x=0$.

Definition 2. Let X be a vector lattice. A nonzero element $x\in X$ is said to be an *atom* if $0\leqslant y$, $z\leqslant |x|$ and $y\wedge z=0$ imply that $y=0$ or $z=0$.

Note that every atom x is either positive or negative since $0\leqslant x^+$, $x^-\leqslant |x|$ and $x^+\wedge x^-=0$. Hence we shall always take atoms to be positive. Note further that for any $x>0$, either $x\geqslant y$ for some atom y (i.e. x dominates y) or there is an infinite sequence $\{y_n\}$ of nonzero elements of X such that $0\leqslant y_n\leqslant x$ and $y_n\wedge y_m=0$ for $n\neq m$. For, if x is an atom, then it dominates an atom. Otherwise, there are $0<y_1, y_2\leqslant x$ with $y_1\wedge y_2=0$. Either one of y_1, y_2 are atoms or they can be further decomposed. Clearly this process either terminates at an atom or yields such a sequence.

Theorem 9. *Let X be a vector lattice with the Archimedean property. Then $x>0$ is an atom if and only if for each $y\in X$ with $|y|\leqslant x$, $y=ax$ for some $a\in\mathbb{R}$. Moreover, for each atom x, the band $B(x)$ generated by x is a projection band.*

Proof. Suppose $0<y\leqslant x$ and put $A=\{a\in\mathbb{R}^+ : ay\leqslant x\}$. Since X has the Archimedean property, A is bounded. Thus $a=\sup A\geqslant 1$. Let $\{a_n\}$ be an increasing sequence in A with $\lim a_n=a$. Then $\inf(a-a_n)y=0$ and $ay=\sup a_n y\leqslant x$. Suppose $x-ay=z>0$. Since $\sup\left(z-\frac{1}{n}y\right)^+=z$, $\left(z-\frac{1}{n}y\right)^+>0$ for some n. Clearly $\left(z-\frac{1}{n}y\right)^+=\left(x-\left(a+\frac{1}{n}\right)y\right)^+$ $\leqslant x\leqslant 2ax$. Also, $\left(x-\left(a+\frac{1}{n}\right)y\right)^->0$ since otherwise $x\geqslant\left(a+\frac{1}{n}\right)y$ which contradicts the definition of a. Moreover, $\left(x-\left(a+\frac{1}{n}\right)y\right)^-$ $=\left(\left(a+\frac{1}{n}\right)y-x\right)^+\leqslant\left(a+\frac{1}{n}\right)x\leqslant 2ax$. But this contradicts the fact that $2ax$ is an atom. Therefore $y=x/a$.

The converse is clear. To show that $B(x)$ is a projection band it suffices to observe that for each $y\in X^+$, $\sup y\wedge nx$ exists in X. To see this, note that $y\wedge nx\in B(x)$ so that $y\wedge nx=a_n x$ for some a_n. Moreover, a_n is increasing and bounded and $\sup y\wedge nx=ax$ where $a=\lim a_n$. ∎

We now show that any finite dimensional vector lattice X has atoms and that, in fact, if X has the Archimedean property, then it has a Hamel basis consisting of atoms. For this we use the following elementary fact. If $x_1,\dots,x_n$ are nonzero elements of X and $x_i\wedge x_j=0$ for $i\neq j$, then $\{x_1,\dots,x_n\}$ is a linearly independent set.

Theorem 10. *Let X be a finite dimensional vector lattice. Then for each $x>0$ in X, x dominates an atom. (If x did not dominate an atom, then there would exist a sequence $\{y_n\}$ such that $0<y_n\leqslant x$ and $y_n \wedge y_m=0$ for $n\neq m$.)* ∎

Theorem 11. *Let X be a vector lattice with the Archimedean property. If each collection of disjoint positive elements in X is finite, then X is finite dimensional and has a Hamel basis consisting of atoms.*

Proof. Clearly X has atoms by arguments above. Let $\{x_1,\dots,x_n\}$ be a maximal set of atoms (obtainable by Zorn's lemma and finite by hypothesis). Since X has the Archimedean property, the band $B(x_i)$ generated by x_i is a projection band. Let $y\in X^+$ and for each i let $a_i\in\mathbb{R}$ be such that $a_i x_i$ is the component of y in $B(x_i)$. Clearly $a_i\geqslant 0$ and $a_1x_1+\dots+a_nx_n\leqslant y$. Let $z=y-(a_1x_1+\dots+a_nx_n)$. Then $0\leqslant z\leqslant y-a_ix_i$ implies $z\in B(x_i)^\perp$. Hence $z\wedge x_i=0$ and if $z\neq 0$, then it dominates some atom x and $\{x,x_1,\dots,x_n\}$ is a collection of atoms which is impossible. Thus $y=a_1x_1+\dots+a_nx_n$ and it follows that $\{x_1,\dots,x_n\}$ is a Hamel basis for X. ∎

We close this section with some remarks on complex vector lattices.

Definition 3. Let X be a complex vector space and suppose M is a real linear subspace of X such that each $x\in X$ can be uniquely written as $x=x_1+ix_2$ with $x_1,x_2\in M$. Let C be a positive cone in M. We say (X,C) is a *complex vector lattice* if (M,C) is a (real) vector lattice and if for each $x\in X$, $\sup\{\mathrm{Re}(e^{i\theta}x):0\leqslant\theta\leqslant 2\pi\}$ exists in M (where $\mathrm{Re}\,x=x_1$ in the above representation).

The requirement that $\sup\{\mathrm{Re}(e^{i\theta}x):0\leqslant\theta\leqslant 2\pi\}$ exists in M is to guarantee an absolute value on X. Note that if $x\in M$, then one has $|x|=\sup\{\mathrm{Re}(e^{i\theta}x):0\leqslant\theta\leqslant 2\pi\}$. Thus we shall write $|x|$ for this supremem for all $x\in X$. (The reader can easily verify that $|x+y|\leqslant|x|+|y|$, $|x|=0$ if and only if $x=0$, and $|x_1|\vee|x_2|\leqslant|x|$ where $x=x_1+ix_2$.)

We extend the rest of the terminology as follows. A *sublattice* of X is a (complex) linear subspace Y such that $\mathrm{Re}\,x\in Y$ whenever $x\in Y$ and $Y\cap M=\mathrm{Re}\,Y$ is a (real) sublattice of (M,C). An *ideal* Y in X is a linear subspace such that if $x\in X$, $y\in Y$ and $|x|\leqslant|y|$, then $x\in Y$. Clearly an ideal is a sublattice. A *band* is an ideal Y such that $\mathrm{Re}\,Y$ is a band in (M,C). For a nonempty subset D of X, $D^\perp=\{x\in X:|x|\wedge|y|=0$ for all $y\in D\}$. Clearly $D^\perp$ is a band in X. As before, Y is said to be a *projection band* if $Y\oplus Y^\perp=X$. The projection of X onto Y with kernel $Y^\perp$ is called a *band projection.*

Projection bands and band projections are characterized by the real case as is seen from the following lemma.

Lemma 3. *Let X be a complex vector lattice. A band Y in X is a projection band if and only if* $\operatorname{Re} Y$ *is a projection band in* $\operatorname{Re} X$.

Proof. Suppose $Y \oplus Y^{\perp} = X$ and let $x \in \operatorname{Re} X$. Then $x = (y_1 + i y_2) + (z_1 + i z_2)$ with $y_1 + i y_2 \in Y$ and $z_1 + i z_2 \in Y^{\perp}$. Clearly $y_2 = -z_2$ and since $y_2 \in Y$ and $z_2 \in Y^{\perp}$, we get that $y_2 = z_2 = 0$. Hence $\operatorname{Re} X = \operatorname{Re} Y \oplus (\operatorname{Re} Y)^{\perp}$ (note: $(\operatorname{Re} Y)^{\perp}$ is taken in $\operatorname{Re} X$).

Conversely, suppose that $\operatorname{Re} X = \operatorname{Re} Y \oplus (\operatorname{Re} Y)^{\perp}$ and $x \in X$. Then $x = x_1 + i x_2$ with $x_1, x_2 \in \operatorname{Re} X$. Thus $x_1 = y_1 + z_1$ and $x_2 = y_2 + z_2$ with $y_1, y_2 \in \operatorname{Re} Y$ and $z_1, z_2 \in (\operatorname{Re} Y)^{\perp}$, and $|y_1 + i y_2| \wedge |z_j| = 0$ for $j = 1, 2$. Hence $|y_1 + i y_2| \wedge |z_1 + i z_2| \leqslant |y_1 + i y_2| \wedge |z_1| + |y_1 + i y_2| \wedge |z_2| = 0$ and $X = Y \oplus Y^{\perp}$. ∎

Note that the above proof shows that if $P: X \to Y$ is the band projection, then $P|\operatorname{Re} X$ is the (real) band projection of $\operatorname{Re} X$ onto $\operatorname{Re} Y$ (and similarly for $I - P$). Consequently, for $x \geqslant 0$, $Px = \sup\{y \in \operatorname{Re} Y : 0 \leqslant y \leqslant x\}$ and $x - Px = \sup\{z \in (\operatorname{Re} Y)^{\perp} : 0 \leqslant z \leqslant x\}$. Conversely, if $\sup\{y \in \operatorname{Re} Y : 0 \leqslant y \leqslant x\}$ exists for all $x \geqslant 0$, then Y is a projection band (see theorem 4). In particular, for $u \in X$, $u^{\perp\perp}$ is a projection band if and only if for each $x \geqslant 0$, $\sup\{x \wedge n|u| : n = 1, 2, \ldots\}$ exists in $\operatorname{Re} X$. Clearly by an *order complete* complex vector lattice X we mean that $\operatorname{Re} X$ is order complete and the Riesz theorem (theorem 5) and its corollary have a complex analogue based on the above remarks.

§ 2. Partially Ordered Normed Linear Spaces

Let X be a real normed linear space and C be a positive cone in X. There is a natural dual wedge to C given by $C^* = \{x^* \in X^* : x^*(x) \geqslant 0$ for all $x \in C\}$. One can easily see that C^* is a cone if and only if $C - C$ is dense in X (if $C - C$ is not dense in X, then by the Hahn-Banach theorem there is a nonzero $x^* \in X^*$ which is zero on $C - C$ and, hence, $x^* \in C^* \cap (-C^*)$). In this section we study properties on C and the norm on X to see what properties they induce on C^* and the norm on X^* (and conversely). The properties we are interested in are those that characterize certain subspaces of the Banach space $C(T, \mathbb{R})$ of all continuous real valued functions on a compact Hausdorff space T (supremum norm and pointwise ordering) and their duals. (See [197] and [241] for more general results.)

The basic tools we use are the Hahn-Banach theorem and the second basic separation theorem (which says that a closed convex set and a compact convex set can be separated by a continuous linear functional when they are disjoint; see the appendix). We say that C^* *determines* C

if $x \in C$ whenever $x^*(x) \geqslant 0$ for all $x^* \in C^*$. If C is closed, then C^* determines C. For, if $x \notin C$; then by the second basic separation theorem there is an $x^* \in X^*$ such that $x^*(x) < \inf_{y \in C} x^*(y)$. Since $0 \in C$, $x^*(x) < 0$ and since C is a cone, it follows that $x^*(y) \geqslant 0$ for all $y \in C$, that is, $x^* \in C^*$.

To motivate the following definition consider the Banach space $L_1(\mu, \mathbb{R}) = L_1(T, \Sigma, \mu, \mathbb{R})$ for some measure space (T, Σ, μ) (with the usual norm and ordering). Then whenever $f, g \in (L_1(\mu, \mathbb{R}))^+$ we have that $\|f+g\| = \|f\| + \|g\|$.

Definition 1. Let X be a normed linear space and C a positive cone in X. The norm on X is said to be *additive* (with respect to C) if $\|x+y\| = \|x\| + \|y\|$ for all $x, y \in C$.

We shall use the following notation throughout the text: $b_0(X) = \{x \in X : \|x\| < 1\}$, $b(X) = \{x \in X : \|x\| \leqslant 1\}$ ($b_0(X)$ (resp. $b(X)$) is called the *open unit ball* (resp. *closed unit ball*) of X).

By a *partially ordered normed linear space X* we mean X is a normed linear space with a partial ordering given by a closed positive cone C.

Theorem 1. *Let X be a partially ordered normed linear space. If the norm is additive on C^*, then $b_0(X)^+$ is directed upwards.*

Proof. First note that C^* is a cone since if $x^* \in C^* \cap (-C^*)$, then $0 = \|x^* - x^*\| = 2\|x^*\|$. Consider the normed linear spaces $X \times \mathbb{R}$ and $X^* \times \mathbb{R}$ where the norms are given by $\|(x,a)\| = \|x\| + |a|$ and $\|(x^*, a)\| = \max(\|x^*\|, |a|)$ respectively. Clearly the operator A defined by $A(x^*, a)(x, b) = x^*(x) + ab$ is a linear isometry of $X^* \times \mathbb{R}$ onto $(X \times \mathbb{R})^*$. Let $x, y \in b_0(X)^+$ and set $H_z = \{(x^*, a) : x^* \in C^*$ and $0 \leqslant a \leqslant x^*(z)\}$ for $z = x$ or $z = y$. Then H_x and H_y are weak* closed cones in $X^* \times \mathbb{R}$ and $H = H_x + H_y$ is also a cone. We wish to show that H is also weak* closed. By the Krein-Šmulian theorem (see the appendix) it is enough to show that $H \cap b(X^* \times \mathbb{R})$ is weak* compact. So, let $\{(x_d^*, a_d)\}$ be a net in $H \cap b(X^* \times \mathbb{R})$. Then $x_d^* = y_d^* + z_d^*$ and $a_d = b_d + c_d$ for some $(y_d^*, b_d) \in H_x$ and $(z_d^*, c_d) \in H_y$. Since the norm on X^* is additive, $\|x_d^*\| = \|y_d^*\| + \|z_d^*\|$. Thus there is a subnet $\{(x_{d'}^*, a_{d'})\}$ such that $\{(y_{d'}^*, b_{d'})\}$ converges to some $(y^*, b) \in H_x$ and $\{(z_{d'}^*, c_{d'})\}$ converges to some $(z^*, c) \in H_y$. Clearly $(y^* + z^*, b + c)$ $H \cap b(X^* \times \mathbb{R})$.

Now let $K = \{(x^*, 1) : \|x^*\| \leqslant 1\}$. Then K is convex and weak* compact. We claim that K and H are disjoint. For, suppose $(x^*, 1) \in K$ and $(x^*, 1) = (y^*, a) + (z^*, b)$ with $(y^*, a) \in H_x$ and $(z^*, b) \in H_y$. Since $\|x\|, \|y\| < 1$ and the norm on X^* is additive, $1 = a + b \leqslant y^*(x) + z^*(y) < \|y^*\| + \|z^*\| = \|x^*\| \leqslant 1$ which is impossible. Hence by the second basic separation theorem there is a weak* continuous linear functional L on $X^* \times \mathbb{R}$ such that $\sup L(H) < \inf L(K)$. Since H is a cone, it follows that $0 = \sup L(H) < \inf L(K)$. Let $b = L(0,1)$. Then since $(0,1) \in K$, we have that $b > 0$. We define x^{**} on X^* by $x^{**}(x^*) = -(1/b)L(x^*,0)$. It is easy to see that x^{**} is weak* continuous since L is. Thus there is a $z \in X$ with $x^{**}(x^*) = x^*(z)$ for all $x^* \in X^*$. Hence $0 = x^*(z)L(0,1) + L(x^*,0) = L(x^*, x^*(z)) = \sup L(H)$ for all $x^* \in X^*$. If $x^* \in C^*$, then $(x^*, x^*(x)) \in H_x$ and $L(x^*, x^*(x)) \leqslant L(x^*, x^*(z))$. Since $b > 0$, $x^*(x) \leqslant x^*(z)$ for all $x^* \in C^*$. Similarly we obtain $x^*(y) \leqslant x^*(z)$ for all $x^* \in C^*$. Since C^* determines C, $x, y \leqslant z$. All that remains is to show that $\|z\| < 1$. By the Hahn-Banach theorem there is a $z^* \in X^*$ with $\|z^*\| = 1$ and $z^*(z) = \|z\|$. Thus $(z^*, 1) \in K$ and $0 = L(z^*, z^*(z)) < L(z^*, 1)$ and it follows that $z^*(z) < 1$. ∎

Clearly $b_0(C(T, \mathbb{R}))$ is directed upwards for any compact Hausdorff space T. For, if $\|f\|, \|g\| < 1$, then $|f| \vee |g| \geqslant f, g$ and $\||f| \vee |g|\| = \max(\|f\|, \|g\|) < 1$. It is also true that the norm on $C(T, \mathbb{R})^*$ is additive. For, note that $\|f\| \leqslant 1$ if and only if $-1 \leqslant f \leqslant 1$. Hence for any positive linear functional x^* on $C(T, \mathbb{R})$, $-x^*(1) \leqslant x^*(f) \leqslant x^*(1)$. In particular, $\|x^*\| = x^*(1)$ and it follows that $\|x^* + y^*\| = (x^* + y^*)(1) = x^*(1) + y^*(1) = \|x^*\| + \|y^*\|$ when $x^*, y^* \geqslant 0$. We now investigate these two properties in an abstract setting (the proofs here are based on ideas in [225]).

Definition 2. Let X be a normed linear space and C be a cone on X. We say that the norm on X is *regular* (with respect to C) if for each $x \in b_0(X)$ there is a $y \in b_0(X)$ with $y \geqslant x, -x$. The norm on X is said to be *monotone* (with respect to C) if $-x \leqslant y \leqslant x$ implies $\|y\| \leqslant \|x\|$.

We shall need the following lemma concerning regularity and monotonicity (additional results are in [78] and [225]).

Lemma 1. *Let (X, C) be a partially ordered normed linear space such that C^* is a cone.*

(1) *If the norm on X is regular, then $0 \leqslant y^* \leqslant x^*$ implies that $\|y^*\| \leqslant \|x^*\|$.*

(2) *If the norm on C^* is monotone, then the norm on C is regular.*

Proof. (1) Let $\varepsilon > 0$ and choose $x \in b_0(X)$ with $y^*(x) > \|y^*\| - \varepsilon$. By the regularity condition there is a $y \in b_0(X)$ with $y \geqslant x, -x$. Now $x^*(y) \geqslant y^*(y) \geqslant y^*(x) > \|y^*\| - \varepsilon$ and hence, $\|x^*\| \geqslant \|y^*\|$.

(2) Let $x \in b_0(X)$ and define $h(x^*) = \sup\{y^*(x): -x^* \leqslant y^* \leqslant x^*\}$ for all $x^* \in C^*$. Then $H = \{(x^*, a): x^* \in C^*,\ a \leqslant h(x^*)\}$ is a weak* closed cone in $X^* \times \mathbb{R}$. Clearly $D = \{(x^*, 1): \|x^*\| \leqslant 1\}$ is a weak* compact convex set in $X^* \times \mathbb{R}$. Moreover, since the norm is monotone on X^*, H and D are disjoint. Thus by the second basic separation theorem there

is a weak* continuous linear functional L on $X^* \times \mathbb{R}$ such that $\sup L(H) < \inf L(D)$. Since H is a cone, it follows that $\sup L(H) = 0$. As before, we put $b = L(0,1)$ and since $(0,1) \in D$, $b > 0$ and we again define $x^{**}(x^*) = -(1/b)\,L(x^*,0)$ for $x^* \in X^*$. Since x^{**} is weak* continuous, there is a $z \in X$ such that $x^{**}(x^*) = x^*(z)$ $(x^* \in X^*)$. Hence $L(x^*, x^*(z)) = 0$ $(x^* \in X^*)$ and since $(x^*, \pm x^*(x)) \in H$ $(x^* \in C^*)$ we have that $L(x^*, \pm x^*(x)) \leqslant L(x^*, x^*(z))$ for all $x^* \in C^*$. That is, $(\pm x^*(x) - x^*(z))\,L(0,1) \leqslant 0$ for all $x^* \in C^*$ and since $L(0,1) > 0$ and C is closed, it follows that $z \geqslant x, -x$. The proof that $\|z\| < 1$ is the same as theorem 1. ∎

We note that if the norm is additive on X, then it is also monotone. For, suppose $-x \leqslant y \leqslant x$. Then $2x = (x+y) + (x-y)$ and $2\|x\| = \|x+y\| + \|x-y\|$. On the other hand, $2y = (y+x) + (y-x)$ and $2\|y\| \leqslant \|y+x\| + \|y-x\| = 2\|x\|$.

Theorem 2. *Let X be a partially ordered normed linear space. Then the norm is additive on X^* if and only if $b_0(X)$ is directed upwards.*

Proof. Let $x^*, y^* \in C^*$ and $x, y \in b_0(X)$. Then since $b_0(X)$ is directed upwards there is a $z \in b_0(X)$ with $z \geqslant x, y$. Thus $x^*(x) + y^*(y) \leqslant (x^* + y^*)(z) < \|x^* + y^*\|$ so that $\|x^*\| + \|y^*\| \leqslant \|x^* + y^*\|$. Since the reverse inequality is always true, we get that $\|x^* + y^*\| = \|x^*\| + \|y^*\|$.

Suppose the norm is additive on X^*. Then it is monotone and by lemma 1, the norm on X is regular. Let $x, y \in b_0(X)$. By regularity there are $x_1, y_1 \in (b_0(X))^+$ with $x_1 \geqslant x, -x$ and $y_1 \geqslant y, -y$. From theorem 1 we obtain a $z \in (b_0(X))^+$ with $z \geqslant x_1, y_1$. Thus $z \geqslant x, y$ and $b_0(X)$ is upwards directed. ∎

Let X be a normed linear space with a positive cone C. An element $e \in X$ is said to be a *strong order unit* (with respect to C) if $\|e\| = 1$ and $e \geqslant x$ for all $x \in b(X)$. The norm on X is said to be a *strong order unit norm* (with respect to e) if $x \in b(X)$ is equivalent to $-e \leqslant x \leqslant e$. Clearly if e is a strong order unit in a space which the norm is monotone, then the norm is a strong order unit norm. Moreover, if the norm is a strong order unit norm, then $\|x\| = \inf\{a > 0\colon -ae \leqslant x \leqslant ae\}$ $(x \in X)$. We noted above that the constant function 1 is a strong order unit in $C(T, \mathbb{R})$ and the norm in $C(T, \mathbb{R})$ is a strong order unit norm. More generally, if X is any linear subspace of $C(T, \mathbb{R})$ containing 1, then 1 is a stronger order unit in X (with respect to the induced order from $C(T, \mathbb{R})$) and the norm on X is a strong order unit norm (we shall see later that any partially ordered normed linear space with a strong order unit norm can be so represented).

Lemma 2. *Let X be a normed linear space with a positive cone C and suppose e is a strong order unit in X and X has strong order unit norm with respect to e. Then*

(1) $X=C-C$,

(2) *any positive linear functional on X is continuous,*

(3) $x^*\in C^*$ *if and only if* $x^*(e)=\|x^*\|$,

(4) *the norm is additive on X^*.*

Proof. (1) For any $x\in X$ with $\|x\|=1$, $-e\leqslant x\leqslant e$. Thus $x=(x+e)-e$ is in $C-C$ and it follows that $X=C-C$.

(2) If $x^*(x)\geqslant 0$ for all $x\in C$, then $-e\leqslant x\leqslant e$ implies $-x^*(e)\leqslant x^*(x)\leqslant x^*(e)$ so that $\|x^*\|=x^*(e)$ and x^* is continuous.

(3) Suppose $x^*\in X^*$ and $x^*(e)=\|x^*\|$. Let $x\in C$ with $\|x\|=1$. Then $-e\leqslant x-e\leqslant e$ so that $\|x-e\|\leqslant 1$. In particular, $|x^*(x)-x^*(e)|\leqslant\|x^*\|$ and it follows that $x^*(x)\geqslant 0$. Hence $x^*\in C^*$. The converse is clear from (2).

(4) If $x^*,y^*\in C^*$, then $\|x^*+y^*\|=x^*(e)+y^*(e)=\|x^*\|+\|y^*\|$. ∎

Definition 3. Let X be a normed linear space and C a positive cone in X. Then C is said to be c *normal* $(c\geqslant 1)$ if $x\leqslant y\leqslant z$ implies that $\|y\|\leqslant c\max(\|x\|,\|z\|)$. The cone C is said to be c *generating* if for each $x\in X$, $x=x_1-x_2$ with $x_1,x_2\in C$ and $c\|x\|\geqslant\|x_1\|+\|x_2\|$.

Theorem 3. *Let X be a normed linear space, C a positive cone in X, e a strong order unit, and suppose X has strong order unit norm. Then C^* is* 1 *generating.*

Proof. Recall that $X=C-C$ so that C^* is a positive cone. Let $Y=X\times X$ and $D=C\times C$. Then D is a positive cone in Y, (e,e) is a strong order unit in Y, and the norm defined by $\|(x,y)\|=\max(\|x\|,\|y\|)$ is a strong order unit norm on Y. Let $Z=\{t(e,e)-(x,-x):t\in\mathbb{R},x\in X\}$. Then the cone $C_0=D\cap Z$ is a positive cone in Z and Z has strong order unit norm with respect to C_0 and (e,e). For $x^*\in X^*$ we define $F(t(e,e)-(x,-x))=t\|x^*\|-x^*(x))$. Then F is a positive linear functional on Z (and hence continuous). By the Hahn-Banach theorem F has a norm preserving extension (called $\tilde{F}$) to Y. Clearly the extension is still positive since $\tilde{F}(e,e)=\|\tilde{F}\|$. Let x_1^*,x_2^* be defined by $x_1^*(x)=\tilde{F}(x,0)$ and $x_2^*(x)=\tilde{F}(0,x)$ for $x\in X$. Then x_1^*,x_2^* are in C^* and $x^*=x_1^*-x_2^*$. Moreover, it is easy to see that $\|x^*\|=\|x_1^*\|+\|x_2^*\|$. ∎

Theorem 4 (Grossberg-Krein). *Let X be a partially ordered normed linear space with positive cone C. Then C is c normal if and only if C^* is c generating.*

Proof. Suppose C^* is c generating and $x\leqslant y\leqslant z$ with $x,z\in b(X)$. By the Hahn-Banach theorem there is an $x^*\in X^*$ with $\|x^*\|=1$ and

$x^*(y)=\|y\|$. Since C^* is c generating, $x^*=x_1^*-x_2^*$ with x_1^*, x_2^* in C^* and $c\|x^*\| \geqslant \|x_1^*\|+\|x_2^*\|$. Now $-\|x_i^*\| \leqslant x_i^*(x) \leqslant x_i^*(y) \leqslant x_i^*(z) \leqslant \|x_i^*\|$ $(i=1,2)$ and, thus, $\|y\|=x^*(y)=x_1^*(y)-x_2^*(y) \leqslant \|x_1^*\|+\|x_2^*\| \leqslant c$.

Suppose C is c normal. Let $C_1=\{(x,t): x\in X, t\in \mathbb{R}^+$, and there is a $y\in X$ such that $\|y\|\leqslant 1$ and $x+ty\in C\}$. Then C_1 is a positive cone in $X\times\mathbb{R}$. Let the norm in $X\times\mathbb{R}$ be defined by $\|(x,t)\|=\inf\{a>0:(0,-a) \leqslant (x,t) \leqslant (0,a)\}$. (The reader can easily check that this indeed defines a norm. The condition $\|(x,t)\|=0$ implies that $(x,t)=(0,0)$ comes from the fact that C is closed.) Note that for $x\in X$, $\pm(x,0)+\|x\|(0,1)\in C_1$ and consequently $\|(x,0)\|\leqslant\|x\|$. On the other hand, for any $\varepsilon>0$ we have that $-(\|(x,0)\|+\varepsilon)(0,1)\leqslant(x,0)\leqslant(\|(x,0)\|+\varepsilon)(0,1)$. Hence there are $y,z\in b_0(X)$ such that $(\|(x,0)\|+\varepsilon)z\leqslant x\leqslant(\|(x,0)\|+\varepsilon)y$. Since C is c normal we have that $\|x\|\leqslant c(\|(x,0)\|+\varepsilon)$ and it follows that $\|(x,0)\|\leqslant\|x\|\leqslant c\|(x,0)\|$. Clearly $X\times\mathbb{R}$ has strong order unit norm with respect to C_1 and $(0,1)$. Let $x^*\in X^*$ and consider y^* defined on $X\times\{0\}$ by $y^*(x,0)=x^*(x)$. Then $\|x^*\|\leqslant\|y^*\|\leqslant c\|x^*\|$ and by the Hahn-Banach theorem y^* has a norm preserving extension F to all of $X\times\mathbb{R}$. By theorem 3, $F=G-H$ where G and H are positive linear functionals and $\|F\|=\|G\|+\|H\|$. Let $x_1^*(x)=G(x,0)$ and $x_2^*(x)=H(x,0)$. Then $x_1^*, x_2^*\in C^*$ and $c\|x^*\| \geqslant\|x_1^*\|+\|x_2^*\|$. ∎

We now wish to give a representation of partially ordered normed linear spaces X with 1 normal positive cone and upwards directed open unit balls. We also give the promised representation of partially ordered normed linear spaces with strong order unit norm. These are based on the notion of affine continuous functions. Let F be a locally convex Hausdorff topological vector space and K be a nonempty compact convex set in F. By $A(K,\mathbb{R})$ we mean the Banach space of all affine continuous real functions on K under the supremum norm. Clearly $A(K,\mathbb{R})$ is a partially ordered Banach space under pointwise ordering and has the constant function 1 as a strong order unit and strong order unit norm. This main tool we use is the approximation lemma for continuous affine functions which says that if $h\in A(K,\mathbb{R})$ and $\varepsilon>0$, then there is an $x^*\in F^*$ and a $c\in\mathbb{R}$ such that $|x^*(x)+c-h(x)|<\varepsilon$ for all $x\in K$ (see the appendix for a proof). It is easy to see that the positive cone is 1 normal in $A(K,\mathbb{R})$ so that the dual cone is 1 generating. Also, since $A(K,\mathbb{R})$ has strong order unit norm, the norm in $A(K,\mathbb{R})^*$ is additive. Finally, by the Krein-Milman theorem we have that $\|h\|=\sup\{|h(x)|: x\in \operatorname{ext} K\}$.

Theorem 5. *Let X be a partially ordered normed linear space with 1 normal positive cone and upwards directed open unit ball. Let $K=\{x^*\in C^*: \|x^*\|\leqslant 1\}$. Then K is weak* compact convex set and there is an order preserving linear isometry ϕ of X onto a dense subspace of*

$A_0(K,\mathbb{R})=\{h\in A(K,\mathbb{R}):h(0)=0\}$. *Furthermore, if X is complete, then* $X=A_0(K,\mathbb{R})$.

Proof. Clearly K is convex and weak* compact ($K=C^*\cap b(X^*)$). Suppose $x^*\in X^*$ and $\|x^*\|=1$. If $x^*\notin C^*\cup(-C^*)$, then since C^* is 1 generating $x^*=x_1^*-x_2^*$ with $x_1^*,x_2^*\in C^*\setminus\{0\}$ and $\|x^*\|=\|x_1^*\|+\|x_2^*\|$. Thus $x^*=\|x_1^*\|(x_1^*/\|x_1^*\|)+\|x_2^*\|(-x_2^*/\|x_2^*\|)$ and it follows that $x^*\notin \operatorname{ext} b(X^*)$. That is, $\operatorname{ext} b(X^*)=\operatorname{ext} K\cup(-\operatorname{ext} K)\setminus\{0\}$. In particular, if $\phi(x)(x^*)=x^*(x)$ for all $x\in X$ and $x^*\in K$, then ϕ is a linear isometry since $\|\phi(x)\|=\sup\{|x^*(x)|:x\in\operatorname{ext}(K)\}=\|x\|$. Since C is closed, ϕ is order preserving (i.e., $\phi(x)\geqslant 0$ if and only if $x\geqslant 0$). The range of ϕ is dense in $A_0(K,\mathbb{R})$ by the approximation lemma for continuous affine functions. For, given $\varepsilon>0$ and $h\in A_0(K,\mathbb{R})$ there is an $x\in X$ and a $k\in\mathbb{R}$ such that $|h(x^*)+k-x^*(x)|<\varepsilon$ for all $x^*\in K$. In particular, $|k|<\varepsilon$ and it follows that $\|h-\phi(x)\|<2\varepsilon$. ∎

Note: In the above proof we applied the approximation lemma to X^* with the weak* topology and used the fact that the dual of this space is X.

Theorem 6. *Let X be a partially ordered normed linear space with positive cone C. Suppose e is a strong order unit for X and X has strong order unit norm with respect to e. Then $K=\{x^*\in X^*:x^*(e)=1=\|x^*\|\}$ is weak* compact and convex and there is an order preserving linear isometry ϕ of X onto a dense subspace of $A(K,\mathbb{R})$. In particular, if X is complete, then $X=A(K,\mathbb{R})$.*

Proof. Clearly K is weak* compact and convex. As in theorem 5, $\operatorname{ext} b(X^*)=\operatorname{ext} K\cup(-\operatorname{ext} K)$ and, so, the operator ϕ defined by $\phi(x)(x^*)=x^*(x)$ $(x\in X, x^*\in K)$ is an order preserving linear isometry of X into $A(K,\mathbb{R})$. (Since $\|x\|=\sup\{|x^*(x)|:x^*\in\operatorname{ext} b(X^*)\}=\sup\{|x^*(x)|:x^*\in\operatorname{ext} K\}=\sup\{|x^*(x)|:x^*\in K\}$.) The approximation lemma for continuous affine functions shows that the range of ϕ is dense in $A(K,\mathbb{R})$. For, if $h\in A(K,\mathbb{R})$ and $\varepsilon>0$, then there is an $x\in X$ and a $c\in\mathbb{R}$ such that $|h(x^*)-x^*(x)-x^*(c\,e)|=|h(x^*)-x^*(x)-c|<\varepsilon$. That is, $\|h-\phi(x+c\,e)\|<\varepsilon$. ∎

Now suppose F is a locally convex Hausdorff space and K is a nonempty compact convex set in F. Let $K_1=\{x^*\in A(K,\mathbb{R})^*:x^*(1)=1=\|x^*\|\}$. Then, as above, K_1 is a weak* compact convex set. Let $\psi:K\to K_1$ be defined by $\psi(x)(h)=h(x)$ for $x\in K$ and $h\in A(K,\mathbb{R})$. Note that since $\psi(x)(1)=1$ and $\|\psi(x)\|\leqslant 1$, we do get that $\psi(x)(1)=1=\|\psi(x)\|$ so that $\psi(x)$ is in K_1. Clearly ψ is affine and since $A(K,\mathbb{R})$ separates the points of K (by the Hahn-Banach theorem applied to F), ψ is one to one. The definition of the weak* topology shows that ψ is continuous. Consequently we need only show that ψ is onto. Suppose $x^*\in K_1\setminus\psi(K)$.

Then by the second basic separation theorem there is an $h \in A(K, \mathbb{R})$ such that $x^*(h) > \sup\{h(x): x \in K\} = a$ which is clearly impossible since $x^* \in K_1$. For, $h \leqslant a \cdot 1$ implies $x^*(h) \leqslant a x^*(1) = a$. Thus K and K_1 are completely identified. Henceforth when we consider $A(K, \mathbb{R})$ we shall always identify K with K_1.

Theorem 7. *Let X be a normed linear space and C a positive cone in X. Suppose that C is 1 generating and the norm on X is additive. Then $e^*(x-y) = \|x\| - \|y\|$ $(x, y \in C)$ defines a strong order unit in X^* with respect to C^* and X^* has strong order unit norm.*

Proof. Note that since $C - C = X$, C^* is a positive cone. Since the norm on X is additive, e^* is a well defined linear operator and since C is one generating, $\|e^*\| = 1$. Moreover, by definition of e^*, $e^* \in C^*$. Now if $\|x^*\| \leqslant 1$ and $x \in C$, than $(e^* \pm x^*)(x) = \|x\| \pm x^*(x) \geqslant 0$ and, thus, $-e^* \leqslant x^* \leqslant e^*$. On the other hand, if $-e^* \leqslant x^* \leqslant e^*$, then for any $x \in X$ with $\|x\| = 1$, $x = x_1 - x_2$ where $x_1, x_2 \in C$ and $1 = \|x_1\| + \|x_2\|$. Thus $e^*(x_1) \geqslant x^*(x_1)$ and $e^*(x_2) \geqslant -x^*(x_2)$ and it follows that $1 = e^*(x_1) + e^*(x_2) \geqslant x^*(x_1) - x^*(x_2) = x^*(x)$. ∎

§ 3. Normed Linear Lattices

Banach lattices play a large role in our development of the theory of classical Banach spaces. In this section we develop some general theory of Banach lattices and apply the results of sections 1 and 2 to obtain Banach lattices of certain types (see [197] and [241] for a more general treatment).

Throughout the text we shall mainly be concerned with Banach lattices which have the property that $\|x+y\|$ is a function of $\|x\|$ and $\|y\|$ when $|x| \wedge |y| = 0$. This is the case, for example, in $C(T, \mathbb{R})$ and $L_p(T, \Sigma, \mu, \mathbb{R})$ $(1 \leqslant p < \infty)$ since in the first case, $\|x+y\| = \max(\|x\|, \|y\|)$ and in the second, $\|x+y\| = [\|x\|^p + \|y\|^p]^{1/p}$ whenever $x \wedge y = 0$. We shall see later a Banach lattice with the above property is either a sublattice of $C(T, R)$ (for some compact Hausdorff space T) or an $L_p(T, \Sigma, \mu, \mathbb{R})$ (for some measure space (T, Σ, μ)). This necessitates some technical results on the relationship between norm and order and we develop two of them here and the rest later in the text.

Definition 1. A normed linear lattice is a *normed linear space* which is also a vector lattice such that $\|x\| \leqslant \|y\|$ whenever $|x| \leqslant |y|$. If X is complete, we call it a *Banach lattice*.

We note that this definition makes sense in both the real and complex case. However, the results of this section are given for the real case only with some comments about the complex case where appropriate.

Lemma 1. *Let X be a normed linear lattice. Then*

(1) $\||x|\|=\|x\|$ *for all* $x\in X$,

(2) *if* $x\wedge y=0$, *then* $\|x-y\|=\|x+y\|$,

(3) *the lattice operations are continuous,*

(4) *the positive cone of X is closed,*

(5) *the norm on X is regular and monotone.*

Proof. (1) is immediate.

(2) Suppose $x\wedge y=0$ and $z=x-y$. Then $z^+=x$ and $z^-=y$ so that $|x-y|=x+y$. Hence $\|x-y\|=\||x-y|\|=\|x+y\|$.

(3) By theorem 1 (11) of section one $|x-y| \geqslant |x\vee z-y\vee z| \vee |x\wedge z-y\wedge z|$. Thus if $x=\lim x_n$ and $y=\lim y_n$, then $|x_n\vee y_n-x\vee y| \leqslant |x_n\vee y_n-x\vee y_n|+|x\vee y_n-x\vee y| \leqslant |x_n-x|+|y_n-y|$ implies that $x\vee y=\lim x_n\vee y_n$ (and similarly, $x\wedge y=\lim x_n\wedge y_n$).

(4) This follows immediately from (3).

(5) Suppose $\|x\|<1$. Then $|x|\geqslant x,-x$ and $\||x|\|=\|x\|<1$ so that the norm is regular. Moreover, if $-x\leqslant y\leqslant x$, then $|y|\leqslant x$ and $\|y\|\leqslant\|x\|$ so the norm is monotone. ∎

The following theorem is due to Riesz.

Theorem 1. *Let X be a normed linear space with a positive cone C such that (X,C) has the decomposition property and the norm on X is regular and monotone. Then (X^*,C^*) is an order complete Banach lattice.*

Proof. First note that $X=C-C$ since for each $x\in X$ there is a $y\in X$ with $y\geqslant x,-x$ (the norm is regular) so that $x=(x+y)-y\in C-C$. To show that (X^*,C^*) is a vector lattice, we need only show that $(x^*)^+$ exists for all $x^*\in X^*$ since in such a case, $x^*\vee y^*=(x^*-y^*)^++y^*$. To define $(x^*)^+$ it suffices to define it on C and show that it is additive and positive homogeneous there. For this, let $(x^*)^+(x)=\sup\{x^*(y):0\leqslant y\leqslant x\}$ for $x\in C$. Since the norm is monotone in X, this supremum is indeed finite. It is easy to see that $(x^*)^+(ax)=a(x^*)^+(x)$ for all $a\in\mathbb{R}^+$ and $x\in C$. If $x_1,x_2\in C$, then by the definition of $(x^*)^+$, $(x^*)^+(x_1+x_2)\geqslant(x^*)^+(x_1)+(x^*)^+(x_2)$. On the other hand, since (X,C) has the decomposition property, if $0\leqslant y\leqslant x_1+x_2$, then $y=y_1+y_2$ with $0\leqslant y_1\leqslant x_1$ and $0\leqslant y_2\leqslant x_2$. Thus $x^*(y)=x^*(y_1)+x^*(y_2)\leqslant(x^*)^+(x_1)+(x^*)^+(x_2)$. Hence, $(x^*)^+$ has a unique linear extension to $C-C=X$ and since the norm is regular, $\|(x^*)^+\|\leqslant\|x^*\|$. For, let $y\in b_0(X)$. Then there is an $x\in b_0(X)$ with $x\geqslant y,-y$. Thus $(x^*)^+(x)\geqslant|(x^*)^+(y)|$ and it follows that $\|(x^*)^+\|=\sup\{(x^*)^+(x):x\in(b_0(X))^+\}=\sup\{x^*(y):y\in(b_0(X))^+\}\leqslant\|x^*\|$. Now clearly $(x^*)^+\geqslant x^*,0$ and if $y^*\geqslant x^*,0$, then for any $x,y\in C$ with $0\leqslant y\leqslant x$, $y^*(x)\geqslant y^*(y)\geqslant x^*(y)$. Thus $y^*\geqslant(x^*)^+$ and $(x^*)^+=x^*\vee 0$ as we need. Hence (X^*,C^*) is a vector lattice.

Suppose $x^*, y^* \in X^*$ and $-x^* \leqslant y^* \leqslant x^*$. Let $x \in b_0(X)$ and choose $y \in b_0(X)$ with $y \geqslant x, -x$. Then $x^*(x)+x^*(y)-y^*(x)-y^*(y) = (x^*-y^*)(x+y) \geqslant 0$ and $-x^*(x)+x^*(y)-y^*(x)+y^*(y)=(x^*+y^*)(y-x) \geqslant 0$. By adding these two inequalities we obtain $y^*(x) \leqslant x^*(y) \leqslant \|x^*\|$. Hence $\|y^*\| \leqslant \|x^*\|$. Thus to show that X^* is a Banach lattice we need only show that $\|x^*\| \geqslant \||x^*|\|$ for all $x^* \in X^*$. For, by the above we have that $\|x^*\| \leqslant \||x^*|\|$ and $\|y^*\| \leqslant \||x^*|\|$ if $|y^*| \leqslant |x^*|$. Let $x^* \in X^*$ and $\varepsilon > 0$ be given. For $y \in C$ let $0 \leqslant z \leqslant y$ be such that $x^*(z) \leqslant (x^*)^+(y) \leqslant x^*(z)+\varepsilon/2$. Since $|x^*| = 2(x^*)^+ - x^*$, we have that $\left||x^*|(y) - x^*(2z-y)\right| < \varepsilon$ and since $-y \leqslant 2z-y \leqslant y$, it follows that $|x^*(2z-y)| \leqslant \|x^*\| \, \|y\|$, that is, $\||x^*|\| \leqslant \|x^*\|$.

Let A be a nonempty upwards directed set in X^* which is bounded from above by x^* (if A is not upwards directed, replace A with the set A' of all finite supremums of elements of A, then $\sup A$ exists if and only if $\sup A'$ exists and in such a case the two are equal). For $y^* \in A$, let $B_{y^*} = \{z^* : y^* \leqslant z^* \leqslant x^*\}$. Since the norm on X^* is monotone and $y^* \leqslant z^* \leqslant x^*$ implies that $-(|y^*| \vee |x^*|) \leqslant z^* \leqslant |y^*| \vee |x^*|$, B_{y^*} is a norm bounded set. Since B_{y^*} is clearly weak* closed, it follows that it is weak* compact. The family $\{B_{y^*} : y^* \in A\}$ has the finite intersection property and $B = \bigcap_{y^* \in A} B_{y^*}$ is a nonempty weak* compact set. Moreover, $z^* \in B$ is and only if $y^* \leqslant z^* \leqslant x^*$ for all $y^* \in A$. In particular, B is closed with respect to taking finite supremums. For $z^* \in B$ let $A_{z^*} = \{w^* \in B : z^* \leqslant w^*\}$. Clearly $A_{z_1^*} \cap \cdots \cap A_{z_n^*} \supset A_{z_1^* \wedge \cdots \wedge z_n^*}$ for any finite set $z_1^*, \ldots, z_n^*$ in B. Since A_{z^*} is weak* compact also, $\bigcap_{z^* \in F} A_{z^*}$ is nonempty. Let w^* be in this intersection. Then since $w^* \in B$, $w^* \geqslant y^*$ for all $y^* \in A$. Moreover, if $v^* \geqslant y^*$ for all $y^* \in A$, then $v^* \wedge x^* \in B$ and it follows that $v^* \wedge x^* \geqslant w^*$ and $w^* = \sup A$. ∎

This theorem has a general converse (see, for example, [15]). We shall only need the following converse which is proved later in the text. If (X, C) is a partially ordered Banach space such that the norm on X^* is additive and 1 generating (with respect to C^*), then (X, C) has the decomposition property.

Note that by theorem 1, if X is a normed linear lattice, then X^{**} is an order complete Banach lattice. We wish to show that the natural map $J : X \to X^{**}$ ($(Jx)(x^*) = x^*(x)$ for $x \in X$ and $x^* \in X^*$) preserves the lattice operations.

Let us further note that from theorem 1 we have that
$(x^* \vee y^*)(x) = \sup\{x^*(y) + y^*(x-y) : 0 \leqslant y \leqslant x\}$ for $x \geqslant 0$ and
$(x^* \wedge y^*)(x) = \inf\{x^*(z) + y^*(x-z) : 0 \leqslant z \leqslant x\}$.

Theorem 2. *Let X be a normed linear lattice. Then the natural map $J : X \to X^{**}$ preserves the lattice operations. In particular, the completion of a normed linear lattice is a Banach lattice.*

Proof. Clearly if $x \geqslant 0$, then $Jx \geqslant 0$. Thus for any $x, y \in X$, $J(x \vee y) \geqslant (Jx) \vee (Jy)$. Suppose $x \wedge y = 0$. Then $(Jx) \vee (Jy)(x^*) = \sup\{x^*(y) + y^*(x-y): 0 \leqslant y^* \leqslant x^*\}$ for $x^* \in C^*$. By theorem 7 of section 1 there is a y^* such that $0 \leqslant y^* \leqslant x^*$, $y^*(x) = x^*(x)$, and $y^*(z) = 0$ if $x \wedge z = 0$. Let $z^* = x^* - y^*$. Then $0 \leqslant z^* \leqslant x^*$, $z^*(y) = x^*(y)$ and $z^*(x) = 0$. Thus $(Jx) \vee (Jy)(x^*) \geqslant x^*(x) + x^*(y) = x^*(x \vee y) = J(x \vee y)(x^*)$. Hence $(Jx) \vee (Jy) = J(x \vee y)$ and by lemma 2 of section 1, J preserves the lattice operations. ∎

If $\{x_d\}$ is a net of elements in a Banach lattice such that $\sup x_d = x$ exists, it is not necessarily true that $x = \lim x_d$. For example, consider the Banach lattice $l\infty(N, \mathbb{R})$ of all bounded real sequences (supremum norm and pointwise ordering). Let $x_n = (1, \ldots, 1, 0, \ldots)$ (1's for the first n coordinates and then 0's). Then $\sup x_n = (1, 1, \ldots)$, but $\{x_n\}$ is not norm Cauchy since $\|x_n - x_m\| = 1$ for $n \neq m$. We wish to investigate when this convergence is valid.

Definition 2. A normed linear lattice X is said to have *order continuous norm* if for each upwards directed family $\{x_d\}$ (with respect to the order on X) which has an upper bound, $\{x_d\}$ is Cauchy in X.

Let us note that if $\{x_d\}$ is an upwards directed (or downwards directed) convergent net, then it must converge to $\sup x_d (\inf x_d)$. For, let $x = \lim x_d$ and $x^* \geqslant 0$. Then $x^*(x) = \lim x^*(x_d) = \sup x^*(x_d)$ which implies that $x \geqslant x_d$ for all d. If $y \geqslant x_d$ for all d, then $x^*(y) \geqslant \sup x^*(x_d) = x^*(x)$ implies that $y \geqslant x$. Thus $x = \sup x_d$ (the proof of the dual case is the same).

Theorem 3. *Let (X, C) be a Banach lattice. Then the norm is order continuous in X if and only if whenever $\{x_d\}$ is a downwards directed set in C with $\inf x_d = 0$, it follows that $\inf \|x_d\| = 0$.*

Proof. Suppose the norm is order continuous. Clearly $\{-x_d\}$ is upwards directed and bounded from above by 0. Thus $\{-x_d\}$ is norm Cauchy which implies $\{x_d\}$ converges. Since $\inf x_d = 0$, $\lim x_d = 0$ and it follows that $\inf \|x_d\| = \lim \|x_d\| = 0$.

Suppose the converse. Let $\{x_d\}$ be an upwards directed set which has an upper bound. Let $\{y_b\}$ be the collection of all upper bounds for $\{x_d\}$. Then $\{y_b\}$ is downwards directed and $z_{bd} = y_b - x_d$ defines a downwards directed family of positive elements such that $\inf z_{bd} = 0$. For, suppose $0 \leqslant x \leqslant z_{bd}$ for all b and d. Then for any b, $x_d \leqslant y_b - x$ for all d so that $y_b - x$ is an upper bound of $\{x_d\}$. In particular, $y_b - nx$ is an upper bound of $\{x_d\}$ for $n = 1, 2, \ldots,$. That is, $y_b - x_d \geqslant nx$ for all n. Since this implies $\|y_b - x_d\| \geqslant n\|x\|$ for all n (b and d fixed), we obtain that $x = 0$. Thus we have that $\inf z_{bd} = 0$. Since the norm preserves infimums $\inf \|y_b - x_d\| = 0$. Let $\varepsilon > 0$ and choose b', d' so that if $y_b \leqslant y_{b'}$ and $x_d \geqslant x_{d'}$, then $\|y_b - x_d\| < \varepsilon$. Hence for $x_d \geqslant x_{d'}$, $\|x_d - x_{d'}\| \leqslant \|x_d - y_b\|$

$+\|y_b - x_{d'}\| < 2\varepsilon$ and it follows that $\{x_d\}$ is norm Cauchy. Thus by the above remarks, $\sup x_d = \lim x_d$. ∎

The norm in $L_p(T, \Sigma, \mu, \mathbb{R}) = L_p(\mu, \mathbb{R})$ is order continuous for all $1 \leqslant p < \infty$. For, first note that for $f, g \geqslant 0$ we have that $\|f+g\|^p \geqslant \|f\|^p + \|g\|^p$ (for $a, b \geqslant 0$, $(a+b)^p \geqslant a^p + b^p$ and, hence, $\int (f+g)^p d\mu \geqslant \int f^p d\mu + \int g^p d\mu$ for $f, g \geqslant 0$). If $\{f_d\}$ is an upwards directed net which has an upper bound, then $\|f_d\|^p$ is an upwards directed convergent net. If $f_d \geqslant f_{d'}$, then $\|f_d\|^p = \|f_d - f_{d'}\|^p \geqslant \|f_d - f_{d'}\|^p + \|f_d\|^p$ so that $\|f_d\|^p - \|f_{d'}\|^p \geqslant \|f_d - d_{d'}\|^p$. From this it is clear that $\{f_d\}$ is a Cauchy net and hence must converge to $\sup f_d$.

We note further that the above definition and theorem are valid for real or complex vector lattices as well as the following theorem. Also, the above remarks are valid for real $L_p(\mu, \mathbb{R})$ and complex $L_p(\mu, \mathbb{C})$ (moreover, the reader can easily see that $L_p(\mu, \mathbb{C})$ is a complex Banach lattice).

Theorem 4 (Ando). *Let X be a Banach lattice. Then the norm is order continuous in X if and only if each closed ideal in X is the range of a positive contractive projection.*

Proof. If the norm is order continuous in X, then clearly X is order complete. Moreover, any closed ideal Y in X is a band since if $x = \sup x_d$ where $\{x_d\}$ is an upwards directed family in Y, then $x = \lim x_d \in Y$. Thus by theorem 4 of section one Y is a projection band. Let P be the band projection of X onto Y. Then by theorem 4 of section one and the monotonicity of the norm it follows readily that P is contractive. For, if $x \in X$, then $|x| \geqslant P|x| \geqslant |Px|$ and, thus, $\|Px\| = \||Px|\| \leqslant \|P|x|\| \leqslant \||x|\| = \|x\|$.

Now suppose the converse. Let Y be a closed ideal in X and P be a positive contractive projection onto Y. For a fixed $x > 0$ in X and $y > 0$ in Y, $y \wedge (x - x \wedge Px) \in Y$ and $y \wedge (x - x \wedge Px) = P(y \wedge (x - x \wedge Px)) \leqslant (x - x \wedge Px) \wedge (Px - x \wedge Px) = 0$. Hence $x - x \wedge Px$ is in $Y^\perp$. Thus if $x > 0$ and $x \in Y^{\perp\perp}$, then $0 = x \wedge (x - x \wedge Px)$, that is, $x = x \wedge Px$ and $x \in Y$. Hence $Y = Y^{\perp\perp}$ and Y is a band.

Let $\mathscr{P}$ be a family of band projections in X which is upwards directed (i.e. $P_1 \leqslant P_2$ if and only if $P_1 x \leqslant P_2 x$ for all $x \geqslant 0$). We shall show that if $x = \sup Px$ $(P \in \mathscr{P})$, then $x = \lim Px$ $(P \in \mathscr{P})$. Let Y be the closed ideal generated by $\{Px : P \in \mathscr{P}\}$. Then by the above Y is a band and, in particular, $x \in Y$. Let $\{u_n\}$ be a sequence in the ideal generated by $\{Px : P \in \mathscr{P}\}$ (Y is the closure of this ideal) with $x = \lim u_n$.

Then $x^+ = \lim u_n^+$ and $x = x \wedge x = \lim x \wedge u_n^+$. Thus we may assume that $0 \leqslant u_n \leqslant x$ for all n. Since $\mathscr{P}$ is upwards directed, for each n there are $a > 1$ and $P_n \in \mathscr{P}$ such that $u_n \leqslant a P_n x$. Moreover, $u_n = u_n \wedge x$

$\leqslant (aP_n x) \wedge x \leqslant P_n x$. Thus for all $P \geqslant P_n$, $0 \leqslant x - Px \leqslant x - P_n x \leqslant x - u_n$ and $\|x - Px\| \leqslant \|x - u_n\|$ and it follows that $x = \lim Px$.

Let Z be a downwards directed subset of positive elements of X such that $\inf Z = 0$. Let $x \in Z$ and without loss of generality we assume that $y \leqslant x$ for all $y \in Z$. Let $\varepsilon > 0$ and let P_y denote the band projection of X onto $\{(\varepsilon x - y)^+\}^{\perp\perp}$. Then since $P_y x = P_y(x - (1/\varepsilon) y) + P_y((1/\varepsilon) y) \geqslant (x - (1/\varepsilon) y)^+$ for $y \in Z$, $x \geqslant \sup\{P_y x : y \in Z\} \geqslant \sup\{(x - (1/\varepsilon) y) : y \in Z\} = (x - (1/\varepsilon) \inf Z) = x$. Hence by the above, $P_y x \to x$ $(y \in Z)$ so there is a $y \in Z$ with $\|x - P_y x\| < \varepsilon$. Thus we have that $y = P_y(y) + (I - P_y)(y) \leqslant P_y(\varepsilon x) - P_y(\varepsilon x - y) + (I - P_y(x)) \leqslant \varepsilon x - (\varepsilon x - y)^+ + x - P_y(x) \leqslant \varepsilon x + x - P_y(x)$ so that $\|y\| \leqslant \varepsilon\|x\| + \varepsilon$ and we have that $\inf\{\|y\| : y \in Z\} = 0$. ∎

We now discuss two dual conditions on the norm first studied by Kakutani in [147] and [148]. For technical reasons we define these conditions in a weak form and then prove that a strengthened form is valid.

Definition 3. Let X be a real Banach lattice. Then X is said to be an *abstract M space* if $\|x + y\| = \max(\|x\|, \|y\|)$ whenever $x \wedge y = 0$.

The first thing we must prove about abstract M spaces is that $\|x \vee y\| = \max(\|x\|, \|y\|)$ for all $x, y \geqslant 0$. This condition is, in fact, Kakutani's original definition of an abstract M space (see [148]). The reason we choose the weaker one is that it expresses $\|x + y\|$ as a function of $\|x\|$ and $\|y\|$ whenever $x \wedge y = 0$. We give two proofs that the above assertion is true. The first one, due to Bohnenblust and Kakutani [50], embeds X into a concrete $C(T, \mathbb{R})$ space as a sublattice and the second one, due to Bernau [41], gives a proof using only the Banach lattice properties of X itself.

Theorem 5. *Let X be an abstract M space. Then $\|x \vee y\| = \max(\|x\|, \|y\|)$ whenever $x, y \geqslant 0$.*

Proof. The idea of Bohnenblust and Kakutani is to embed X into $C(T, \mathbb{R})$ for some compact Hausdorff space T by a linear isometric mapping ϕ which preserves the lattice operations. Thus the range of ϕ is a sublattice of $C(T, \mathbb{R})$ and clearly has the desired property and, thus, so does X.

To this end let $K = C^* \cap b(X^*)$. Then, as before, K is a weak* compact convex set (and $K \neq \{0\}$ whenever $X \neq \{0\}$). Let $x^* \in K$ with $\|x^*\| = 1$. And let y^* be defined by $y^*(z) = \sup\{x^*(z \wedge nx) : n = 1, 2, \ldots\}$ for $z \geqslant 0$. By theorem 7 of section 1, $0 \leqslant y^* \leqslant x^*$ and $y^*(x) = x$ and $y^*(z) = 0$ whenever $z \wedge x = 0$. Clearly $\|y^*\| \leqslant \|x^*\|$. Let $z^* = x^* - y^*$. Then $1 \leqslant \|y^*\| + \|z^*\|$. We wish to show that $1 = \|y^*\| + \|z^*\|$. Let $\varepsilon > 0$ be given and choose $x_1 \in X$ with $\|x_1\| = 1$ and $y^*(x_1) > \|y^*\| - \varepsilon$ (since $y^*(|x_1|) \geqslant y^*(x_1)$ we may assume that $x_1 \geqslant 0$). By definition of y^* there is an n_1 such that $x^*(x_1 \wedge n_1 x) \geqslant y^*(x_1) - \varepsilon \geqslant \|y^*\| - 2\varepsilon$. Put $x_2 = x_1 \wedge n_1 x$. Then $\|x_2\| \leqslant 1$ and $y^*(x_2) = x^*(x_2) > \|y^*\| - 2\varepsilon$ and in particular, $z^*(x_2) = 0$. Choose $y_1 \geqslant 0$ with $\|y_1\| = 1$ so that $z^*(y_1) > \|z^*\| - \varepsilon$ and choose n_2 so that $1 < \varepsilon(n_2 + 1)$. For $y_2 = (y_1 - n_2 x_2)^+$, $0 \leqslant y_2 \leqslant y_1$ and $\|y_2\| \leqslant 1$. Moreover,

$$z^*(y_2) = z^*(y_1 \vee n_2 x_2) - n_2 z^*(x_2) = z^*(y_1 \vee n_2 x_2) \geqslant z^*(y_1) > \|z^*\| - \varepsilon.$$

Now $x_2 \wedge y_2 = (x_2 \wedge (y_1 - n_2 x_2))^+$, $n_2[x_2 \wedge (y_1 - n_2 x_2)] \leqslant n_2 x_2$, and $x_2 \wedge (y_1 - n_2 x_2) \leqslant y_1 - n_2 x_2$. By adding we get $(n_2 + 1)(x_2 \wedge (y_1 - n_2 x_2)) \leqslant y_1$ and $0 \leqslant x_2 \wedge y_2 \leqslant (y_1/n_2 + 1) \leqslant \varepsilon y_1$. Put $x_3 = x_2 - x_2 \wedge y_2$ and $y_3 = y_2 - x_2 \wedge y_2$. Then $x_3 \wedge y_3 = 0$ and $\|x_3\|, \|y_3\| \leqslant 1$. Since $0 \leqslant z^*(x_2 \wedge y_2) \leqslant z^*(x_2) = 0$, we get $z^*(x_3) = 0$ and $z^*(y_3) = z^*(y_2) > \|z^*\| - \varepsilon$. Also, $y^*(x_3) = y^*(x_2) - y^*(x_2 \wedge y_2) \geqslant \|y^*\| - 2\varepsilon$ and $y^*(y_3) = y^*(y_2) - y^*(x_2 \wedge y_2) \geqslant -\varepsilon$. Thus $1 \geqslant x^*(x_3 + y_3) = y^*(x_3) + y^*(y_3) + z^*(x_3) + z^*(y_3) \geqslant \|y^*\| + \|z^*\| - 4\varepsilon$ and it follows that $1 \geqslant \|y^*\| + \|z^*\|$.

Let $T_0 = \operatorname{ext} K \setminus \{0\}$. By the above, if $x^* \in T_0$ and $x \wedge y = 0$, then $x^*(x) x^*(y) = 0$. If this is not true, then by the above $x^* = y^* + z^*$ where $0 \leqslant y^* \leqslant x^*$, $y^*(x) = x^*(x) \neq 0$ and $x^*(y) = z^*(y) \neq 0$ and $1 = \|y^*\| + \|z^*\|$. Thus $x^* = \|y^*\| (y^*/\|y^*\|) + \|z^*\| (z^*/\|z^*\|)$ which contradicts the fact that x^* is an extreme point of K.

Let T be the weak* closure of T_0. Clearly the elements of T also have the property that $x \wedge y = 0$ implies $x^*(x) x^*(y) = 0$. The operator $\phi: X \to C(T, \mathbb{R})$ is defined by $\phi(x)(x^*) = x^*(x)$ for all $x^* \in T$ and $x \in X$. Clearly ϕ is linear. Moreover, $x \wedge y = 0$ implies that $\phi(x \vee y)(x^*) = x^*(x + y) = \phi(x) x^* + \phi(y) x^* = (\phi(x) \vee \phi(y))(x^*)$ (since $\phi(x) \wedge \phi(y) = 0$). Thus by lemma 2 of section 1 ϕ preserves the lattice operations. In particular, $\phi(x^+) = \phi(x)^+$ and $\phi(x^-) = \phi(x)^-$ so that $\|x\| = \max(\|x^+\|, \|x^-\|) = \max(\|\phi(x^+)\|, \|\phi(x^-)\|) = \max(\|\phi(x)^+\|, \|\phi(x)^-\|) = \|\phi(x)\|$ (to see that $\|\phi(z)\| = \|z\|$ when $z \geqslant 0$ note that for any $x^* \in b(X^*)$, $|x^*| \in K$ and $|x^*| \geqslant x^*$ so that $|x^*|(z) \geqslant x^*(z)$ and $\|z\| = \sup\{x^*(z): x^* \in b(X^*)\} = \sup\{x^*(z): z \in K\} = \sup\{x^*(z): x^* \in T_0\} = \sup\{x^*(z) = x^* \in T\}$). Thus, if $x, y \geqslant 0$, then

$$\|x \vee y\| = \|\phi(x) \vee \phi(y)\| = \max(\|\phi(x)\|, \|\phi(y)\|) = \max(\|x\|, \|y\|). \quad \blacksquare$$

Corollary. *A Banach lattice is an abstract M space if and only if it is linearly isometric and lattice isomorphic to a closed sublattice of $C(T, \mathbb{R})$ for some compact Hausdorff space T.* ∎

For the proof due to Bernau we need some notation and two lemmas. For $x, y \geqslant 0$ and $n = 1, 2, \ldots$, let $u(n) = \{(n+1)[(n-1)(x \vee y) - n(x \wedge y)]^+\} \wedge (x \vee y)$ and $v(n) = \{(n+1)[(n-1)(x \vee y) - n(x \wedge y)]^-\} \wedge (x \wedge y)$.

Lemma 2. $v(n) + u(n+1) \geqslant (n/n+1)(x \vee y)$.

Proof. First observe that $v(n)+u(n+1)=\{(n+1)[(n-1)(x\vee y)-n(x\wedge y)]^- + u(n+1)\}\wedge\{x\wedge y+u(n+1)\}$. Now

$(n+1)[(n-1)(x\vee y)-n(x\wedge y)]^- + u(n+1)$
$=\{(n+1)[(n-1)(x\vee y)-n(x\wedge y)]^- + (n+2)[n(x\vee y)-(n+1)(x\wedge y)]^+\}$
$\wedge (x\vee y)$; $(n+1)[(n-1)(x\vee y)-n(x\wedge y)]^- + u(n+1)$
$\geqslant\{(n+1)[(n-1)(x\vee y)-n(x\wedge y)]^- + (n+2)[n(x\vee y)-(n+1)(x\wedge y)]^+\}$
$\wedge (x\vee y)$; and $(n+1)[(n-1)(x\vee y)-n(x\wedge y)]^-$
$=[n^2(x\vee y)-(n^2+n)(x\wedge y)-x\vee y]^- \geqslant x\vee y - n[n(x\vee y)$
$-(n+1)(x\wedge y)]^+$.

Hence $(n+1)[(n-1)(x\vee y)-n(x\wedge y)]^- + u(n+1)\geqslant x\vee y$. Moreover, $x\wedge y=(n/n+1)(x\vee y)-(1/n+1)[n(x\vee y)-(n+1)(x\wedge y)]$ and $u(n+1)\geqslant\{(1/n+1)[n(x\vee y)-(n+1)(x\wedge y)]^+\}\wedge(x\vee y)=(1/n+1)[n(x\vee y)-(n+1)(x\wedge y)]^+$. Thus $x\wedge y+u(n+1)\geqslant(n/n+1)(x\vee y)$ and $v(n)+u(n+1)\geqslant(n/n+1)(x\vee y)$. ∎

Lemma 3. $\|u(n)+v(n)\|\leqslant\max(\|x\|,\|y\|)$.

Proof. $[(n-1)(x\vee y)-n(x\wedge y)]^+=[n(x\vee y-x\wedge y)-x\vee y]^+$
$=[n(x-x\wedge y)\vee(y-x\wedge y)-x\vee y]^+=[n(x-x\wedge y)-x\vee y]^+$
$\vee[n(y-x\wedge y)-x\vee y]^+$. Thus $u(n)=[(n+1)(n(x-x\wedge y)-x\vee y)^+ \wedge(x\vee y)]\vee[(n+1)(n(y-x\wedge y)-x\vee y)^+\wedge(x\vee y)]$. Moreover, $\{(n+1)[n(x-x\wedge y)-x\vee y]^+\}\wedge(x\vee y)\leqslant[(n+1)n(x-x\wedge y)^+]$
$\wedge[x+(y-x\wedge y)]\leqslant[n(n+1)(x-x\wedge y)]\wedge x+n(n+1)(x-x\wedge y)$
$\wedge(y-x\wedge y)\leqslant x+0=x$.

Similarly, $\{(n+1)[n(y-x\wedge y)-x\vee y]^+\}\wedge(x\vee y)\leqslant y$. Since $[n(x-x\wedge y)-x\vee y]^+\wedge[n(y-x\wedge y)-x\vee y]^+\leqslant n(x-x\wedge y)\wedge(y-x\wedge y)=0$, it follows that $\|u(n)\|\leqslant\max\{\|x\|,\|y\|\}$ and since $0\leqslant v(n)\leqslant x\wedge y$, it follows that $\|v(n)\|\leqslant\max\{\|x\|,\|y\|\}$. Since $u(n)\wedge v(n)=0$, we have that $\|u(n)+v(n)\|=\max\{\|u(n)\|,\|v(n)\|\}\leqslant\max\{\|x\|,\|y\|\}$. ∎

By lemma 2, $z(n)=v(n)+u(n+1)+\cdots+(v(2n)+u(2n+1)]\geqslant(n/n+1)(x\vee y)+\cdots+(2n/2n+1)(x\vee y)\geqslant n(x\vee y)$ and by lemma 3, $\|z(n)\|\leqslant\|v(n)\|+\|u(n+1)+v(n+1)\|+\cdots+\|u(2n)+v(2n)\|+\|u(2n+1)\|\leqslant(n+2)\max\{\|x\|,\|y\|\}$. Thus $\|x\vee y\|\leqslant(1/n)\|z(n)\|\leqslant(n+2/n)\max\{\|x\|,\|y\|\}$ and by taking limits we get that $\|x\vee y\|\leqslant\max\{\|x\|,\|y\|\}$. ∎

Definition 4. Let X be a real Banach lattice. Then X is said to be an *abstract L_1 space* if $\|x+y\|=\|x\|+\|y\|$ whenever $x\wedge y=0$.

Kakutani's original definition is simply that the norm is additive on X [147]. We shall prove that this weaker condition implies additivity of the norm below. Before we do, let us note that if X is an abstract M space, then the norm is additive in X^*. For, clearly by the above $b_0(X)$

is directed upwards and theorem 2 of section 2 applies. Moreover, since the positive cone of X is 1 normal, X is identified with $A_0(K,\mathbb{R})$ $(K=b(X^*)\cap C^*)$ by theorem 5 of section two.

Theorem 6. *Let X be an abstract L_1 space. Then X^* is an abstract M space with strong order unit and strong order unit norm. In particular, the norm on X^{**} is additive and, hence, the norm on X is additive.*

Proof. Let x^*, y^* be in X^* with $x^* \wedge y^* = 0$. Clearly $\|x^*+y^*\| \geqslant \max(\|x^*\|,\|y^*\|)$ by the monotonicity of the norm. Let $\varepsilon>0$ be given. There is an $x \geqslant 0$ in X with $\|x\|=1$ such that $\|x^*+y^*\| \leqslant (x^*+y^*)(x)+\varepsilon$. Since $0=(x^*\wedge y^*)(x)=\inf\{x^*(y)+y^*(x-y): 0\leqslant y\leqslant x\}$, there is a $0\leqslant y\leqslant x$ with $x^*(y)+y^*(x-y)<\varepsilon$. Thus $(x^*+y^*)(x)=x^*(x)-x^*(y)+x^*(y)+y^*(x)-y^*(y)+y^*(y)\leqslant x^*(x-y)+y^*(y)+\varepsilon\leqslant x^*[(x-y)-(x-y)\wedge y]+y^*[y-(x-y)\wedge y]+2\varepsilon\leqslant \|x^*\|\,\|(x-y)-(x-y)\wedge y\|+\|y^*\|\,\|y-(x-y)\wedge y)\|+2\varepsilon\leqslant\max(\|x^*\|,\|y^*\|)[\|(x-y)-(x-y)\wedge y\|+\|y-(x-y)\wedge y\|]+2\varepsilon$. Now $[(x-y)-(x-y)\wedge y]\wedge[y-(x-y)\wedge y]=0$ and, thus, $\|(x-y)-(x-y)\wedge y\|+\|(y-(x-y)\wedge y)\|=\|x-y-(x-y)\wedge y+y-(x-y)\wedge y\|=\|x-2(x-y)\wedge y\|\leqslant\|x\|$. Hence we have that $\|x^*+y^*\|\leqslant\max(\|x^*\|,\|y^*\|)$.

From this it follows that X^* is an abstract M space and by the remarks preceeding the theorem, the norm is additive in X^{**}. Thus by theorem 2, the norm is additive in X and since the norm in X is clearly one generating $(x=x^+-x^-$ and $\|x\|=\|\,|x|\,\|=\|x^++x^-\|=\|x^+\|+\|x^-\|)$, by theorem 7 of section two X^* has a strong order unit and strong order unit norm. ∎

A summary of the above results is given in the next theorem.

Theorem 7. *Let X be a Banach lattice.*

(a) *X is an abstract M space if and only if X^* is an abstract L_1 space.*

(b) *X is an abstract L_1 space if and only if X^* is an abstract M space.* ∎

Let Γ be a nonempty set. A real or complex valued function f on Γ is said to be *p summable* $(1\leqslant p<\infty)$ if $\sup\{|f(\gamma)|^p: F\subset\Gamma$ is a finite set$\}<\infty$. By $l_p(\Gamma,\mathbb{R})$ (resp. $l_p(\Gamma,\mathbb{C})$) we mean the Banach lattice of all real valued (resp. complex valued) p summable functions on Γ. The norm is the pth root of the above supremum. Clearly $f\in l_p(\Gamma,\mathbb{R})$ (resp. $l_p(\Gamma,\mathbb{C})$) if and only if there is a countable set $\{\gamma_n\}\subset\Gamma$ such that $f(\gamma)=0$ if $\gamma\notin\{\gamma_n\}$ and $\Sigma|f(\gamma_n)|^p<\infty$. The Banach lattice $c_0(\Gamma,\mathbb{R})$ (resp. $c_0(\Gamma,\mathbb{C})$) is the collection of all real valued (resp. complex valued) functions f on Γ such that for each $\varepsilon>0$, $\{\gamma: |f(\gamma)|\geqslant\varepsilon\}$ is finite. Clearly $f\in c_0(\Gamma,\mathbb{R})$ (resp. $c_0(\Gamma,\mathbb{C})$) if and only if there is a countable set $\{\gamma_n\}\subset\Gamma$ such that $f(\gamma)=0$ if $\gamma\notin\{\gamma_n\}$ and $\lim f(\gamma_n)=0$. The norm on $c_0(\Gamma,\mathbb{R})$ is the supremum norm. By $l\infty(\Gamma,\mathbb{R})$ (resp. $l\infty(\Gamma,\mathbb{C})$) we mean the Banach lattice of all bounded real valued (resp. complex valued) functions on Γ under the supremum norm. The ordering taken in each case is the natural pointwise ordering.

Exercises. 1. Let X be a real or complex Banach lattice in which the norm is order continuous and let $u \in X^+$. Show that $u^{\perp\perp}$ is the norm closure of $\{x \in X : |x| \leqslant nu$ for some $n = 1, 2, \ldots\}$. Conclude that the linear span of $\{y : y \wedge (u-y) = 0\}$ is dense in $u^{\perp\perp}$.

2. Prove that the norm is order continuous in an abstract L_1 space and in $c_0(\Gamma, \mathbb{R})$ for any Γ.

3. Let X be a Banach lattice and $x^* \in X^*$. Show that for any $x \geqslant 0$, $x^*(x) = \sup\{|x^*(y)| : |y| \leqslant x\}$.

The following exercises are stated for real spaces. They are also (in general) true for complex spaces.

4. For $(1/p)+(1/q)=1$, $l_p(\Gamma, \mathbb{R})^*$ is linearly isometric and order isomorphic to $l_q(\Gamma, \mathbb{R})$ $(1 \leqslant p < \infty)$ and $c_0(\Gamma, \mathbb{R})^*$ is linearly isometric and order isomorphic to $l_1(\Gamma, \mathbb{R})$.

5. Prove that $l\infty(\Gamma, \mathbb{R})$ is order complete but the norm is not order continuous if Γ is infinite.

6. Prove that any closed sublattice of $l_p(\Gamma, \mathbb{R})$ or $c_0(\Gamma, \mathbb{R})$ $(1 \leqslant p < \infty)$ is linearly isometric to $l_p(\Delta, \mathbb{R})$ or $c_0(\Delta, \mathbb{R})$ respectively for some set Δ.

7. Find an infinite dimensional closed sublattice of $l\infty(\Gamma, \mathbb{R})$ not linearly isometric and order isomorphic to $l\infty(\Gamma, \mathbb{R})$ when Γ is countably infinite.

8. Prove that there is a closed sublattice of $L_1([0,1], \mathbb{R})$ linearly isometric and order isomorphic to $l_1(\mathbb{N}, \mathbb{R})$, where $\mathbb{N}$ is the set of positive integers.

9. Let $\{r_n\}$ be an enumeration of the rationals in $[0,1]$ and let $T = \{(r_n, 1/m) : m \geqslant n\} \cup [0,1]$ with the induced topology from the plane. Let $X = \{f \in C(T, R) : f \mid [0,1]$ is affine$\}$. Show that X is a partially ordered Banach space with the decomposition property and strong order unit norm, but that X is not a vector lattice. Let $K = \{x^* \in X^* : x^*(1) = 1 = \|x^*\|\}$. Show that $\operatorname{ext} K$ is countable and that the weak* closure of $\operatorname{ext} K$ is homomorphic to T. Prove that $X_0 = \{f \in X : f(0) = 0\}$ is an abstract M space.

10. Let $c(\mathbb{N}, \mathbb{R})$ denote the Banach lattice of all real valued convergent sequences. Show that for any positive sequence of norm one $\{b_n\}$ in $l_1(\mathbb{N}, \mathbb{R})$, $X = \{\{a_n\} : \Sigma a_n b_n = \lim a_n\}$ is a partially ordered Banach space with the decomposition property and strong order unit norm. When is X a vector lattice (i.e. an abstract M space)?

Chapter 2

Some Aspects of Topology and Regular Borel Measures

We assume a basic knowledge of general topology and integration theory (as found, for example, in [91], [104], [130], and [253]). The purpose of this chapter is to present some special results which are not necessarily found in general references. In section 4 we prove an interpolation theorem and investigate when compact Hausdorff spaces can be mapped continously onto the closed unit interval $[0,1]$. A brief development of dispersed spaces and their relationship to spaces of ordinal numbers is given in section 5. Section 6 is devoted to a study of the Cantor set and section 7 is concerned with extremally disconnected compact Hausdorff spaces and their role as projectives (in the category of compact Hausdorff spaces and continuous maps). In section eight we briefly develop the theory of regular Borel measures and prove representation theorems for $C(T,\mathbb{R})^*$ (and $C(T,\mathbb{C})^*$).

§ 4. Existence Theorems for Continuous Functions

We prove theorems concerning the existence of continuous functions which will be used throughout the text. The first theorem is an interpolation theorem due to Tong [270] from which we can derive the Tietze extension theorem and Urysohn's lemma.

Theorem 1 (Tong). *Let T be a normal topological space, $u\colon T\to\mathbb{R}$ be a bounded upper semicontinuous function and $l\colon T\to\mathbb{R}$ be a bounded lower semicontinuous function and suppose $u\leqslant l$. Then there is a continuous function $f\colon T\to\mathbb{R}$ such that $u\leqslant f\leqslant l$.*

We shall need the following lemma. Recall that an $\mathscr{F}_\sigma$ set is a countable union of closed sets and a G_δ set is a countable intersection of open sets. $\operatorname{Int}(A)$ is the *interior* of A.

Lemma 1. *Let T be a normal topological space and let A,B be subsets of T such that A is an $\mathscr{F}_\sigma$ set, B is a G_δ set, $\bar{A}\subset B$ and $A\subset\operatorname{int}(B)$. Then there is an open $\mathscr{F}_\sigma$ set W such that $A\subset W\subset\bar{W}\subset B$.*

Proof. Let $A=\bigcup_{n=1}^{\infty} A_n$ where each A_n is closed and $B=\bigcap_{n=1}^{\infty} B_n$ where each B_n is open. Since $A_1\subset \operatorname{int}(B)$, there is an open set U_1 with $A_1\subset U_1\subset \bar{U}_1\subset \operatorname{int}(B)$. Now $\bar{A}\cup \bar{U}_1\subset B_1$ and there is an open set V_1 with $\bar{A}\cup \bar{U}_1\subset V_1\subset \bar{V}_1\subset B_1$. Now suppose $U_0,\dots,U_n$ and $V_0,\dots,V_n$ have been chosen so that $A_{k+1}\cup \bar{U}_k\subset U_{k+1}\subset \bar{U}_{k+1}\subset \operatorname{int}(B)\cap V_k$ for $k=0,\dots,n-1$ (where $V_0=T$ and $U_0=\emptyset$) and $\bar{A}\cup \bar{U}_k\subset V_k\subset \bar{V}_k\subset B_k\cap V_{k-1}$ for $k=1,\dots,n$. Then, in particular, $\bar{U}_n\cup A_{n+1}\subset \operatorname{int}(B)$, $A_{n+1}\subset \bar{A}\subset V_n$ and $\bar{U}_n\subset V_n$. Thus $\bar{U}_n\cup A_{n+1}\subset \operatorname{int}(B)\cap V_n$ and there is an open set U_{n+1} such that $\bar{U}_n\cup A_{n+1}\subset U_{n+1}\subset \bar{U}_{n+1}\subset \operatorname{int}(B)\cap V_n$. Thus $\bar{A}\cup \bar{U}_{n+1}\subset B_{n+1}\cap V_n$ and there is an open set V_{n+1} such that $\bar{A}\cup \bar{U}_{n+1}\subset V_{n+1}\subset \bar{V}_{n+1}\subset B_{n+1}\cap V_n$.

By induction there are sequences $\{U_n\}$ and $\{V_n\}$ of open sets such that (1) $A_n\subset U_n\subset \bar{U}_n\subset V_n$; (2) $B_n\supset \bar{V}_n\supset V_{n+1}$; and (3) $\bar{U}_n\subset U_{n+1}$ for all n. Let $W=\bigcup_{n=1}^{\infty} \bar{U}_n$. Then $A\subset W$, W is an $\mathscr{F}_\sigma$ set and $\bar{W}\subset \bigcap_{n=1}^{\infty} \bar{V}_n\subset B$. ∎

Proof of Theorem 1. For $a\in \mathbb{R}$, let $L(a)=\{t\in T: l(t)<a\}$ and $U(a)=\{t\in T: u(t)\leqslant a\}$. Then $L(a)$ is an $\mathscr{F}_\sigma$ set, $U(a)$ is a G_δ set, $\overline{L(a)}\subset U(a)$, $L(a)\subset \operatorname{int} U(a)$, and if $a\leqslant b$, then $L(a)\subset L(b)$ and $U(a)\subset U(b)$.

Now let $\{r_n\}$ be an enumeration of the rationals. We shall construct a function W from the rationals to the open sets in T such that if $a<b$, then $\overline{W(a)}\subset W(b)$ and $L(a)\subset W(a)\subset U(a)$.

Let $W(r_1)$ be an open set such that $L(r_1)\subset W(r_1)\subset \overline{W(r_1)}\subset U(r_1)$ and suppose $W(r_1),\dots,W(r_n)$ have been chosen so that if $r_i<r_j$, then $\overline{W(r_i)}\subset W(r_j)$ and $L(r_i)\subset W(r_i)\subset \overline{W(r_i)}\subset U(r_i)$ for $i,j=1,\dots,n$.

Set $A=L(r_{n+1})\cup \bigcup\{\overline{W(r_i)}: 1\leqslant i\leqslant n \text{ and } r_i<r_{n+1}\}$ and $B=U(r_{n+1})\cap \bigcap\{W(r_i): 1\leqslant i\leqslant n \text{ and } r_i>r_{n+1}\}$. Then $\bar{A}\subset \{t: l(t)\leqslant r_{n+1}\}\cup \bigcup\{\overline{W(r_i)}: 1\leqslant i\leqslant n \text{ and } r_i<r_{n+1}\}\subset B$ and $\operatorname{int}(B)\supset \{t: u(t)<r_{n+1}\}\cap \bigcap\{W(r_i): 1\leqslant i\leqslant n \text{ and } r_i>r_{n+1}\}\supset A$. Let $W(r_{n+1})$ be an open set such that $A\subset W(r_{n+1})\subset \overline{W(r_{n+1})}\subset B$.

If we define $f: T\to R$ by $f(t)=\inf\{r_i: t\in W(r_i)\}$ for all $t\in T$, then it is easy to check that f is continuous and $u\leqslant f\leqslant l$. ∎

Corollary 1 (Urysohn's lemma). *Let T be a normal topological space, F_1, F_2 two closed nonempty disjoint sets in T. Then there is a continuous map f of T into $[0,1]$ such that $f=0$ on F_1 and $f=1$ on F_2.* ∎

Corollary 2 (Tietze Extension theorem). *Let T be a normal topological space and F a closed nonempty set in T. If $f: F\to R$ is a continuous function, then there is a continuous function $g: T\to R$ such that $g|F=f$ and if f is bounded, so is g and $\sup_{t\in T}|f(t)|=\sup_{t\in T}|g(t)|$.* ∎

The above proof was shown to the author by G.L. Seever. We leave the proof of the two corollaries to the reader.

The next few theorems are concerned with existence of continuous functions whose range is all of $[0,1]$.

Lemma 2. *Let S and T be compact Hausdorff spaces and f a continuous onto map from S to T. Then there is a minimal closed set $F \subset S$ such that $f(F)=T$.*

Proof. Let $\mathscr{K}$ be the class of all closed $K \subset S$ such that $f(K)=T$. Then $\mathscr{K}$ is a partially ordered set under subset inclusion. If $\mathscr{C}$ is a chain (a totally ordered subset) in $\mathscr{K}$, then $\mathscr{C}$ has the finite intersection property and hence $\emptyset \neq \bigcap \mathscr{C} = K_0$. Let $t \in T$ and for each $K \in \mathscr{C}$ let $s_K \in K$ with $f(s_K)=t$. Then $\{s_K\}$ is a net and has a subset which converges to some $s \in K_0$. Clearly $f(s)=t$ and it follows that K_0 is a lower bound to $\mathscr{C}$. Thus by Zorn's lemma, $\mathscr{K}$ has a minimal element. ▮

A *perfect set* P in a topological space T is a nonempty closed set each point of which is a limit point of P.

Lemma 3. *Let S and T be compact Hausdorff spaces and suppose T contains a perfect set. If $f: S \to T$ is a continuous onto map, then S has a perfect set.*

Proof. Let $P \subset T$ be a perfect set. By lemma 2 there is a minimal closed set $F \subset X$ such that $f(F)=P$. It is easy to check that F is perfect. ▮

Lemma 4. *Let T be a perfect compact Hausdorff space. If F is a closed set in X and U is an open set in X with $F \subsetneqq U$, then there is an open set V with $F \subsetneqq V \subset \overline{V} \subsetneqq U$.*

Proof. Suppose $F \subsetneqq U$. Let t_1, t_2 be two distinct points in $U \setminus F$ and for each $t \in F$ choose an open set V_t containing t, t_2 with $V_t \subset \overline{V}_t \subset U \setminus \{t_1\}$. Then there are finitely many $V_{t_3}, \ldots, V_{t_n}$ which cover F and $V = V_{t_3} \cup \cdots \cup V_{t_n}$ will suffice. ▮

Lemma 5. *Let T be a compact Hausdorff space, D a dense set in $(0,1)$, $\{U_r\}$ $(r \in D)$ a family of open sets in T such that if r, s are in D with $r < s$, then $\overline{U}_r \subsetneqq U_s$. Then there is a continuous function f from T onto $[0,1]$.*

Proof. Let $f(t) = \sup\{r \in D; t \notin U_r\}$ for $t \in T$ (where $\sup \emptyset = 0$). ▮

The following theorem was proved in [235] (in a different manner).

Theorem 2. *Let T be a compact Hausdorff space. Then T has a perfect set if and only if there is a continuous map of T onto $[0,1]$.*

Proof. The necessity is lemma 3. By the Tietze Extension Theorem it suffices to assume that T is perfect for the sufficiency. The result then follows from lemmas 4 and 5 using the dyadic rationals as D. ▮

The reader should note that the proof of lemma 5 is just a modification of the usual proof of Urysohn's lemma (see, for example [131]). The set of *dyadic rationals* is the set of all rationals numbers of the form $m/2^n$ where $1 \leqslant m < 2^n$ and $n = 1, 2, 3, \ldots$.

§ 5. Dispersed Compact Hausdorff Spaces

A compact Hausdorff space T is said to be *dispersed* if it does not contain any perfect sets (recall that perfect sets are defined to be nonempty). We shall briefly develop some general results about such spaces and their relationship to spaces of ordinal numbers. In the development we shall use without explanation the language of ordinal numbers (the reader can check [256] for the relevant terminology).

The first lemma and theorem follows immediately from results in section four.

Lemma 1. *Let S, T be compact Hausdorff spaces and suppose $f: S \to T$ is a continuous onto function. If S is dispersed, then so is T.*

Proof. By lemma 3 of section 4, if T contains a perfect set, then so does S. ∎

Theorem 1. *Let T be a compact Hausdorff space. Then T is dispersed if and only if every $f \in C(T, \mathbb{C})$ has countable range.*

Proof. First note that a compact set $D \subset \mathbb{C}$ is dispersed if and only if it is countable. For, by theorem 2 of section 4, every countable space is dispersed. If D is dispersed, then it is countable since, otherwise, the set of condensation points in D is a nonempty perfect set (recall that x is a *condensation point* of D if and only if each neighborhood of x intersects D in an uncountable set).

If T is dispersed, then so is $f(T)$ for all $f \in C(T, \mathbb{C})$, and hence f has countable range. On the other hand, if T is not dispersed, then there is a continuous function f from T onto $[0, 1]$. ∎

If ξ is an ordinal number, then $\Gamma(\xi)$ denotes the set of all ordinals $\leqslant \xi$. For a, b in $\Gamma(\xi)$ let $L(a) = \{\eta : \eta < a\}$ and $U(b) = \{\eta \in \Gamma(\xi) : \eta > b\}$. Then the collection of all $L(a)$'s and $U(b)$'s forms a subbase for the *order topology* on $\Gamma(\xi)$. It is easily seen that $\Gamma(\xi)$ is compact and Hausdorff in this topology since every set in $\Gamma(\xi)$ has both a supremum and an infimum in $\Gamma(\xi)$.

The *ordinal rays* $\Gamma(\xi)$ play a significant role in the theory of dispersed spaces as shall be seen below. We shall adopt the following notation. If T is a compact Hausdorff space, the *derivative* of a set $A \subset T$ is defined to be the set of limit points of A. Thus a transfinite inductive sequence is defined as follows: $T^{(0)} = T, T^{(1)}$ is the derivative of $T, T^{(2)}$ is the derivative of $T^{(1)}$, in general suppose $T^{(\alpha)}$ has been defined for all ordinals $\alpha<\beta$, if $\beta=\gamma+1$, then $T^{(\beta)}$ is the derivative of $T^{(\gamma)}$, otherwise $T^{(\beta)}=\bigcap_{\alpha<\beta} T^{(\alpha)}$.

Lemma 2. *Let T be an infinite dispersed compact Hausdorff space. Then there is a nonlimit ordinal* $\beta=\gamma+1$ *such that* $T^{(\beta)}=\emptyset$, $T^{(\gamma)}$ *is finite, and for* $\xi<\eta\leqslant\beta$, $T^{(\xi)}$ *is a proper subset of* $T^{(\eta)}$.

Proof. Since T is dispersed, $T^{(0)}\supsetneqq T^{(1)}\supsetneqq T^{(2)}\supsetneqq\cdots\supsetneqq T^{(\xi)}\supsetneqq T^{(\eta)}\supsetneqq\cdots$ for all $\xi<\eta$. By the finite intersection property if $T^{(\alpha)}\neq\emptyset$ for all $\alpha<\beta$ and β is a limit ordinal, then $T^{(\beta)}\neq\emptyset$. Since the above inclusions are strict, $T^{(\beta)}=\emptyset$ for some β (clearly β is less than or equal to the initial ordinal whose cardinality is that of T), which is the smallest such ordinal. Then since β is not a limit ordinal, $\beta=\gamma+1$ and since $T^{(\gamma)}$ is compact and $T^{(\gamma+1)}$ is empty, $T^{(\gamma)}$ is finite. ∎

Note that if T is a compact Hausdorff space with a perfect set P, then $T^{(\alpha)}\supset P$ for all ordinals α. Hence a compact Hausdorff space T is dispersed if and only if $T^{(\beta)}=\emptyset$ for some β. From this it is easy to see that $\Gamma(\xi)$ is dispersed since $\Gamma(\xi)^{(\xi+1)}=\emptyset$ for all ξ.

Let T be a topological space. If $T^{(\xi)}$ is finite and contains exactly n points, the pair (ξ,n) is called the *characteristic system* of T. The set of ordinals $<\xi$ with the order topology will be denoted by $\Gamma_0(\xi)$. Hence $\Gamma(\xi)=\Gamma_0(\xi+1)$.

A characterization, due to Baker [23], of dispersed spaces which are of the type $\Gamma(\xi)$ is now given. It will follow from the lemmas and corollaries below. As usual, we use ω to denote the ordinal whose order type is that of the set of positive integers.

Lemma 3. *For every ordinal* λ, $\Gamma(\omega^\lambda)^{(\lambda)}=\{\omega^\lambda\}$. *In particular, if n is a natural number, then* card $\Gamma(\xi)^{(\lambda)}\geqslant n$ *if and only if* $\xi\geqslant\omega^\lambda\cdot n$.

Proof. This is clear for $\lambda=1$. Suppose the lemma is true for all $\lambda<\gamma$. If γ is not a limit ordinal, then $[\omega^{\gamma-1}\cdot(n-1),\,\omega^{\gamma-1}\cdot n]^{(\gamma-1)}=\{\omega^{\gamma-1}\cdot n\}$ and $\Gamma(\omega^\gamma)=\left(\bigcup_{n=1}^{\infty}[\omega^{\gamma-1}\cdot(n-1),\,\omega^{\gamma-1}\cdot n]\cup\{\omega^\gamma\}\right)$. Therefore $\Gamma(\omega^\gamma)^{(\gamma-1)}$ $=\bigcup_{n=1}^{\infty}\{\omega^{\gamma-1}\cdot n\}\cup\{\omega^\gamma\}$ and $\Gamma(\omega^\gamma)^{(\gamma)}=\{\omega^\gamma\}$. If γ is a limit ordinal, and $\mu<\omega^\gamma$, then $\mu\in\Gamma(\omega^\alpha)$ for some $\alpha<\gamma$. Now $\Gamma(\omega^\alpha)$ is open in $\Gamma(\omega^\gamma)$ and $\Gamma(\omega^\alpha)^{(\alpha+1)}=\emptyset$ and since $\alpha+1<\gamma$, $\mu\notin\Gamma(\omega^\gamma)^{(\gamma)}$. Therefore $\Gamma(\omega^\gamma)^{(\gamma)}$ $\subset\{\omega^\gamma\}$. Since $\Gamma(\omega^\gamma)^{(\alpha)}\neq\emptyset$ for all $\alpha<\gamma$, it follows that $\Gamma(\omega^\gamma)^{(\gamma)}\neq\phi$ and, hence, $\Gamma(\omega^\gamma)^{(\gamma)}=\{\omega^\gamma\}$. ∎

In the lemma below, γ is an ordinal number, $\{U_\xi\}_{\xi<\gamma}$ is a (transfinite) sequence of sets such that if $\xi<\eta$, then $U_\xi\supset U_\eta$.

Lemma 4. *Suppose T is a totally disconnected compact space and $t \in T$. If t has a neighborhood base consisting of a decreasing sequence $\{U_\xi\}_{\xi<\gamma}$ of sets, then $\{U_\xi\}_{\xi<\gamma}$ can also be selected with each U_ξ closed and open.*

Proof. Clearly, it may be assumed that t is not an isolated point so that γ is a limit ordinal. Using transfinite induction, a family $\{W_\xi\}_{\xi<\gamma}$ is selected such that if $\tau<\gamma$, then

(1) For each $\alpha<\tau$, either $W_\alpha=U_\xi$ for some $\xi\geqslant\alpha$ or $W_\alpha=W_\tau=\{t\}$

(2) If $\alpha<\beta<\tau$ and $W_\alpha\neq\{t\}$, then $\overline{W}_\beta\subset\operatorname{int}(W_\alpha)$.

First, let $W_1=U_1$. Suppose that $\tau<\gamma$ and W_α has been selected for all $\alpha<\tau$. Let $S=\{\xi<\gamma: U_\xi\subset W_\alpha \text{ for all } \alpha<\tau\}$. If S is empty, define $W_\tau=\{t\}$. If S is nonempty, let μ be its first ordinal number. Then $U_\mu\subset W_\alpha$ for all $\alpha<\tau$. There exists $\sigma>\tau$ such that $\overline{U}_\sigma\subset\operatorname{int}(U_\mu)$. If W_τ is defined by $W_\tau=U_\sigma$, it is easy to see that $\{W_\xi\}_{\xi<\tau}$ satisfies (1) and (2).

Let $L=\{\alpha:\alpha=\gamma \text{ or } W_\alpha=\{t\}\}$ and λ be the first element of L. Then λ is a limit ordinal and $\{W_\xi\}_{\xi<\lambda}$ is a neighborhood base for t. For each $\alpha<\lambda$, $\alpha+1<\lambda$ and $\overline{W}_{\alpha+1}\subset\operatorname{int}(W_\alpha)$. For each s in $\overline{W}_{\alpha+1}$, there is a closed and open neighborhood V_s of s contained in $\operatorname{int}(W_\alpha)$. Let $V_{s_1}, V_{s_2}, \ldots, V_{s_n}$ be a finite subcover of this cover. Define $G_\alpha=\bigcup_{i=1}^{n} V_{s_i}$. Since $\overline{W}_{\alpha+1}\subset G_\alpha\subset\operatorname{int}(W_\alpha)$, $\{G_\alpha\}_{\alpha<\lambda}$ is a decreasing sequence of closed and open sets which form a neighborhood base for t. ∎

Theorem 2. *Let T be a compact, dispersed space with characteristic system (λ, n). If each point t in T has a neighborhood base consisting of a decreasing sequence $\{U_\alpha\}_{\alpha<\tau}$ of sets, then there is a map of T onto $\Gamma(\omega^\lambda\cdot n)$.*

Proof. First note that if the theorem is true for $(\lambda,1)$, it is also true for (λ,n) for each positive integer n. For suppose that $T^{(\lambda)}$ has exactly n points say $t_1,\ldots,t_n$ and that the theorem has been established for $(\lambda,1)$. Then T can be partitioned into disjoint closed and open sets $U_1, U_2, \ldots, U_n$ so that $t_i\in U_i$ for each i. But there exists a map f_i of U_i onto $(\omega^\lambda\cdot i, \omega^\lambda(i+1)]$ and the map f defined by $f|U_i=f_i$ for $i=1,\ldots,n$ is a map of T onto $\Gamma(\omega^\lambda\cdot n)$.

If S is a closed subspace of T with characteristic system $(0,1)$, then S is homeomorphic to $\Gamma(1)$ and the theorem statement is valid for S. Suppose the theorem has been established for each closed subset of T with characteristic system $(\gamma,1)$ where $\gamma<\lambda$ and $\lambda\geqslant 1$. Then by the preceding paragraph, it may also be assumed that the theorem has been proved for each closed subspace with characteristic system (γ,m) where $\gamma<\lambda$ and m is a positive integer. Let S be a closed subset of T with characteristic system $(\lambda,1)$ and let s_0 be the one point in $T^{(\lambda)}$. There is a decreasing sequence $\{U_\alpha\}_{\alpha<\tau}$ of sets in S which form a neighborhood base for s_0 and by lemma 4 it can be assumed that each U_α is closed and open. It is convenient to assign $U_0 = S$ and $U_\tau = \emptyset$. Since $\lambda \geqslant 1$, τ is a limit ordinal. Suppose $W_\alpha = U_\alpha \setminus U_{\alpha+1}$ has characteristic system $(\lambda_\alpha, n_\alpha)$. Since $W_\alpha \subset (S \setminus S^{(\lambda)})$, $\lambda_\alpha < \lambda$ and, by hypothesis, there exists a map f_α of W_α onto $\Gamma(\omega^{\lambda_\alpha} \cdot n_\alpha)$. Therefore, there is a map g_α of W_α onto

$$\Big(\sum_{\sigma<\alpha} \cdot \omega^{\lambda_\sigma} \cdot n_\sigma, \sum_{\sigma \leqslant \alpha} \omega^{\lambda_\sigma} \cdot n_\sigma\Big].$$

We define a map h_α of $S \setminus U_\alpha$ onto $\Gamma\Big(\sum_{\sigma<\alpha} \omega^\alpha \cdot n_\sigma\Big)$ for each $\alpha < \tau$ with the property that $\alpha < \beta$ implies $h_\beta | S \setminus U_\alpha = h_\alpha$. Let $h_1 = g_0$. Suppose h_α has been defined for each $\alpha < \beta$ where β is a fixed ordinal and $\beta \leqslant \tau$. If β is a nonlimit ordinal, then $h_{\beta-1}$ exists and we define h_β by $h_\beta | S \setminus U_{\beta-1} = h_{\beta-1}$ and $h_\beta | W_{\beta-1} = g_{\beta-1}$. Clearly, h_β is the desired map since

$$S \setminus U_\beta = (S \setminus U_{\beta-1}) \cup W_{\beta-1}$$

and $S \setminus U_{\beta-1}$ and $W_{\beta-1}$ are closed and open disjoint subsets of S.

Next, suppose β is a limit ordinal. Since $\mu < \alpha < \beta$ implies $h_\alpha | S \setminus U_\mu = h_\mu$, the map g defined by $g | S \setminus U_\alpha = h_\alpha$ is a map of $S_\beta = \bigcup_{\alpha<\beta} (S \setminus U_\alpha)$ onto $\Gamma_0\Big(\sum_{\alpha<\beta} \omega^{\lambda_\alpha} \cdot \eta_\alpha\Big)$. For if $s \in S_\beta$, there exists $\xi < \beta$ such that $s \in S \setminus U_\xi$ and

$$g(s) = h_\xi(s) \in \Gamma\Big(\sum_{\alpha<\xi} \omega^{\lambda_\alpha} \cdot n_\alpha\Big) \subseteq \Gamma_0\Big(\sum_{\alpha<\beta} \omega^{\lambda_\alpha} \cdot n_\alpha\Big).$$

Moreover, if $\delta \in \Gamma_0\Big(\sum_{\alpha<\beta} \omega^{\lambda_\alpha} \cdot n_\alpha\Big)$, then $\delta \in \Gamma\Big(\sum_{\alpha<\xi} \omega^{\lambda_\alpha} \cdot n_\alpha\Big)$ for some $\xi < \beta$ and there exists $s \in (S \setminus U_\xi) \subset S_\beta$ such that $g(s) = h_\xi(s) = \delta$.

Let h_β be defined by

$$h_\beta(s) = \left\{\begin{array}{ll} g(s), & \text{if } s \in S_\beta = S \setminus \Big(\bigcap_{\alpha<\beta} U_\alpha\Big) \\ \sum_{\alpha<\beta} \omega^{\lambda_\alpha} \cdot n_\alpha, & \text{if } s \in \Big(\bigcap_{\alpha<\beta} U_\alpha\Big) \setminus U_\beta \end{array}\right\}.$$

The domain of h_β is $S \setminus U_\beta$, and it is obvious that h_β is continuous at each point in $S \setminus \Big(\bigcap_{\alpha<\beta} U_\alpha\Big)$. Suppose $s \in \Big(\bigcap_{\alpha<\beta} U_\alpha\Big) \setminus U_\beta$ and $\{s_\mu\}_{\mu \in M}$ is a net in $S \setminus U_\beta$ such that $s_\mu \to s$. Let $\delta < \sum_{\alpha<\beta} \omega^{\lambda_\alpha} \cdot n_\alpha$. There exists $\xi < \beta$ such that $\delta < \sum_{\alpha<\xi} \omega^{\lambda_\alpha} \cdot n_\alpha$. Thus, $\delta < h_\beta(t) < \sum_{\alpha<\beta} \omega^{\lambda_\alpha} \cdot n_\alpha$ for $t \in \Big[U_\xi \setminus \Big(\bigcap_{\alpha<\beta} U_\alpha\Big)\Big]$. But for $t \in \Big(\bigcap_{\alpha<\beta} U_\alpha\Big) \setminus U_\beta$, $h_\beta(t) = \sum_{\alpha<\beta} \omega^{\lambda_\alpha} \cdot n_\alpha$. Since $s_\mu \to s$, there exists a μ_0 such that $s_\mu \in U_\xi \setminus U_\beta$ for $\mu \geqslant \mu_0$. Therefore,

$$h_\beta(s_\mu) \in \Big[\delta, \sum_{\alpha<\beta} \omega^{\lambda_\alpha} \cdot n_\alpha\Big]$$

for all $\mu \geqslant \mu_0$, so $h(s_\mu) \to h(s)$. This shows h_β is continuous; consequently, h_α can be defined for each $\alpha \leqslant \tau$.

Thus, h_τ is a map of S onto $\Gamma\left(\sum_{\alpha<\tau}\omega^{\lambda_\alpha}\cdot n_\alpha\right)$. Now, $\sum_{\alpha<\tau}\omega^{\lambda_\alpha}\cdot n_\alpha \geqslant \omega^\lambda$. If λ is a nonlimit ordinal, then $\lambda_\alpha=\lambda-1$ for infinitely many values of α and, by induction on n,

$$\sum_{\alpha<\tau}\omega^{\lambda_\alpha}\cdot n_\alpha=\lim_{\gamma<\tau}\left(\sum_{\alpha<\gamma}\omega^{\lambda_\alpha}\cdot n_\alpha\right)\geqslant\sum_{i=1}^{n}\omega^{\lambda-1}$$

for each natural number n. Thus,

$$\sum_{\alpha<\tau}\omega^{\lambda_\alpha}\cdot n_\alpha\geqslant\sum_{i=1}^{\infty}\omega^{\lambda-1}=\omega^\lambda .$$

If λ is a limit ordinal and $\gamma<\lambda$, there exists $\sigma<\tau$ such that $\lambda_\sigma>\gamma$. Therefore, $\sum_{\alpha<\tau}\omega^{\lambda_\alpha}\cdot n_\alpha\geqslant\omega^\lambda\geqslant\omega^\gamma$. Since this is true for all $\gamma<\lambda$, $\sum_{\alpha<\tau}\omega^{\lambda_\alpha}\cdot\eta_\alpha\geqslant\omega^\lambda$ in both cases. If $\sum_{\alpha<\tau}\omega^{\lambda_\alpha}\cdot\eta_\alpha>\omega^\lambda$, there exists $\gamma<\tau$ such that $\sum_{\alpha<\gamma}\omega^{\lambda_\alpha}\cdot\eta_\alpha>\omega^\lambda$. Since h_γ is a mapping of $S\setminus U_\gamma$ onto $\Gamma\left(\sum_{\alpha<\gamma}\omega^{\lambda_\alpha}\cdot n_\alpha\right)$, $(S\setminus U_\gamma)^{(\gamma)}$ is nonempty by lemma 3, which is impossible. Hence $\sum_{\sigma<\tau}\omega^{\lambda_\sigma}\cdot n_\sigma=\omega^\lambda$. Thus h_τ is a map of S onto $\Gamma(\omega^\lambda)$. It follows by transfinite induction and by the statement in the first paragraph of this proof that if S is a closed subspace of T with characteristic system (λ,n), there exists a map of S onto $\Gamma(\omega^\lambda\cdot n)$. This completes the proof. ∎

A point t in T satisfies (D) in T if T has a neighborhood base consisting of a decreasing sequence $\{U_\alpha\}_{\alpha<\tau}$ of closed and open sets with the additional property that $\left(\bigcap_{\alpha<\beta}U_\alpha\right)\setminus U_\beta$ contains at most one point for each limit ordinal β with $\beta<\tau$. If each point in T satisfies (D), then we say T has *property* (D). It should be noted that every first countable, regular space and every set of ordinals satisfies (D). Theorem 3 below gives a complete characterization of compact, dispersed, (Hausdorff) spaces with property (D). In particular, it characterizes sets of ordinals which are homeomorphic.

Theorem 3. *Let T be a compact, dispersed space with characteristic (λ,n). If T has property (D), T is homeomorphic to $\Gamma(\omega^\lambda\cdot n)$.*

Proof. The proof of this theorem is identical to the proof of Theorem 2 except "continuous map" is replaced with "homeomorphism" throughout the proof. Note that in such a case h_β is a one-to-one since $\left(\bigcap_{\alpha<\beta}U_\alpha\right)\setminus U_\beta$ contains exactly one point. ∎

A subset T of $\Gamma(\xi)=\Gamma_0(\xi+1)$ is well ordered and has order type β for some $\beta \leqslant \xi+1$. Therefore, there is a one-to-one order preserving map ϕ of $\Gamma_0(\beta)$ onto T. If T is a closed subspace of $\Gamma(\xi)$, ϕ is also a homeomorphism. Thus, a closed subspace of $\Gamma(\xi)$ is homeomorphic to $\Gamma(\eta)$ for some $\eta \leqslant \xi$; the proof that ϕ is a homeomorphism is omitted since the following corollary is an easy consequence of Theorem 3.

Corollary 1. *A closed subspace of* $\Gamma(\xi)$ is *homeomorphic to* $\Gamma(\eta)$ *for some* $\eta \leqslant \xi$. ∎

Theorem 3 is a generalization of a theorem due to Mazurkiewicz and Sierpiński [202]. In the following the ordinal Ω is the initial ordinal whose cardinal number is uncountable.

Corollary 2 (Semadeni). *A first countable, dispersed, compact space T is metrizable. In fact, T is homeomorphic to $\Gamma(\omega^\alpha \cdot n)$ where (α, n) is the characteristic system of T and $\alpha < \Omega$.*

Proof. Since T satisfies (D), it is homeomorphic to $\Gamma(\omega^\alpha \cdot n)$ by Theorem 3. Now $\Omega \notin \Gamma(\omega^\alpha \cdot n)$ because T is first countable. Therefore, T is denumerable and second countable, and by the Urysohn Metrization theorem, T is metrizable. ∎

The reader can check [256] for further results on dispersed spaces including a discussion of the noncompact case.

§ 6. The Cantor Set

The Cantor set plays a significant role in analysis since it is characterized as the only perfect totally disconnected metrizable compact space, and it can be embedded homeomorphically into any uncountable separable complete metric space, and it can be mapped continuously onto any compact metric space. We shall give proofs of these three assertions in this section.

Definition 1. The *Cantor set* is the compact metric space Δ which is the set of all sequences $\{a_n\}$ with $a_n \in \{-1, 1\}$ for all n. The topology on Δ is taken to be the product topology induced by the discrete topology on $\{-1, 1\}$.

There are, of course, many ways to define a metric on Δ which is compatible with the product topology. One such is clearly given by $d(\{a_n\}, \{b_n\}) = \sum_{n=1}^{\infty} \frac{|a_n - b_n|}{2^n}$. Also Δ can be identified with the classical Cantor set in $[0, 1]$ constructed by removing "middle thirds" as follows.

Let $\Delta'=\left\{\sum_{n=1}^{\infty}\frac{b_n}{3^n}: b_n\in\{0,2\}\right\}$. Then Δ' is exactly the set in $[0,1]$ constructed by removing the middle third $(\frac{1}{3},\frac{2}{3})$ from $[0,1]$. The middle third $(\frac{1}{9},\frac{2}{9})$ from $[0,\frac{1}{3}]$ and the middle third $(\frac{7}{9},\frac{8}{9})$ from $[\frac{2}{3},1]$, and so on. Moreover, if $b_n, c_n\in\{0,2\}$ and $\sum_{n=1}^{\infty}\frac{b_n}{3^n}=\sum_{n=1}^{\infty}\frac{c_n}{3^n}$, then $b_n=c_n$ for all n. It is easy to check that the map $f:\Delta'\to\Delta$ defined by $f\left(\sum_{n=1}^{\infty}\frac{b_n}{3^n}\right)=\{b_n-1\}$ is a homeomorphism of Δ' onto Δ.

We shall use the following notation: If $\{(A_n,B_n)\}$ is a sequence of pairs of sets, $1\cdot A_n=A_n$ and $-1\cdot A_n=B_n$.

Definition 2. Let T be a compact Hausdorff space. A sequence $\{(A_n,B_n)\}$ of pairs of closed sets A_n and B_n in T is said to be *interlocking* if A_n and B_n are disjoint and for each finite set $F\subset N$, $\bigcap_{i\in F}\varepsilon_i A_i\neq\emptyset$ where $\varepsilon_i\in\{-1,1\}$ for all $i\in F$. It is said to be a *separation* if for each pair x,y of distinct points in T there is an n such that $x\in A_n$ and $y\in B_n$ (or visa versa).

Lemma 1. *Let T be a compact Hausdorff space. Then T is homeomorphic to Δ if and only if there is a sequence $\{(A_n,B_n)\}$ of pairs of closed sets in T which is both interlocking and a separation and $A_n\cup B_n=T$ for all n.*

Proof. The necessity is immediate by putting $A_n=\{\{a_k\}:a_n=-1\}$ and $B_n=\{\{a_k\}:a_n=1\}$.

For the sufficiency, note that for each $\{a_n\}$ in Δ, $\bigcap_{n=1}^{\infty}a_n\cdot A_n$ is nonempty since each finite intersection $\bigcap_{k=1}^{p}a_{n_k}\cdot A_{n_k}$ is nonempty. We define $f:T\to\Delta$ by $f(t)=\{a_n\}$ where $a_n=-1$ if $t\in B_n$ and $a_n=1$ if $t\in A_n$. By the above remark, f is onto. Since we have the separation property, f is one to one. To see that f is continuous it is enough to check that for each n, $f(t)\to a_n$ is continuous. But, $t\in B_n$ implies that each net $\{x_d\}$ which converges to t is eventually in B_n (since B_n is open). Consequently, the n*th* coordinate of $f(x_d)$ must eventually be -1 (clearly the same argument applies if $t\in A_n$). It follows that f is continuous and, since T is compact, that f is also a homeomorphism. ∎

Recall that the *diameter* of a set A in a metric space is given by $d(A)=\sup\{d(x,y):x,y\in A\}$.

Lemma 2. *Let T be a perfect totally disconnected metrizable compact space (with metric d). Then for each $\varepsilon>0$ there is an N such that for all $n\geqslant N$ there are pairwise disjoint closed and open sets $A_1,\ldots,A_n$ in T such that $d(A_i)<\varepsilon$ and $T=\bigcup_{i=1}^{n}A_i$.*

Proof. Since T is compact and totally disconnected there are nonempty closed and open sets $B_1, \ldots, B_k$ in T with $d(B_j) < \varepsilon$ for $j = 1, \ldots, k$ and $T = \bigcup_{j=1}^{k} B_j$. Let $A_1 = B_1$ and for $1 < j \leq k$ let $A_j = B_j \setminus \bigcup_{i=1}^{j-1} B_i$. Then $A_1, \ldots, A_k$ is a pairwise disjoint collection of closed and open sets whose union is T and with $d(A_i) < \varepsilon$ for all $i = 1, \ldots, k$. By discarding the empty ones we can assume each A_i is nonempty. Since T is perfect and each A_i is open, it follows that each A_i is infinite. Thus by suitably decomposing the A_i's we can obtain such decompositions of T. ∎

We establish the following theorem using lemmas 1 and 2.

Theorem 1. *Let T be a perfect totally disconnected compact metric space. Then T is homeomorphic to Δ.*

Proof. By lemma 1 we need only build an interlocking sequence $\{(A_n, B_n)\}$ with the separation property such that A_n, B_n are disjoint closed sets and $T = A_n \cup B_n$. We outline how to do this. By lemma 2 we obtain an N_1 and pairwise disjoint closed and open sets $D_1, \ldots, D_{2^{N_1}}$ each of diameter less than 1/2. We now define $A_1, B_1, \ldots, A_{N_1}, B_{N_1}$ as follows. Let $A_1 = D_1 \cup \cdots \cup D_{(1/2)2^{N_1}}$ and $B_1 = T \setminus A_1$, $A_2 = D_1 \cup \cdots \cup D_{(1/4)2^{N_1}} \cup D_{(1/2)2^{N_1}} \cup \cdots \cup D_{(3/4)2^{N_1}}$, and $B_2 = T \setminus A_2, \ldots, A_{N_1} = \bigcup \{D_i : 1 \leq i \leq 2^{N_1}, i \text{ odd}\}$ and $B_{N_1} = T \setminus A_{N_1}$. By lemma 2 again we obtain an N_2 such that for each $1 \leq i \leq 2^{N_2}$ there are pairwise disjoint closed and open sets D_{ij} $(j = 1, \ldots, 2^{N_2})$ each of diameter less than 1/4 such that $D_i = \bigcup_{j=1}^{2^{N_2}} D_{ij}$. We decompose T again in the above fashion to obtain disjoint pairs $A_{N_1+1}, B_{N_1+1}, \ldots, A_{N_1+N_2}, B_{N_1+N_2}$. The sequence so constructed is an interlocking sequence with the separation property such that $T = A_n \cup B_n$ for each n. ∎

We shall need the following fact about complete metric spaces in the next theorem. The reader may consult [104] for a proof of it. *Let T be a complete metric space and S a subset of T. Then it is possible to define a metric d on S (compatible with the topology of S) such that (S, d) is a complete metric space if and only if S is a G_δ set in T.*

The following theorem is due to Kuratowski (see [158]).

Theorem 2. *Let T be a complete separable metric space and S be an uncountable G_δ set in T. Then S contains a subspace homeomorphic to Δ.*

Proof. By the above remarks it suffices to consider that $S = T$. Since T is uncountable it contains a perfect set. For, let A denote the set of all *condensation points* of T (that is, $x \in A$ if and only if each neighborhood of x is uncountable). Then $T \setminus A$ is countable. For, let $\{U_n\}$ be a basis for the topology of T. Then for each $x \in T \setminus A$ there is an n such that $x \in U_n$ and U_n is countable. Hence $T \setminus A$ is contained in a countable union of countable sets and is itself countable. Hence A is nonempty and it follows readily that A is perfect. Thus we may assume that T itself is perfect.

Let A_1, B_1 be two nonempty open sets in T such that $\bar{A}_1, \bar{B}_1$ are disjoint and have diameters less than one. Since T is perfect, A_1 and B_1 are infinite. Hence there are nonempty open sets $U_1 \cup V_1 \subset A_1$ and $U_2 \cup V_2 \subset B_1$ such that $\bar{U}_1, \bar{V}_1$ and $\bar{U}_2, \bar{V}_2$ are disjoint and all have diameters less than 1/2. Let $A_2 = U_1 \cup U_2$ and $B_2 = V_1 \cup V_2$. Again we can choose nonempty open sets $U_{11} \cup V_{11} \subset U_1$, $U_{12} \cup V_{12} \subset V_1$, $U_{21} \cup V_{21} \subset U_2$, and $U_{22} \cup V_{22} \subset V_2$ so that $\bar{U}_{11}, \bar{V}_{11}, \bar{U}_{12}, \bar{V}_{12}, \bar{U}_{21}, \bar{V}_{21}$, and $\bar{U}_{22}, \bar{V}_{22}$ are pairwise disjoint and all have diameters less than 1/4. We let $A_3 = U_{11} \cup U_{12} \cup U_{21} \cup U_{22}$ and $B_3 = V_{11} \cup V_{12} \cup V_{21} \cup V_{22}$. We continue this process to obtain a sequence $\{(A_n, B_n)\}$. Let $D = \bigcap_{n=1}^{\infty} (\bar{A}_n \cup \bar{B}_n)$. Using the fact that T is complete it is easy to see that D is nonempty and sequentially compact (and thus compact). Moreover, if $C_n = \bar{A}_n \cap D$ and $D_n = \bar{B}_n \cap D$, then the sequence $\{(C_n, D_n)\}$ forms an interlocking sequence with the separation property and $C_n \cup D_n = D$ for all n. Thus by lemma 1 D is homeomorphic to Δ. ∎

Let us note that $\Delta^{\mathbb{N}} = \Delta \times \Delta \times \cdots$ in the product topology is homeomorphic to Δ. For, clearly $\Delta^{\mathbb{N}}$ is a compact perfect totally disconnected metric space so by theorem 1 it is homeomorphic to Δ.

Lemma 3. *Let T be a separable metric space. Then there is a homeomorphic embedding of T into $H = [0,1] \times [0,1] \times \cdots$ (with the product topology).*

Proof. Clearly we can assume T has a metric d (compatible with the topology of T) such that $d(x,y) \leqslant 1$ for all x, y.

Let $\{x_n\}$ be a dense set in T and define $\phi: T \to H$ by $\phi(x) = \{d(x, x_n)\}$. It is easy to check that ϕ is a homeomorphic embedding of T into H. ∎

Lemma 4. *There is a continuous mapping of Δ onto $[0,1]$ and, consequently, there is a continuous mapping of Δ onto H.*

Proof. We consider Δ to be the sequences whose terms lie in $\{-1, 1\}$. Let $f(\{a_n\}) = \sum_{n=1}^{\infty} \frac{a_n + 1}{2^{n+1}}$. Then f is continuous and since each element of $[0,1]$ has a binary expansion, f is onto.

Let $g: \Delta^{\mathbb{N}} \to H$ be defined by $g(\{x_n\}) = \{f(x_n)\}$. Clearly g is continuous and onto H. Since $\Delta^{\mathbb{N}}$ is homeomorphic to Δ, we get a continuous mapping of Δ onto H. ∎

In the following lemma we use the metric d on Δ defined by $d(\{a_n\},\{b_n\})=\sum_{n=1}^{\infty}\frac{|a_n-b_n|}{10^n}$. Note that d is compatible with the topology of Δ and has the property that if $d(\{a_n\},\{b_n\})=d(\{a_n\},\{c_n\})$, then $\{b_n\}=\{c_n\}$ (by the uniqueness of decimal expansions when the terms do not terminate in repeated zeros).

Lemma 5. *Let F be a closed subset of Δ. Then there is a continuous mapping of Δ onto F.*

Proof. For each $x\in\Delta$ there is a $y\in F$ such that $d(x,y)=d(x,F)=\inf\{d(x,z):z\in F\}$. By the above remarks such a y is unique (with respect to this metric).

Thus we let $f(x)=y$ where $d(x,y)=d(x,F)$. Note that $f(x)=x$ when $x\in F$ so that f is onto. It is easy to check that f is continuous since if $\lim x_n=x$, then $\lim d(x_n,F)=d(x,F)=\lim d(x_n,y_n)$. ∎

Putting the above results together we have the following theorem.

Theorem 3. *Let T be a compact metric space. Then there is a continuous mapping of Δ onto T.*

Proof. By lemma 3 there is a homeomorphic embedding ϕ of T into H. By lemma 4 there is a continuous mapping f of Δ onto H. For $F=f^{-1}(\phi(T))$, by lemma 5 there is a continuous mapping g of Δ onto F. Thus $h=\phi^{-1}\circ f\circ g$ is the required mapping. ∎

Let I be an index set and for a topological space T, let T^I denote topological space of all functions from I to T with the product topology (i.e. T^I is the product of T with itself the cardinality of I times with the corresponding product topology). The *weight* of a topological space T is the smallest cardinal m for which there is a base for the topology of T with cardinality m. The reader can easily check that if I is an infinite set, then the weight of $\{-1,1\}^I$ and $[0,1]^I$ is the cardinality of I. Moreover, if T is a completely regular Hausdorff space of weight m, then T can be nomeomorphically embedded into $[0,1]^I$ where the cardinality of I is m (see [104] for a proof of this last statement, and note that it is a generalization of lemma 3).

In general it is not true that a compact Hausdorff space T (of weight m) is the continuous image of $\{-1,1\}^I$ (where card $I=m$). Thus theorem 3 does not have a full generalization. However, there are such spaces (called *dydadic spaces*). Reference [104] has considerable information about these spaces.

§ 7. Extremally Disconnected Compact Hausdorff Spaces

The theory of extremally disconnected compact Hausdorff spaces is needed because such spaces are used in the characterization of Banach spaces X which have the property that whenever $X \subset Y$ where Y is a Banach space, then X is the range of a contractive projection on Y.

We shall need the concept of the Stone-Čech compactification of completely regular Hausdorff spaces. The development given here makes use of the weak* topology in Banach spaces.

Definition 1. Let T be a completely regular Hausdorff space. A pair $(\beta T, \beta)$ is said to be the *Stone-Čech Compactification* of T provided that βT is a compact Hausdorff space, β is a homeomorphism of T onto a dense set in βT, and each bounded continuous real valued function f on T has a (unique) continuous extension βf on βT such that $\sup_{t \in T} |f(t)| = \sup_{t \in \beta T} |\beta f(t)|$.

It is well known that the pair $(\beta T, \beta)$ is unique in the sense that if $(\beta' T, \beta')$ is any other pair satisfying the definition, then there is a homeomorphism ϕ of βT onto $\beta' T$ such that $\phi \circ \beta = \beta'$. For a proof of this fact and other details concerning the Stone-Čech compactification the reader may consult reference [104].

Let T be a completely regular Hausdorff space and X be the Banach space of all bounded continuous real-valued functions on T with the supremum norm. Let $\beta: T \to b(X^*)$ be defined by $\beta(t)(f) = f(t)$ for all $t \in T$ and $f \in X$. Clearly β is continuous relative to the weak* topology on $b(X^*)$. Moreover, the definition of complete regularity yields that β is one to one. To see that β is a homeomorphism, suppose $\{t_d\}$ is a net in T and $\{t_d\}$ does not converge to t. We shall show that $\{\beta(t_d)\}$ does not converge to $\beta(t)$ which proves that β^{-1} is continuous. There is an open set U containing t such that for each d there is a $j(d) \geqslant d$ such that $t_{j(d)} \notin U$. Since T is completely regular, there is a $f \in X$ such that $f(t) = 1$ and $f(s) = 0$ for all $s \in T \setminus U$. Thus $f(t_{j(d)}) = 0$ for all d and it follows that $\{f(t_d)\}$ does not converge to $f(t)$, that is, $\{\beta(t_d)\}$ does not converge to $\beta(t)$.

Theorem 1. *Let βT be the weak* closure of $\beta(T)$ in $b(X^*)$. Then $(\beta T, \beta)$ is the Stone-Čech compactification of T.*

Proof. For $f \in X$ let βf be defined on βT by $(\beta f)(x^*) = x^*(f)$ for all $x^* \in \beta T$. Clearly by the definition of the weak* topology, βf is continuous on βT. Moreover, $|(\beta f)(x^*)| = |x^*(f)| \leqslant \|x^*\| \, \|f\| \leqslant \|f\|$ implies that $\|f\| = \|\beta f\|$. Thus $(\beta T, \beta)$ is the Stone-Čech compactification of T. ∎

By $\beta\mathbb{N}$ we mean the Stone-Čech compactification of the set of positive integers with the discrete topology.

Definition 2. A compact Hausdorff space Ω is said to be *extremally disconnected* if the closure of each open set in Ω is again open in Ω.

We shall also need the following concepts. A compact Hausdorff space will be called *free* if it is the Stone-Čech compactification of the set of its isolated points (an *isolated point* being one which is both closed and open as a set).

A free space is clearly extremally disconnected. For, let U be an open set in the free space T and D the set of isolated points of T. Then $U \cap D$ is a closed and open set in D so that the characteristic function f of $U \cap D$ is continuous on D. Hence its extension, βf, is the characteristic function of $\bar{U}$ in T, that is, $\bar{U}$ is open.

A compact Hausdorff space P is said to be *projective* if for each pair S, T of compact Hausdorff spaces and each continuous onto map f from S to T and any continuous map g of P to T there is a continuous map h of P to S such that $f \circ h = g$. Our purpose is to prove that a compact Hausdorff space is extremally disconnected if and only if it is projective. This result is due to Gleason [112]. The development here follows Rainwater [242].

Lemma 1. *Let T be a compact Hausdorff space and $f: T \to T$ a continuous map which is not the identity. Then there is a proper closed set $F \subset T$ such that $F \cup f^{-1}(F) = T$.*

Proof. Let $t \in T$ with $f(t) \neq t$ and U, V disjoint open sets in T with $t \in U$, $f(t) \in V$. Then $F = T \setminus [U \cap f^{-1}(V)]$ is such a set. ∎

Lemma 2. *Every free compact Hausdorff space is projective and every compact Hausdorff space is the continuous image of a free space.*

Proof. Let P be a free space, S, T be compact Hausdorff spaces, and $f: S \to T$, $g: P \to T$ be continuous maps with f onto. For each isolated point $p \in P$ let $s_p \in S$ with $f(s_p) = g(p)$. Then $h_0: P_0 \to S$ defined by $h_0(p) = s_p$ is clearly continuous, where P_0 is the set of isolated points in P. Hence the continuous mapping $h = \beta h_0$ is such that $f \circ h = g$ and it follows that P is projective.

To see that every compact Hausdorff space is the continuous image of a free space let T be a compact Hausdorff space and T_0 a dense set in T. Consider T_0 as a discrete space. Then the extension of the map $f: T_0 \to T$ defined by $f(t) = t$ for all $t \in T_0$ is a map of βT_0 onto T. (In particular, any separable space is the continuous image of $\beta\mathbb{N}$.). ∎

Recall that a topological space S is said to be a *retract* of a topological space T if there are continuous maps $r: T \to S$ and $s: S \to T$ such that rs is the identity and s is a homeomorphic embedding.

Lemma 3. *A compact Hausdorff space P is projective if and only if it is the retract of a free space.*

Proof. Suppose P is the retract of a projective space Q with maps r and s. Let S, T be compact Hausdorff spaces and $f: S \to T$, $g: P \to T$ continuous maps with f onto. Then $gr: Q \to T$ and since Q is projective there is a continuous map $h: Q \to S$ with $fh = gr$. Now $hs: P \to T$ and $fhs = grs = g$ and it follows that P is projective.

Suppose P is projective. By lemma 2 there is a free space βT, where T is a discrete set, and there is a continuous map r of βT onto P. Since P is projective there is a continuous map $s: P \to \beta T$ such that $rs =$ identity. Hence s is one to one and since P is compact, s is a homeomorphic embedding. ∎

Let P, T be compact Hausdorff spaces and f a continuous mapping of P onto T. Then f is said to be *minimal* if for each closed subset $F \neq P$, $f(F) \neq T$.

Theorem 2. *For each compact Hausdorff space T there is a projective space P and a minimal map f of P onto T. Moreover, P is unique in the sense that if P' is projective and f' is a minimal map of P' onto T, then there is a homeomorphism $g: P \to P'$ such that $f'g = f$.*

Proof. Let S be a discrete space such that there is a continuous onto map f_0 from βS to T. Let P be a minimal closed subspace of βS such that $f_0(P) = T$ and let $f = f_0 | P$. Then f is a minimal map of P onto T and it remains to show that P is projective. Since βS is projective there is a continuous map $r: \beta S \to P$ such that $fr = f_0$. Let $s: P \to \beta T$ be the natural embedding. If $rs \neq$ the identity, then by lemma 1 there is a proper closed set $F \subset P$ such that $F \cup (rs)^{-1}(F) = P$. Since $rs(rs)^{-1}(F) \subset F$, $f(F) \supset frs(rs)^{-1}(F) = f(rs)^{-1}(F)$ and $f(F) = T$ which is a contradiction. Hence P is a retract of βT and is therefore projective.

The uniqueness is clear from the condition of projectivity. ∎

Note that in the course of the proof it was shown that if P is projective and $f: P \to T$ is continuous and onto and F is a minimal closed set with $f(F) = T$, then F is projective. The proofs of the following two lemmas are left to the reader.

Lemma 4. *Let S, T be compact Hausdorff spaces and $f: S \to T$ a minimal onto continuous map. Then for any open set $U \subset S$, $f(U) \subset \overline{T \setminus f(S \setminus U)}$.* ∎

Lemma 5. *Let T be an extremally disconnected compact Hausdorff space. Then the closures of any two disjoint open sets in T are again disjoint.* ∎

Lemma 6. *Let S, T be compact Hausdorff spaces with T extremally disconnected. If $f: S \to T$ is a minimal onto continuous map, then it is a homeomorphism.*

Proof. By compactness it suffices to show that f is one to one. Suppose $s_1 \neq s_2$ and $f(s_1) = f(s_2)$. Let U_1, U_2 be two disjoint open neighborhoods of s_1, s_2 respectively. Now $T \setminus f(S \setminus U_1)$ and $T \setminus f(S \setminus U_2)$ are open and disjoint and hence their closures are disjoint. But, by lemma 4, $f(s_1) = f(s_2)$ is in both closures. ∎

The stated characterization of extremally disconnected spaces now follows from the above results.

Theorem 3 (Gleason). *Let P be a compact Hausdorff space. Then P is projective if and only if it is extremally disconnected.*

Proof. Suppose P is extremally disconnected. By theorem 2 there is a projective Q and a minimal map of Q onto P. By lemma 6 this map is a homeomorphism.

Suppose P is projective and $U \subset P$ is open. Let $S = P \times \{a, b\}$ and $T = [(S \setminus U) \times \{a\}] \cup [U \times \{b\}]$. If $s: T \to S$ is the natural embedding and $\pi: S \to P$ the natural projection, then there is a continuous map $f: P \to T$ such that $\pi \circ s \circ f$ is the identity. Since $\pi \circ s$ is one-to-one on $U \times \{b\}$, $f(s) = (s, b)$ for all $s \in \bar{U}$ and by continuity $f(s) = (s, b)$ for all $s \in \bar{U}$. Similarly for $s \in S \setminus \bar{U}$, $f(s) = (s, a)$ and $\bar{U} = f^{-1}(\bar{U} \times \{b\})$. It follows that $\bar{U} \times \{b\}$ is open in T and $\bar{U}$ is open in P. ∎

A remark which is worth noting is the following. If T is an extremally disconnected space of weight $m \geqslant \aleph_0$ and S is a compact Hausdorff space which can be mapped continuously onto $[0,1]^I$, $(\operatorname{card} I = m)$ then S contains a subspace homeomorphic to T. For, T can be embedded into $[0,1]^I$ and there is a minimal subspace F of S which can be mapped onto the homeomorphic image of T in $[0,1]^I$ by the map of S onto $[0,1]^I$. Thus by lemma 6, F is homeomorphic to T.

§ 8. Regular Borel Measures

Let T be a compact Hausdorff space and μ be a regular Borel measure on T. Then $x^*(f) = \int f \, d\mu$ defines a positive linear functional on $C(T, \mathbb{R})$. The Riesz representation theorem states that every positive linear functional on $C(T, \mathbb{R})$ is uniquely represented by a regular Borel measure in the above manner. This theorem plays a central role in this book. We shall prove it and some other facts about regular Borel measures in this section.

First we recall several standard results about finite valued measures. An excellent source for this material is [131] (starting at page 304).

Let T be a nonempty set and Σ be a σ algebra of subsets of T. We denote by $m(T,\Sigma,\mathbb{C})$ the vector space of all countably additive set functions $\mu:\Sigma\to\mathbb{C}$ (and by $m(T,\Sigma,\mathbb{R})$ the real linear subspace of such set functions whose range is contained in $\mathbb{R}$). The vector space operations are defined by $(\mu+v)(A)=\mu(A)+v(A)$ and $(a\mu)(A)=a\mu(A)$ for all $\mu,v\in m(T,\Sigma,\mathbb{C})$, $A\in\Sigma$, and $a\in\mathbb{C}$. There is a natural partial ordering on $m(T,\Sigma,\mathbb{R})$ defined by $\mu\leqslant v$ if and only if $\mu(A)\leqslant v(A)$ for all $A\in\Sigma$. The positive cone relative to this ordering is denoted by $m^+(T,\Sigma,\mathbb{R})$ and its elements are call *positive measures.*

Let $\mu\in m(T,\Sigma,\mathbb{R})$. A pair (P,P') (where $P\in\Sigma$ and $P'=T\setminus P$) is said to be the *Hahn decomposition* of μ if $\mu(A\cap P)\geqslant 0$ and $\mu(A\cap P')\leqslant 0$ for all $A\in\Sigma$. It is well known that each μ has a unique Hahn decomposition in the sense that if (Q,Q') is another pair with the same properties, then $\mu(A\cap Q)=\mu(A\cap P)$ and $\mu(A\cap Q')=\mu(A\cap P')$ for all $A\in\Sigma$.

We can use the Hahn decomposition of μ to define μ^+ (which is the supremum of μ and 0) as follows. For each $A\in\Sigma$, $\mu^+(A)=\mu(A\cap P)$. Clearly $\mu^+\in m^+(T,\Sigma,\mathbb{R})$ and $\mu^+\geqslant\mu,0$. The reader can easily check that if $v\geqslant\mu,0$, then $v\geqslant\mu^+$. Thus we obtain immediately that $m(T,\Sigma,\mathbb{R})$ is a vector lattice since we can define $\mu\vee v$ by $\mu\vee v=(\mu-v)^+ + v$ and $\mu\wedge v$ by $\mu\wedge v=-[(-\mu)\wedge(-v)]$.

Clearly any $\mu\in m(T,\Sigma,\mathbb{C})$ can be expressed uniquely as $\mu=v_1+iv_2$ where $v_1,v_2\in m(T,\Sigma,\mathbb{R})$. In order for $m(T,\Sigma,\mathbb{C})$ to be a complex vector lattice we need to be able to define the modulus function on it. This is accomplished as follows. For $\mu\in m(T,\Sigma,\mathbb{C})$, $|\mu|(A)=\sup\left\{\sum_{i=1}^{n}|\mu(A_i)|: A_i\in\Sigma,\ A_i\cap A_j=\emptyset \text{ for } i\neq j,\ A=\bigcup_{i=1}^{n}A_i\right\}$. The modulus has the following properties: (1) $|\mu|\in m(T,\Sigma,\mathbb{R})$; (2) $|\mu|=\mu^+-\mu^-$ if $\mu\in m(T,\Sigma,\mathbb{R})$; (3) $|\mu+v|\leqslant|\mu|+|v|$; (4) $|a\mu|=|a||\mu|$; (5) $|\mu|=0$ if and only if $\mu=0$; and (6) $|\mu|=\sup\{\operatorname{Re}(e^{i\theta}\mu):0\leqslant\theta\leqslant 2\pi\}$ (properties 1—5 are proved in [131] and property 6 is left as an exercise).

The modulus can also be used to define a norm on $m(T,\Sigma,\mathbb{C})$ as follows. For each $\mu\in m(T,\Sigma,\mathbb{C})$, $\|\mu\|=|\mu|(T)$. From the properties of modulus it is clear that this is a norm and that $|\mu|\leqslant|v|$ implies $\|\mu\|\leqslant\|v\|$. Thus to show that $m(T,\Sigma,\mathbb{C})$ is a Banach lattice we need only show completeness. Let $\{\mu_n\}$ be a Cauchy sequence in $m(T,\Sigma,\mathbb{C})$. Then for any $A\in\Sigma$, $|\mu_n(A)-\mu_m(A)|\leqslant\|\mu_n-\mu_m\|$ so that $\mu(A)=\lim\mu_n(A)$ exists for all $A\in\Sigma$. Clearly $\mu\in m(T,\Sigma,\mathbb{C})$, we shall show that $\mu=\lim\mu_n$. Let $\varepsilon>0$

be given and choose N so that if $n, m \geqslant N$, then $\|\mu_n - \mu_m\| < \varepsilon$. Let $T = \bigcup_{i=1}^{k} A_i$ with $A_i \in \Sigma$ and A_i's pairwise disjoint. Choose $m \geqslant N$ so that $|\mu_m(A_i) - \mu(A_i)| < \varepsilon/k$ for $i = 1, \ldots, k$. Then for $n \geqslant N$, $\sum_{i=1}^{k} |\mu_n(A_i) - \mu(A_i)| \leqslant \sum_{i=1}^{k} |\mu_n(A_i) - \mu_m(A_i)| + \sum_{i=1}^{k} |\mu_m(A_i) - \mu(A_i)| < 2\varepsilon$, that is, $\|\mu_n - \mu\| \leqslant 2\varepsilon$. Thus $\mu = \lim \mu_n$ and, in particular, if $\mu_n \in m(T, \Sigma, \mathbb{R})$, then so is μ. Finally note that if $\mu, \nu \geqslant 0$, then $\|\mu + \nu\| = |\mu + \nu|(T) = (\mu + \nu)(T) = \mu(T) + \nu(T) = \|\mu\| + \|\nu\|$.

We summarize the above remarks in the following theorem (see section 3).

Theorem 1. *$m(T, \Sigma, \mathbb{C})$ and $m(T, \Sigma, \mathbb{R})$ are Banach lattices in which the norm is additive. In particular, the norm is order continuous and each closed ideal is a projection band.* ∎

We now wish to discuss ideals of a particular type in $m(T, \Sigma, \mathbb{C})$. Let $\mu \in m^+(T, \Sigma, \mathbb{R})$ and recall that $\nu \in m(T, \Sigma, \mathbb{C})$ is said to be *absolutely continuous* with respect to μ if $|\nu|(A) = 0$ whenever $\mu(A) = 0$. Clearly if $f \in L_1(T, \Sigma, \mu, \mathbb{C})$ and $\nu(A) = \int_A f\, d\mu$ for all $A \in \Sigma$, then ν is absolutely continuous with respect to μ. The classical Radon-Nikodym theorem asserts that conversely, if ν is absolutely continuous with respect to μ, then there is a μ-integrable complex valued function f such that $\nu(A) = \int_A f\, d\mu$ for all $A \in \Sigma$. Moreover, f is unique in the sense that if g is another function with the same properties, then $f = g$ almost everywhere (see [131, p. 176]).

Theorem 2. *Let $\mu \in m^+(T, \Sigma, \mathbb{R})$. Then $\mu^{\perp\perp}$ is linearly isometric and lattice isomorphic to $L_1(T, \Sigma, \mu, \mathbb{C})$. In particular, $\mu^{\perp\perp} = \{\nu \in m(T, \Sigma, \mathbb{C}) : \nu$ is absolutely continuous with respect to $\mu\}$.*

Proof. First note that $Y = \{\nu : \nu$ is absolutely continuous with respect to $\mu\}$ is a closed ideal in $m(T, \Sigma, \mathbb{C})$. Moreover, mapping $\phi : L_1(T, \Sigma, \mu, \mathbb{C}) \to Y$ given by $\phi(f)(A) = \int_A f\, d\mu$ for $f \in L(T, \Sigma, \mu, \mathbb{C})$ and $A \in \Sigma$ is onto, a linear isometry, and $\phi(|f|) = |\phi(f)|$ for all $f \in L_1(T, \Sigma, \mu, \mathbb{C})$ (see [131], p. 323). Clearly $\{f : |\phi(f)| \leqslant n\mu$ for some $n\}$ is dense in $L_1(T, \Sigma, \mu, \mathbb{C})$ since the simple functions are. Thus $\{\nu \in Y : |\nu| \leqslant n\mu$ for some $n\}$ is dense in Y. But, this last set is also dense in $\mu^{\perp\perp}$. ∎

We recall that when T is a compact Hausdorff space that the *Borel sets* of T are the members of the smallest σ algebra containing the open sets of T. The class of Borel sets is denoted by $\mathscr{B}$. Since $\mathscr{B}$ will be understood, so for example, we denote $m(T, \mathscr{B}, \mathbb{C})$ by $m(T, \mathbb{C})$, and $L_1(T, \mathscr{B}, \mu, \mathbb{C})$ by $L_1(T, \mu, \mathbb{C})$ (or $L_1(\mu, \mathbb{C})$ if T is understood).

Definition 1. Let $\mu \in m^+(T,\mathbb{R})$. Then μ is said to be a *regular Borel measure* if $\mu(A) = \sup\{\mu(F): F \subset A,\ F \text{ is compact}\}$ for all Borel sets A.

It is immediate that μ is regular if and only if $\mu(A) = \inf\{\mu(U): A \subset U,\ U \text{ is open}\}$ (i.e. μ is regular if and only if it is either inner regular or outer regular).

Clearly the class $M^+(T,\mathbb{R})$ of all regular Borel measures is a cone in $m(T,\mathbb{R})$. We put $M(T,\mathbb{R}) = M^+(T,\mathbb{R}) - M^+(T,\mathbb{R})$ and $M(T,\mathbb{C}) = M(T,\mathbb{R}) + iM(T,\mathbb{R})$. Thus $M(T,\mathbb{R})$ and $M(T,\mathbb{C})$ are respectively real and complex vector lattices (we shall see below that $\mu \in M(T,\mathbb{C})$ if and only if $|\mu| \in M^+(T,\mathbb{R})$.

Suppose $\mu \in M^+(T,\mathbb{R})$ and $v \in m^+(T,\mathbb{R})$ is absolutely continuous with respect to μ. Then $v \in M^+(T,\mathbb{R})$. For, $\varepsilon > 0$ is given, then there is a $\delta > 0$ such that if $B \in \mathscr{B}$ and $\mu(B) < \delta$, then $v(B) < \varepsilon$. Let $A \in \mathscr{B}$ and choose $F \subset A$ such that F is compact and $\mu(A \setminus F) < \delta$ so that $v(A \setminus F) < \varepsilon$. Hence $v(F) > v(A) - \varepsilon$ and it follows that v is regular. Now suppose $v \in m(T,\mathbb{C})$ so that $v = v_1 + iv_2$ with $v_1, v_2 \in m(T,\mathbb{R})$. If $|v| \leqslant \mu$, then $v_1^+ \vee v_1^- \vee v_2^+ \vee v_2^- \leqslant \mu$ and, hence, $v_1^+, v_1^-, v_2^+, v_2^-$ are regular, that is, $v \in M(T,\mathbb{C})$. In particular $v \in M(T,\mathbb{C})$ if and only if $|v| \in M^+(T,\mathbb{R})$.

Theorem 3. *Let T be a compact Hausdorff space. Then $M(T,\mathbb{R})$ and $M(T,\mathbb{C})$ are bands in $m(T,\mathbb{R})$ and $m(T,\mathbb{C})$ respectively. In particular, the norm is order continuous in $M(T,\mathbb{R})$ and $M(T,\mathbb{C})$ and every closed ideal is a projection band.*

Proof. It clearly suffices to show that $M(T,\mathbb{R})$ and $M(T,\mathbb{C})$ are closed in $m(T,\mathbb{R})$ and $m(T,\mathbb{C})$ respectively since we have already shown that they are ideals. Moreover, by the continuity of the lattice operations it suffices to show that $M^+(T,\mathbb{R})$ is closed. Suppose $\mu_n \in M^+(T,\mathbb{R})$ and $\mu = \lim \mu_n$ (recall that $\mu \geqslant 0$). Let $A \in \mathscr{B}$ and $\varepsilon > 0$ be given. There is an n such that $\|\mu_n - \mu\| < \varepsilon$. By regularity there is a compact set $F \subset A$ such that $\mu_n(A \setminus F) < \varepsilon$. Thus $\mu(A \setminus F) \leqslant |\mu_n(A \setminus F) - \mu(A \setminus F)| + \mu_n(A \setminus F) < 2\varepsilon$ and it follows that μ is regular. ∎

The following theorem is the one we refer to as the *Riesz representation theorem*. We put $S(f) = \overline{\{t: f(t) \neq 0\}}$ ($S(f)$ is called the *closed support* of f).

Theorem 4. *Let T be a compact Hausdorff space and x^* be a positive linear functional on $C(T,\mathbb{R})$ of norm one. Then there is a unique regular Borel measure μ on T such that $x^*(f) = \int f\,d\mu$ for all $f \in C(T,\mathbb{R})$.*

Proof. We define a set function μ^* on the class of all subsets of T, show that it is a Caratheodory outer measure, that the measure μ induced by μ^* is regular, and that $x^*(f) = \int f\,d\mu$ for all $f \in C(T,\mathbb{R})$.

The set function μ^* is defined on open sets U as follows: $\mu^*(U)=\sup\{x^*(f): f\in C(T,\mathbb{R}), 0\leqslant f\leqslant 1, S(f)\subset U\}$ if $U\neq\emptyset$ and $\mu(\emptyset)=0$. Suppose U, V are open sets. Let $\varepsilon>0$ and choose f so $0\leqslant f\leqslant 1$ and $x^*(f)>\mu^*(U\cup V)-\varepsilon$. Let $F=\{t: f(t)\geqslant\varepsilon\}$. Then F is a compact subset of $U\cup V$. Let W be an open set such that $F\setminus U\subset W\subset\overline{W}\subset V$. Thus $F_1=F\cap\overline{W}$ and $F_2=F\setminus W$ are compact sets and $F=F_1\cup F_2$. By Urysohn's lemma there are $f_1, f_2\in C(T,\mathbb{R})$ with $0\leqslant f_1$, $f_2\leqslant 1$, $f_1=1$ on F_1, $f_1=0$ on $T\setminus V$ and $f_2=1$ on F_2, $f_2=0$ on $T\setminus U$. Moreover, $f_1+f_2+\varepsilon\geqslant f$. Thus $x^*(f_1)+x^*(f_2)+\varepsilon\geqslant x^*(f)>\mu^*(U\cup V)-\varepsilon$. But, $\mu^*(U)+\mu^*(V)+\varepsilon\geqslant x^*(f_1)+x^*(f_2)+\varepsilon\geqslant\mu^*(U\cup V)-\varepsilon$ and it follows that $\mu^*(U)+\mu^*(V)\geqslant\mu^*(U\cup V)$. Clearly for any finite collection $U_1,\ldots,U_n$ of open sets, $\mu^*\left(\bigcup_{i=1}^{n} U_i\right)\leqslant\sum_{i=1}^{n}\mu^*(U_i)$. Now suppose $\{U_n\}$ is a sequence of open sets. Since for any f with $S(f)\subset\bigcup_{n=1}^{\infty} U_n$ it follows that $S(f)\subset\bigcup_{i=1}^{n} S(f_i)$ for some n (by compactness of $S(f)$), it follows immediately that $\mu^*\left(\bigcup_{n=1}^{\infty} U_n\right)\leqslant\sum_{n=1}^{\infty}\mu^*(U_n)$. It is easy to see that if the U_n's are pairwise disjoint, then we have that $\mu^*\left(\bigcup_{n=1}^{\infty} U_n\right)=\sum_{n=1}^{\infty}\mu^*(U_n)$. (We used the fact that if $0\leqslant f\leqslant 1$ and $f(t)=0$ for $t\notin U$, then $x^*(f)\leqslant\mu^*(U)$. This is proved below.)

Let us further note that for any open set U and any $\varepsilon>0$ there is an open set $W\subset\overline{W}\subset U$ such that $\mu^*(W)>\mu^*(U)-\varepsilon$. For, choose $f\in C(T,\mathbb{R})$ with $0\leqslant f\leqslant 1$ and $S(f)\subset U$ with $x^*(f)>\mu^*(U)-\varepsilon$. Let W be an open set with $S(f)\subset W\subset\overline{W}\subset U$. Then clearly $\mu^*(W)\geqslant x^*(f)>\mu^*(U)-\varepsilon$.

We define the set function μ^* on any subset of A of T as follows: $\mu^*(A)=\inf\{\mu^*(U): A\subset U,\ U \text{ is open}\}$. It is routine to check that μ^* is indeed a Caratheodory outer measure. Recall that a set A is called measurable if and only if $\mu^*(B)=\mu^*(A\cap B)+\mu^*(A'\cap B)$ for any $B\subset T$ (where $A'=T\setminus A$), and that the collection Σ of all measurable sets is a σ algebra and $\mu^*|\Sigma$ is a measure [131, p. 126f.].

To see that any open set is measurable we first show that if U and V are open sets, then $\mu^*(U)\geqslant\mu^*(U\cap V)+\mu^*(U\cap V')$. For, let $\varepsilon>0$ be given. By the above there is an open set $W\subset\overline{W}\subset U\cap V$ such that $\mu^*(W)>\mu^*(U\cap V)-\varepsilon$. Now, $\mu^*(U\cap V)+\mu^*(U\cap V')\leqslant\mu^*(U\cap V)+\mu^*(U\cap(\overline{W})')\leqslant\mu^*(W)+\mu^*(U\cap(\overline{W})')+\varepsilon=\mu^*(W\cup(U\cap(\overline{W})'))+\varepsilon\leqslant\mu^*(U)+\varepsilon$.

Suppose $A, U\subset T$ with U open and let $\varepsilon>0$ be given. There is an open set $V\supset A$ with $\mu^*(V)<\mu^*(A)+\varepsilon$. Now $V=(V\cap U)\cup(V\cap U')$ and $\mu^*(V)\geqslant\mu^*(V\cap U)+\mu^*(V\cap U')\geqslant\mu^*(A\cap U)+\mu^*(A\cap U')$. Thus it follows that $\mu^*(A) \geqslant \mu^*(A \cap U) + \mu^*(A \cap U')$. Since the reverse inequality is clearly valid, we obtain that U is measurable. Hence $\mu = \mu^* | \mathscr{B}$ is a measure.

To see that μ is regular note that the collection of all Borel sets B such that $\mu(B) = \sup\{\mu(F): F \subset B,\ F \text{ is compact}\}$ is a σ algebra. Since it contains the open sets, it must be all of $\mathscr{B}$.

Let U be an open set and suppose f is a continuous function such that $0 \leqslant f \leqslant 1$ and $f(t)=0$ for all $t \in T \setminus U$. Then $x^*(f) \leqslant \mu(U)$. For, let $\varepsilon > 0$ be given and put $F = \{t: f(t) \geqslant \varepsilon\} \subset U$. Then F is closed and there is a continuous function g such that $0 \leqslant g \leqslant f$, $g = f$ on F, and $S(g) \subset U$. Thus $f - g \leqslant \varepsilon \cdot 1$ and $x^*(f) - x^*(g) \leqslant \varepsilon$. Hence $x^*(f) \leqslant x^*(g) + \varepsilon \leqslant \mu(U) + \varepsilon$ and it follows that $x^*(f) \leqslant \mu(U)$. We shall show that $x^*(f) \leqslant \int f \, d\mu$ for all $f \in C(T, \mathbb{R})$. Since this implies that $-x^*(f) = x^*(-f) \leqslant \int -f \, d\mu = -\int f \, d\mu$, we obtain that $x^*(f) = \int f \, d\mu$ for all $f \in C(T, \mathbb{R})$.

Suppose f is a continuous function and $\varepsilon > 0$ is given. Let $-(\|f\|+1) = a_0 < \cdots < a_n = (\|f\|+1)$ be such that $a_i - a_{i-1} < \varepsilon$ for $i = 1, \ldots, n$. Put $A_i = \{t: a_{i-1} < f(t) \leqslant a_i\}$ $(i=1, \ldots, n)$ and $B_i = \{t: f(t) > a_i\}$ $(i = 0, \ldots, n)$. Clearly $A_i = B_{i-1} \setminus B_i$ and B_i is open. We define f_i by

$$f_i(t) = \begin{cases} 1 & \text{if } t \in B_i \\ \dfrac{f(t) - a_{i-1}}{a_i - a_{i-1}} & \text{if } t \in A_i. \\ 0 & \text{if } t \in T \setminus B_{i-1} \end{cases}$$

Then each f_i is continuous and $f = a_0 + \sum_{i=1}^{n} (a_i - a_{i-1}) f_i$. Thus

$$x^*(f) = a_0 + \sum_{i=1}^{n} (a_i - a_{i-1}) x^*(f_i) \leqslant a_0 + \sum_{i=1}^{n} (a_i - a_{i-1}) \mu(B_{i-1}) = \sum_{i=1}^{n} a_i \mu(A_i)$$
$$\leqslant \sum_{i=1}^{n} a_{i-1} \mu(A_i) + \varepsilon \leqslant \int f \, d\mu + \varepsilon.$$

To see that μ is unique observe that if $x^*(f) = \int f \, d\nu$ for all $f \in C(T, \mathbb{R})$, then $\nu \geqslant 0$, $\nu(T) = 1$, and $\nu(U) = \mu(U)$ for all open sets ν. For, if $\nu(A) < 0$ for some A, then by regularity $\nu(U) < 0$ for some open set $U \supset A$. Let $F \subset U$ be a compact set with $\nu(F) \neq 0$ and choose $f \in C(T, \mathbb{R})$ such that $f = 1$ on F, $0 \leqslant f \leqslant 1$, and $f = 0$ on $T \setminus U$. Then $x^*(f) = \int f \, d\nu < 0$ which is a contradiction. A similiar argument shows that $\nu(U) = \mu(U)$ for all open sets U. Thus by regularity it follows that $\nu = \mu$. ∎

Theorem 5. *Let T be a compact Hausdorff space. For $\mu \in M(T, \mathbb{C})$ and $f \in C(T, \mathbb{C})$ let $(L\mu)(f) = \int f \, d\mu$. Then L is a linear isometry of $M(T, \mathbb{C})$ onto $C(T, C)^*$. In particular, $L_0(\mu) = L(\mu) | C(T, \mathbb{R})$ for $\mu \in M(T, \mathbb{R})$ is an order preserving linear isometry of $M(T, \mathbb{R})$ onto $C(T, \mathbb{R})^*$.*

Proof. Clearly if $\mu \geqslant 0$, then $L_0(\mu) \geqslant 0$. Suppose $\mu \in M(T, \mathbb{R})$ and $L_0(\mu) \geqslant 0$. If $\mu^- \neq 0$, then there is a compact set F such that $\mu^-(F) > 0$.

Let U be an open set containing F such that $|\mu|(U \setminus F) < \frac{1}{2}\mu^-(F)$. By Urysohn's lemma there is an $f \in C(T, \mathbb{R})$ such that $0 \leqslant f \leqslant 1$, $f = 1$ on F and $f = 0$ on $T \setminus U$. Now $0 \leqslant \int f\,d\mu = \int_F f\,d\mu + \int_{U \setminus F} f\,d\mu$. But, $\int_F f\,d\mu = -\mu^-(F)$ and $|\int_{U \setminus F} f\,d\mu| \leqslant \int_{U \setminus F} |f|\,d|\mu| \leqslant |\mu|(U \setminus F) < \frac{1}{2}\mu^-(F)$. Hence it follows that $\mu^- = 0$ and L_0 is order preserving. It follows immediately from theorem 4 that L_0 is onto $C(T, \mathbb{R})^*$. Moreover, for $x^* \in C(T, \mathbb{R})^*$, $x^* = (x^*)^+ - (x^*)^-$ and $\|x^*\| = \|(x^*)^+\| + \|(x^*)^-\|$. If $L_0(\mu) = (x^*)^+$ and $L_0(\nu) = (x^*)^-$, then since L_0 is order preserving, $\mu, \nu \geqslant 0$ and $\mu \wedge \nu = 0$. Thus $L_0(\mu - \nu) = x^*$ and $\|x^*\| = \|\mu\| + \|\nu\| = \|L_0(\mu - \nu)\|$ and it follows that L_0 is a linear isometry.

We now prove that L is a linear isometry. If $\mu \in M(T, \mathbb{C})$ and $f \in C(T, \mathbb{C})$, then $|L(\mu)(f)| = |\int f\,d\mu| \leqslant \int |f|\,d|\mu| \leqslant \|f\|\,\|\mu\|$ so that $\|L(\mu)\| \leqslant \|\mu\|$.

Let $\varepsilon > 0$ be given. There are pairwise disjoint Borel sets $A_1, \ldots, A_n$ in T such that $T = \bigcup_{i=1}^{n} A_i$ and $\sum_{i=1}^{n} |\mu(A_i)| > \|\mu\| - \varepsilon$. Let $F_i \subset A_i$ be a compact set such that $|\mu|(A_i \setminus F_i) < \varepsilon/n$ for $i = 1, \ldots, n$. Let $V_i \supset F_i$ be an open set such that $V_1, \ldots, V_n$ are pairwise disjoint and $|\mu|(V_i \setminus F_i) < \varepsilon/n$ for $i = 1, \ldots, n$ choose $f_i \in C(T, \mathbb{C})$ with $0 \leqslant f_i \leqslant 1$, $S(f_i) \subset V_i$ and $f_i = 1$ on F_i for $i = 1, \ldots, n$. Let $a_i \in \mathbb{C}$ be such that $|a_i| = 1$ and $a_i \mu(A_i) = |\mu(A_i)|$ for $i = 1, \ldots, n$. If $f = \sum_{i=1}^{n} a_i f_i$, then $\left|\int f\,d\mu - \|\mu\|\right| \leqslant \left|\sum_{i=1}^{n} a_i \int f_i\,d\mu - \sum_{i=1}^{n} a_i \mu(V_i)\right| + \left|\sum_{i=1}^{n} a_i \mu(V_i) - \sum_{i=1}^{n} a_i \mu(A_i)\right| + \sum_{i=1}^{n} a_i \mu(A_i) - \|\mu\| \leqslant \varepsilon + 2\varepsilon + \varepsilon = 4\varepsilon$. Thus $\|L(\mu)\| = \|\mu\|$.

To see that L is onto, let $x^* \in C(T, \mathbb{C})^*$ and define y^*, z^* on $C(T, \mathbb{R})$ by $y^*(f) = \operatorname{Re} x^*(f)$ and $z^*(f) = -\operatorname{Re} x^*(i f)$ for $f \in C(T, \mathbb{R})$. Then clearly y^*, z^* are linear functionals and $\|y^*\|, \|z^*\| \leqslant \|x^*\|$. By the first part $y^*(f) = \int f\,d\mu$ and $z^*(f) = \int f\,d\nu$ for some $\mu, \nu \in M(T, \mathbb{R})$. Moreover, $x^*(f + i g) = (y^*(f) - z^*(g)) + i(y^*(g) + z^*(f))$ for $f, g \in C(T, \mathbb{R})$ so that $x^*(f + i g) = \int (f + i g)\,d(\mu + i\nu)$. ∎

In [252] Royden gave another integral representation for $C(T, \mathbb{C})^*$ and $C(T, \mathbb{R})^*$ which we now present. The proof is given for the complex case only and will work in the real case with obvious modifications. If $g \in C(T, \mathbb{C})$, $\bar{g}$ denotes the complex conjugate of g.

Theorem 6. *Let T be a compact Hausdorff space and x^* be a nonzero bounded linear functional on $C(T, \mathbb{C})$. Then there is a positive regular Borel measure μ on T and a Borel measurable function h on T such that $|h| = 1$ (μ almost everywhere) and $x^*(f) = \int f h\,d\mu$ for all $f \in C(T, \mathbb{C})$. Moreover, $\|x^*\| = \|\mu\|$.*

Proof. By theorem 5 we can assume that $x^*=(\mu_1-\mu_2)+i(\mu_3-\mu_4)$ where μ_i is a positive regular Borel measure for $i=1,2,3,4$. Let $\nu=\mu_1+\mu_2+\mu_3+\mu_4$. Then ν is a positive Borel measure and each μ_i is absolutely continuous with respect to ν. By the Radon-Nikodym theorem there are Borel measurable functions k_i on T such that $\int f\,d\mu_i=\int f k_i\,d\nu$ for all $f\in C(T,\mathbb{C})$. Now $k=|k_1-k_2+i(k_3-k_4)|$ is integrable and μ defined by, $\mu(B)=\int_B k\,d\nu$ is a regular Borel measure on T. Moreover, if $h=(1/k)[(k_1-k_2)+i(k_3-k_4)]$ where $k\neq 0$ and $h=1$ otherwise, then h is Borel measurable, $|h|=1$ and $x^*(f)=\int fh\,d\mu$ for all $f\in C(T,\mathbb{C})$.

Now, $|x^*(f)|\leqslant\int|f|\,d\mu\leqslant\|f\|\,\|\mu\|$ for all $f\in C(T,\mathbb{C})$. Hence $\|x^*\|\leqslant\|\mu\|$. On the other hand, let $\varepsilon>0$ be given. Then by Lusin's theorem there is a $f\in C(T,\mathbb{C})$ such that $\mu\{t:|f(t)-h(t)|>\varepsilon\}<\varepsilon$. If $g(t)=f(t)$ for $|f(t)|\leqslant 1$ and $g(t)=f(t)/|f(t)|$ for $|f(t)|>1$, then $|g|\leqslant 1$ and since $|h|=1$, $\mu\{t:|g(t)-h(t)|>\varepsilon\}<\varepsilon$. But $\|x^*\|\geqslant|x^*(\bar{g})|$ and $|\int\bar{g}h\,d\mu-\mu(T)|\leqslant\int|\bar{g}h-1|\,d\mu=\int|\bar{g}-\bar{h}|\,d\mu\leqslant 2\varepsilon+\varepsilon\mu(T)$.

Thus $\|x^*\|\geqslant|\int\bar{g}h\,d\mu|\geqslant\mu(T)-\varepsilon[1+\mu(T)]$ and it follows that $\|x^*\|\geqslant\|\mu\|$. ∎

We shall need the following result when we discuss the extreme points of the unit ball of $C(T,\mathbb{C})$. A nonzero bounded linear functional x^* on a Banach space X is said to *attain its norm* if there is an $x\in X$ with $x^*(x)=\|x^*\|$. It is clear that x^* attains its norm if and only if ax^* attains it norm for each scalar $a\neq 0$. Also recall that the *support* of a positive regular Borel measure μ on T is the smallest closed set F such that $\mu(T)=\mu(F)$.

Theorem 7. *Let T be a compact Hausdorff space, x^* a nonzero bounded linear functional on $C(T,\mathbb{C})$, μ a positive regular Borel measure on T, and h a Borel measurable function on T with $|h|=1$ and $x^*(f)=\int fh\,d\mu$ for all $f\in C(T,\mathbb{C})$ and $\|x^*\|=\|\mu\|$. Then x^* attains its norm if and only if there is a $g\in C(T,\mathbb{C})$ with $|g|=1$ on the support $S(\mu)$ of μ, $\|g\|\leqslant 1$, and $x^*(f)=\int fg\,d\mu$ for all $f\in C(T,\mathbb{C})$.*

Proof. Clearly if such a g exists, then x^* attains its norm at $\bar{g}$.

Conversely, suppose x^* attains its norm. Then there is an $f\in C(T,\mathbb{C})$ with $\|f\|\leqslant 1$ and $x^*(f)=\int fh\,d\mu=\int 1\,d\mu$. Thus $fh=1$ almost everywhere on $S(\mu)$. That is, $h=\bar{f}$ almost everywhere on an $S(\mu)$. Hence $g=\bar{f}$ is the required function. ∎

Reflexive spaces are the only Banach spaces X with the property that every $x^*\in X^*$ attains its norm (see [139]). However, the collection of functionals which do attain their norms is always dense in X^* (see

[45]). We give a proof of this for $C(T,\mathbb{C})$ based on the above characterization. That is, for any x^* in $C(T,\mathbb{C})^*$ and any $\varepsilon>0$ there is a y^* in $C(T,\mathbb{C})^*$ which attains its norm and $\|x^*-y^*\|<\varepsilon$. For, let x^*, μ, h be as above and $\varepsilon>0$. There is a continuous function g on T with $\|g\|\leqslant 1$ and a closed set $F\subset S(\mu)$ such that $g|F=h|F$ and $\mu(T\setminus F)<\varepsilon$. If y^* is defined by $y^*(f)=\int_F fg\,d\mu$, then y^* attains its norm at $\bar{g}$ and $\|y^*-x^*\|<2\varepsilon\|x^*\|$.

Corollary. *Let T be a compact Hausdorff space, $U=b(C(T,\mathbb{C}))$ and $D\subset U$. If the closed convex hull of D is not U, then there is a positive regular Borel measure μ on T with $\|\mu\|=1$ and an $f\in C(T,\mathbb{C})$ with on $S(\mu)$ and* $\sup\limits_{g\in D}\operatorname{Re}\int gf\,d\mu<1$.

Proof. If the closed convex hull K of D is not U, then by the second basic separation theorem if $h\in U\setminus K$, there is an x^* in $C(T,\mathbb{C})^*$ such that $\|x^*\|=1$ and $\sup\limits_{g\in D}\operatorname{Re}x^*(g)<\operatorname{Re}x^*(h)\leqslant 1$. Let $\varepsilon>0$ be such that $1-\varepsilon>\sup\limits_{g\in D}\operatorname{Re}x^*(g)$ and choose y^* in $C(T,\mathbb{C})^*$ which attains its norm, $\|y^*\|=1$, and $\|x^*-y^*\|<\varepsilon/2$. Then $\sup\limits_{g\in D}\operatorname{Re}y^*(g)<1$ and μ and f as in theorem 7 for y^* have the required properties. ∎

We now give a brief discussion of atoms in measures. Let $\mu\in M^+(T,\mathbb{R})$. A Borel set A is said to be an *atom* for μ if $\mu(B)=0$ or $\mu(B)=A$ for all Borel sets $B\subset A$. The measure μ is said to be *purely atomic* if every set of positive measure contains a subset which is an atom. On the other hand, μ is said to be *purely nonatomic* if there are no atoms for μ. We shall call $\mu\in M(T,\mathbb{C})$ *purely atomic* if $|\mu|$ is purely atomic.

The following theorem is an easy consequence of Zorn's lemma and compactness and is left to the reader.

Theorem 8. *Let $\mu\in M^+(T,\mathbb{R})$. If A is an atom for μ, then there is a $t\in A$ such that $\mu(\{t\})=\mu(A)$.* ∎

Let $t\in T$. By ε_t we mean the regular Borel measure given by $\varepsilon_t(A)=1$ if $t\in A$ and $\varepsilon_t(A)=0$ if $t\notin A$. Clearly ε_t is purely atomic and $\{t\}$ is an atom for ε_t. For each $t\in T$ let $a_t\in\mathbb{C}$ and suppose $\sum\limits_{t\in T}|a_t|<\infty$ (see chapter 1), that is, $\{t: a_t\neq 0\}=\{t_n\}$ is countable and $\sum|a_{t_n}|<\infty$. Thus $\mu=\sum a_t\varepsilon_t$ exists in $M(T,\mathbb{C})$ and $|\mu|=\sum|a_t|\varepsilon_t$ is purely atomic. We denote the subspace of all purely atomic measures by $l_1(T,\mathbb{C})$. The above theorem yields that every element of $l_1(T,\mathbb{C})$ is of the form $\Sigma a_t\varepsilon_t$ with $\sum|a_t|<\infty$. For, suppose μ is purely atomic and let $\{A_i\}_{i\in I}$ be the collection of all distinct atoms for $|\mu|$. By theorem 8 for each A_i there is a unique $t_i \in A_i$ with $|\mu|(A_i)=|\mu|(\{t_i\})$. Since $|\mu|$ is finite, I is a countable set. We put $a_i=\mu(\{t_i\})$ for all $i \in I$. Clearly $\sum |a_i| < \infty$ and $\mu=\sum a_i \varepsilon_i = \sum a_t \varepsilon_t$ where $a_t=a_i$ if $t=t_i$ and $a_t=0$ otherwise. For, if $F \subset I$ is a finite set, then $\sum_{i \in F} |a_i| \leqslant \sum_{i \in F} |\mu|(\{t_i\}) + \mu(T \setminus \{t_i : i \in F\}) \leqslant \|\mu\|$ so that $\sum |a_i| \leqslant \|\mu\|$. Now for any Borel set A, $\mu(A)=\sum\{a_i : \mu(A \cap A_i) \neq 0\}$. For, since μ is purely atomic, $A=\bigcup\{A \cap A_i : \mu(A \cap A_i) \neq 0\}$ (μ almost everywhere) and $\mu(A \cap A_i) \neq 0$ implies $\mu(A \cap A_i)=a_i$. In particular, $\sum |a_i| = \|\mu\|$ and $\mu=\sum a_i \varepsilon_i$.

We leave the proof of the following theorem to the reader.

Theorem 9. *Let T be a compact Hausdorff space. Then $l_1(T,\mathbb{C})$ is a closed ideal in $M(T,\mathbb{C})$ (resp., $l_1(T,\mathbb{R})$ is a closed ideal in $M(T,\mathbb{R})$). In particular, $M(T,\mathbb{C})=l_1(T,\mathbb{C}) \oplus l_1(T,\mathbb{C})^{\perp}$ (resp., $M(T,\mathbb{R})=l_1(T,\mathbb{R}) \oplus l_1(T,\mathbb{R})^{\perp}$). Moreover, $\mu \in l_1(T,\mathbb{C})^{\perp}$ if and only if $|\mu|$ is purely nonatomic.* ∎

Clearly purely atomic measures always exists. It may be however that $l_1(T,C)^{\perp}=\{0\}$. We investigate exactly when we are guaranteed the existence of a nonzero purely nonatomic measure in the theorem below. The theorem and proof are due to Pełczyński and Semadeni [235]. Alternate proofs are suggested in the exercises. We use λ to denote Lebesgue measure on $[0,1]$.

Theorem 10. *Let T be a compact Hausdorff space. Then there is a purely nonatomic measure on T if and only if T is not dispersed.*

Proof. Suppose μ is a positive purely nonatomic measure on T with $\mu \neq 0$. Then $P=S(\mu)$ is perfect and compact. For, suppose P is not perfect, then there is an open set U in T with $U \cap P=\{t\}$ for some $t \in P$. Now $\mu(\{t\})=0$ and $P \setminus \{t\}$ is a smaller closed set than P such that $\mu(P)=\mu(T)$ which is a contradiction to $P=S(\mu)$.

Now suppose T contains a perfect set. Then there is a continuous map g of T onto $[0,1]$.

Let $L: C([0,1],\mathbb{R}) \to C(T,\mathbb{R})$ be defined by $Lf=f \circ g$. Then L is a linear operator and $\|Lf\|=\|f\|$ for all $f \in C([0,1],\mathbb{R})$. Let x^* be a bounded linear functional on $C(T,\mathbb{R})$ such that $x^*(Lf)=\int_0^1 f(t)\,dt$ for all $f \in C([0,1],\mathbb{R})$ and $\|x^*\|=x^*(1)=1$. Then by the Riesz representation theorem there is a positive regular Borel measure μ on T such that $\int f\,d\mu=x^*(f)$ for all $f \in C(T,\mathbb{R})$. Suppose μ is purely atomic. Then $\mu=\sum_{t \in T} a_t \varepsilon_t$ for some summable family $\{a_t\}$. In particular, for $f \in C([0,1],\mathbb{R})$, $\int f \circ g\,d\mu=\sum_{t \in T} a_t f(g(t))=\int_0^1 f(t)\,dt$. That is $\lambda=\sum_{s \in [0,1]} b_s \varepsilon_s$ where $b_s=\sum_{t \in g^{-1}(s)} a_t$ which is clearly impossible. ∎

The extreme points of the unit sphere of $C(T,\mathbb{C})^*$ play an important role in the analysis of T, $C(T,\mathbb{C})$, and $C(T,\mathbb{C})^*$. The complete identification and characterization of these extreme points in, presented now. The same theorem and proof are valid for $C(T,\mathbb{R})$.

Theorem 11. *Let T be a compact Hausdorff space and x^* be in $C(T,\mathbb{C})^*$ with $\|x^*\| \leqslant 1$. Then the following are equivalent.*

(1) *$x^* = \varepsilon_t$ for some $t \in T$,*

(2) *$x^*(f) \geqslant 0$ for all $f \geqslant 0$ and x^* is an extreme point of the unit sphere of $C(T,\mathbb{C})^*$,*

(3) *$x^*(fg) = x^*(f)x^*(g)$ for all f,g in $C(T,\mathbb{C})$,*

(4) *$x^*(f \vee g) = \max(x^*(f), x^*(g))$ for all f,g in $C(T,\mathbb{R})$,*

(5) *$x^*(f \wedge g) = \min(x^*(f), x^*(g))$ for all f,g in $C(T,\mathbb{R})$.*

Proof. (1) implies (2). Suppose $\varepsilon_t = a\mu + (1-a)\nu$ for μ, ν in $M(T,\mathbb{C})$ with $\|\mu\| \leqslant 1$, $\|\nu\| \leqslant 1$ and $0 \leqslant a \leqslant 1$. Then $1 = a\mu(T) + (1-a)\nu(T)$ implies $\mu(T) = 1 = \nu(T)$. Thus μ and ν are positive measures of norm 1. Let U be an open set containing t and f be in $C(T,\mathbb{C})$ with $f=0$ on $T \setminus U$, $0 \leqslant f \leqslant 1$, and $f(t)=1$. Then $1 = a\int f\,d\mu + (1-a)\int f\,d\nu$ and it follows that $\int f\,d\mu = \int f\,d\nu = 1$. Thus $\mu(U) = \nu(U) = 1$ and it follows that $\mu(\{t\}) = \nu(\{t\}) = 1$. Hence $\mu = \nu = \varepsilon_t$.

(2) implies (1). Suppose $\mu \neq \varepsilon_t$ for any $t \in T$. Then there are two disjoint Borel sets A and B in T with $a = \mu(A)$, $b = \mu(B)$, $0 < a$, $b < 1$, and $a + b = 1$. Let ν_1 be defined by $\nu_1(E) = (1/a)\mu(E \cap A)$ and ν_2 be defined by $\nu_2(E) = (1/b)\mu(E \cap B)$ for all Borel sets $E \subset T$. Then ν_1, ν_2 are regular Borel measures, $\nu_1(T) = \nu_2(T) = 1$, and $\mu = a\nu_1 + b\nu_2$ with $\nu_1 \neq \nu_2$. Thus μ is not an extreme point of the unit sphere of $C(T,\mathbb{C})^*$.

Clearly (1) implies (3), (4), and (5). Moreover, (4) and (5) are obviously equivalent.

(3) implies (1). Let $\mu \neq 0$ be a regular Borel measure on T such that $\int fg\,d\mu = (\int f\,d\mu)(\int g\,d\mu)$ for all f,g in $C(T,\mathbb{C})$. Then $\int f^2\,d\mu = [\int f\,d\mu]^2$ for all $f \in C(T,\mathbb{C})$. Hence μ is a positive measure. Now, $\int 1\,d\mu = [\int 1\,d\mu]^2$ implies $\mu(T) = 1$. Suppose $\mu \neq \varepsilon_t$ for any $t \in T$. Then there are closed disjoint sets A and B in T with $0 < \mu(A)$, $\mu(B) < 1$. Let U and V be disjoint open sets with $A \subset U$ and $B \subset V$. Choose f,g in $C(T,\mathbb{C})$ with $0 \leqslant f$, $g \leqslant 1$ and $f=1$ on A, $f=0$ on $T \setminus U$, $g=1$ on B, and $g=0$ on $T \setminus V$. Then $\int f\,d\mu \geqslant \mu(A)$, $\int g\,d\mu \geqslant \mu(B)$, but $fg=0$ and hence $\int fg\,d\mu = 0$. The same type of argument shows that (4) implies (1). ∎

Corollary. *The set of extreme points of the closed ball of $C(T,\mathbb{C})^*$ is $\{a\varepsilon_t : |a| = 1, t \in T\}$.*

Proof. Let U^* be the closed unit ball of $C(T,\mathbb{C})^*$. Since it is weak* compact, by the Krein-Milman theorem it is the closed convex hull of its extreme points. Suppose U^* is not the weak* closed convex hull of $\{a\varepsilon_t : |a|=1, t\in T\}$. Then there is a $\mu\in U^*$ and an $f\in C(T,\mathbb{C})$ with $\|f\|=1$ and $\operatorname{Re}\int f\,d\mu > \sup_{\substack{|a|=1\\ t\in T}} \operatorname{Re} af(t) = \|f\| = 1$ by the second basic separation theorem, which is impossible. Thus $\{a\varepsilon_t : |a|=1, t\in T\}$ is precisely the set of extreme points of U^*. ∎

Finally, we note that an easy test for when a bounded linear functional x^* on $C(T,\mathbb{C})$ is positive is the requirement that $x^*(1)=\|x^*\|$. It is clearly necessary. Conversely, suppose $f\geqslant 0$ is in $C(T,\mathbb{C})$ and $x^*(1)=\|x^*\|$. Let K be a closed sphere in $\mathbb{C}$ with center d and radius $t>0$ which contains $f(T)$. Then $\|f-d\cdot 1\|\leqslant t$ and $t\geqslant |x^*(f-d\cdot 1)| = |x^*(f)-d|$ and $x^*(f)\in K$. Thus $x^*(f)$ is the closed convex hull of $f(T)$ which is contained in the nonnegative real axis.

Exercises. 1. Show that for each cardinal $m>\aleph_0$ there is a compact Hausdorff dispersed space which is not homeomorphic to a ray of ordinals.

2. Given an example of a nondispersed compact metric space in which the set of isolated points is dense.

3. (Pełczyński-Semadeni). Let T be a compact Hausdorff space. Show that T is dispersed if and only if T is totally disconnected and the Cantor set is not a continuous image of T.

4. (Morris). Prove that a compact Hausdorff space T contains a perfect set if and only if it contains an interlocking sequence $\{(A_n, B_n)\}$ of closed sets.

5. (Lacey-Cohen). Let T be a compact Hausdorff space. A projective space P with a minimal onto continuous map from P to T is called the *Gleason space* of T. Show that if T_1 and T_2 are perfect compact metric spaces, then they have the same Gleason space.

6. (Lacey-Cohen). Show that there are exactly five Gleason spaces for the set of all compact metric spaces.

7. Let S, T be compact Hausdorff spaces and suppose there is a minimal onto continuous mapping f from S to T. Then

(i) if $A\subset T$ is closed and nowhere dense, $f^{-1}(A)$ is closed and nowhere dense.

(ii) if S is totally disconnected and $B\subset S$ is closed and nowhere dense, then $f(B)$ is closed and nowhere dense,

(iii) for each nonempty open set $U\subset S$, $f(U)$ has nonempty interior in T.

8. Let P, T be compact Hausdorff spaces with P a projective space and f a minimal continuous mapping of P onto T. Then for any nonempty closed and open set $U\subset P$, $\operatorname{int} f^{-1}[f(U)]=U$ and $\overline{f^{-1}(\operatorname{int} f(U))}=U$.

If $U \subset T$ is an open set such that $U = \operatorname{int} \bar{U}$, then $\overline{f^{-1}(U)} = \operatorname{int} f^{-1}(\bar{U})$ and $\overline{f^{-1}(U)} \cap \overline{f^{-1}(T \setminus U)} = \emptyset$.

9. Let S and T be compact Hausdorff spaces and $f: S \to T$ an onto continuous function. Show that if μ is a positive regular Borel measure on S, then there is a positive regular Borel measure v on T such that $v(B) = \mu(f^{-1}(B))$ for all Borel sets $B \subset T$ and $\int g \, dv = \int g f \, d\mu$ for all $g \in C(T, \mathbb{C})$.

10. (Hardy-Lacey). Let S, T, f be as in 9. Show that if v is a positive regular Borel measure on T, then there is a positive regular Borel measure μ on S such that $\mu(f^{-1}(B)) = v(B)$ for all Borel sets $B \subset T$.

11. (Hardy-Lacey). Use 10 to show that if S is a compact Hausdorff space with a perfect set, then S admits a purely nonatomic regular Borel measure.

12. (Knowles). Let T be a perfect compact Hausdorff space and K be the set of positive regular Borel measures on T of norm one. Then in the weak* topology from $C(S, \mathbb{R})$, K is compact. Show that for each $a > 0$, $\{v \in K : v(t) \geqslant a$ for some $t \in T\}$ is a closed and nowhere dense set in K. Use this to prove that there is a purely nonatomic element in K.

13. Let K be a compact set in some locally convex Hausdorff topological vector space and T a closed subset of K containing the set of extreme points of K. Show that for each $x \in K$ there is a positive regular Borel measure of norm one μ on T such that $h(x) = \int h \, d\mu$ for all $h \in A(K, \mathbb{R})$.

14. Let K and T be as in 13. Show that for each positive regular Borel measure of norm one μ on T there is a unique $x \in K$ such that $h(x) = \int h \, d\mu$ for all $h \in A(K, \mathbb{R})$.

15. (Darst). Let T be a compact Hausdorff space and μ a purely nonatomic positive regular Borel measure on T. Show that for each Borel set $B \subset T$ with $\mu(B) > 0$ and each a with $0 \leqslant a < \mu(B)$ there is a perfect set $P \subset B$ with $\mu(P) = a$.

16. Let T be a nondispersed compact metric space. Show that there is an uncountable family $\{\mu_i\}$ of positive purely nonatomic regular Borel measures on T of norm one such that $\mu_i \wedge \mu_j = 0$ for all $i \neq j$.

17. If $T = [0, 1]$, show that the family in 16 can be chosen so that the support of each μ_i is $[0, 1]$.

18. (Hebert-Lacey). Let T be a compact metric space. Show that there is a measure whose support is T. Show that if T is perfect, then there is a purely nonatomic measure whose support is T.

19. Let T be a compact Hausdorff space and $F \subset C(T, \mathbb{C})$ a family of functions directed downwards under pointwise ordering and bounded from below. Let $\mu \in M(T, \mathbb{R})$ be positive and $\bar{f} = \inf\{f : f \in F\}$. Show that $\int \bar{f} \, d\mu = \inf\{\int f \, d\mu : f \in F\}$.

20. Let $X \subset C(T, \mathbb{C})$ be a closed linear subspace. Show that for any extreme point x^* of the closed unit ball of X^* there is a $t \in T$ and an $a \in C$ with $|a| = 1$ such that $x^*(f) = af(t)$ for all $f \in X$. (Hint: Show that $\{\mu \in M(T, \mathbb{C}) : \|\mu\| = 1 \text{ and } \mu|X = x^*\}$ has an extreme point and that any extreme point of this set is an extreme point of the closed unit ball in $M(T, \mathbb{C})$.)

Chapter 3

Characterizations of Banach Spaces of Continuous Functions

In this chapter we give various characterizations of $C(T,\mathbb{R})$ and $C(T,\mathbb{C})$. In section 9 we present the Banach lattice and Banach algebra characterizations of these spaces and include a discussion of abstract M spaces and the Banach spaces of continuous functions which vanish at infinity on a locally compact Hausdorff space. Section 10 takes up the study of $C(T,\mathbb{R})$ as a Banach space. This means that geometric conditions are given on a Banach space X and its dual X^* in order to characterize when X is linearly isometric to $C(T,\mathbb{R})$ for some compact Hausdorff space T. In section 11 we characterize those Banach spaces which have the property that they are complemented with respect to a contractive projection in each Banach space which contains them.

§ 9. Lattice and Algebraic Characterizations of Spaces of Continuous Functions

The central tool used in this study is the notion of an abstract M space (as discussed in section 3). We show that an abstract M space with a strong order unit is linearly isometric and order isomorphic to some $C(T,\mathbb{R})$ and that Banach algebras of a certain type are linearly isometric and algebraically isomorphic to $C(T,\mathbb{R})$. We then apply this to obtain similar characterizations of $C(T,\mathbb{C})$ and of $C_0(T,\mathbb{R})$ (resp., $C_0(T,\mathbb{C})$), the Banach space of real (resp., complex) valued functions which vanish at infinity on a locally compact Hausdorff space T. In this study and in the one in section ten we need various approximation theorems. The one we use here is the classical Stone-Weierstrass theorem. The lattice and algebraic versions are presented below.

By an *algebra* we mean a linear space X with a multiplication such that the addition of X and this multiplication make X into a commutative ring and for any $x,y\in X$ and any pair of scalars a and b, $(ax)y = x(ay) = a(xy)$ and $(ab)x = a(bx)$.

Recall that a subset M of $C(T,\mathbb{C})$ is said to *separate the points* of T if whenever t_1, t_2 are distinct points in T, there is an $f\in M$ with $f(t_1)\neq f(t_2)$.

Theorem 1 *(Real Case—lattice form). Let T be a compact Hausdorff space and let $M \subset C(T, \mathbb{R})$ be a sublattice of $C(T, \mathbb{R})$ which contains the constant function 1 and separates the points of T. Then M is dense in $C(T, \mathbb{R})$.*

Proof. For each pair of distinct points s and t in T and every $f \in C(T, \mathbb{R})$, there is an $f_{s,t} \in M$ such that $f_{s,t}(t) = f(t)$ and $f_{s,t}(s) = f(s)$. For, let $h \in M$ be such that $h(t) \neq h(s)$ and put $f_{s,t} = (f(t) - f(s)) \left[\frac{h - h(s) \cdot 1}{h(t) - h(s)}\right] + f(s) \cdot 1$. Let $\varepsilon > 0$ be given and put $V_{s,t} = \{t' : f_{s,t}(t') < f(t') + \varepsilon\}$ and $V_{s,t} = \{t' : f_{s,t}(t') > f(t') - \varepsilon\}$. If t is fixed, choose a cover $V_{s_1,t}, \ldots, V_{s_n,t}$ of T and let $f_t = \min f_{s_i,t}$. Now $f_t(t') < f(t') + \varepsilon$ for all $t' \in T$ and $f_t(t') > f(t') - \varepsilon$ for all $t' \in V_t = \bigcap V_{s_i,t}$. Let $V_{t_1}, \ldots, V_{t_m}$ be a cover of T and put $g = \max f_{t_j}$. Then we have that $\|f - g\| < \varepsilon$ and it follows that M is dense in $C(T, \mathbb{R})$. ∎

Theorem 2 *(Real Case—algebraic form). Let T be a compact Hausdorff space and let A be a subalgebra of $C(T, \mathbb{R})$ which contains the constant function 1 and separates the points of T. Then A is dense in $C(T, \mathbb{R})$.*

Proof. Clearly $\bar{A}$ is a subalgebra of $C(T, \mathbb{R})$. Thus we shall assume that A is closed and show that it is a sublattice of $C(T, \mathbb{R})$.

Let $f \in A$ and $\|f\| \leqslant 1$. Furthermore, let $\{p_n\}$ be a sequence of polynomials in the real variable x which converge uniformly to $\sqrt{x}$ on $[0, 1]$. Then $g_n(t) = p_n(f^2(t))$ defines an element of A and $\{g_n\}$ converges uniformly to $|f|$. Thus $|f|$ is in A for each $f \in A$ and it follows that A is a sublattice of $C(T, \mathbb{R})$. ∎

A linear subspace M of $C(T, \mathbb{C})$ is said to be *self adjoint* if $\bar{f} \in M$ when $f \in M$. Clearly M is self adjoint if and only if $\operatorname{Re} f$ is in M when f is. In particular, if M is self adjoint, then $M = M_1 + i M_1$ where $M_1 = \{\operatorname{Re} f : f \in M\}$.

Theorem 3 *(Complex Case). Let T be a compact Hausdorff space and let M be a self adjoint linear subspace of $C(T, \mathbb{C})$ containing the constant function 1 and separating the points of T. If M is a subalgebra of $C(T, \mathbb{C})$ or if M_1 is a sublattice of $C(T, \mathbb{R})$, then M is dense in $C(T, \mathbb{C})$.*

Proof. In either case M_1 is dense in $C(T, \mathbb{R})$. Thus, M is dense in $C(T, \mathbb{C})$. ∎

Theorem 5 in section three showed that a real Banach lattice X is an abstract M space if and only if it is linearly isometric and lattice isomorphic to a sublattice of $C(T, \mathbb{R})$ for some compact Hausdorff space T (actually theorem 5 only states the necessity, the sufficiency being obvious). Moreover, the space T constructed in theorem 5 is naturally dependent upon X and the image of X in $C(T,\mathbb{R})$ clearly separates the points of T (since they are linear functionals on X). In particular, applying the Stone-Weierstrass theorem we obtain the following characterization due to Kakutani [148].

Theorem 4. *Let X be a real Banach lattice. Then X is linearly isometric and order isomorphic to $C(T,\mathbb{R})$ for some compact space T if and only if X is an abstract M space with a strong order unit.*

Proof. The necessity is clear.

If X is an abstract M space with a strong order unit e, then X has the strong order unit norm with respect to e. Let $K_0=\{x^*\in(X^*)^+:\|x^*\|\leqslant 1\}$ and $K=\{x^*\in(X^*)^+:\|x^*\|=1\}=\{x^*\in X^*:x^*(e)=1=\|x^*\|\}$. Then K_0 and K are weak* compact and clearly K contains the nonzero extreme points of K_0. Thus K contains the weak* closure T of the set of nonzero extreme points of K_0. By theorem 5 of section 3 the mapping $\phi:X\to C(T,\mathbb{R})$ defined by $\phi(x)(x^*)=x^*(x)$ for $x\in X$ and $x^*\in T$ is a linear isometry and preserves the lattice operations. Clearly $\phi(e)=1$, the constant function one on T and $\phi(X)$ separates the points of T. Thus by theorem 1, ϕ is onto $C(T,\mathbb{R})$. ∎

We now investigate the embedding of an abstract M space into $C(T,\mathbb{R})$ more closely.

Lemma 1. *Let X be a real vector lattice and let x^* be a nonzero linear functional on X. Then the following are equivalent.*

(1) $x^*(x\vee y)=\max(x^*(x),x^*(y))$ *for all* $x,y\in X$.

(2) $x^*\geqslant 0$ *and if* $0\leqslant y^*\leqslant x^*$, *then* $y^*=ax^*$ *for some* $a\geqslant 0$.

Proof. (1) implies (2). If $x\geqslant 0$, then $x^*(x)=x^*(x\vee 0)=\max(x^*(x),0)\geqslant 0$. Hence x^* is a positive linear functional. Suppose $0\leqslant y^*\leqslant x^*$. For any $x\in X$, $0=\min(x^*(x^+),x^*(x^-))$. Hence if $x^*(x)=0$, then $|y^*(x)|\leqslant y^*(|x|)\leqslant x^*(|x|)=0$ and it follows that $y^*=ax^*$ for some $a\in\mathbb{R}$. Since $y^*\geqslant 0$, we obtain that $a\geqslant 0$.

(2) implies (1). It suffices to show that $x^*(x+y)=\max(x^*(x),x^*(y))$ if $x\wedge y=0$. Clearly we may assume that $x^*(x)\neq 0$. Let y^* be defined by $y^*(z)=\sup\{x^*(z\wedge nx):n=1,2,\ldots\}$ for all $z\geqslant 0$. Then by theorem 7 of section 1, y^* is a positive linear functional, $0\leqslant y^*\leqslant x^*$, and $y^*(z)=0$ if $z\wedge x=0$. By condition (2), $y^*=ax^*$ for some $a\in\mathbb{R}$ and since $y^*(x)=x^*(x)$, we obtain that $a=1$ and $y^*=x^*$. Hence $x^*(y)=0$ and it follows that $x^*(x+y)=x^*(x)=\max(x^*(x),x^*(y))$. ∎

Corollary. *Let X be an abstract M space and $K=\{x^*\in X^*:x^*\geqslant 0,\|x^*\|\leqslant 1\}$. Then x^* is an extreme point of K if and only if x^* is a lattice homomorphism.*

Proof. Since each x^* in K is positive and has norm 1, if x^* is an extreme point of K, then it satisfies condition (2) above. For, suppose $0 \leqslant y^* \leqslant x^*$ with $y^* \neq 0$, $y^* \neq x^*$. Then $\|y^*\| + \|x^* - y^*\| = 1$ and

$$x^* = \|y^*\| \left(\frac{y^*}{\|y^*\|} \right) + \|x^* - y^*\| \left(\frac{x^* - y^*}{\|x^* - y^*\|} \right).$$

Thus $y^*/\|y^*\| = x^*$.

If x^* is a lattice homomorphism and $\|x^*\| = 1$, then whenever $x^* = a\,y^* + (1-a)z^*$ with $0 < a < 1$ and y^*, z^* in K then y^* and z^* have norm one and it follows that $0 \leqslant a\,y^* \leqslant x^*$ and $0 \leqslant (1-a)z^* \leqslant x^*$. Hence $x^* = b\,a\,y^* = c(1-a)z^*$ for some $b, c > 0$. Therefore $b\,a = 1 = c(1-a)$ and $x^* = y^* = z^*$. ∎

Let T be a compact Hausdorff space and let I be a nonempty index set. For each $i \in I$ let $t_i, t_i' \in T$ (not necessarily distinct) and $a_i \in \mathbb{R}^+$. Then $\{t_i, t_i', a_i\}$ $(i \in I)$ is said to be a family of *relations* on T and $X = \{f \in C(T, \mathbb{R}) : f(t_i) = a_i f(t_i') \text{ for all } i \in I\}$ is said to be the family of continuous functions which satisfy these relations. Clearly X is an abstract M space since it is a closed sublattice of $C(T, \mathbb{R})$. Kakutani proved that any abstract M space can be so represented [148].

Theorem 5. *Let X be an abstract M space. Then there is a compact Hausdorff space T and a set of relations $\{t_i, t_i', a_i\}_{i \in I}$ such that $0 \leqslant a_i < 1$ and X is linearly isometric and lattice isomorphic to the set of all functions $f \in C(T, \mathbb{R})$ such that $f(t_i) = a_i f(t_i')$ for all $i \in I$.*

Proof. Let $K = \{x^* \in (X^*)^+ : \|x^*\| \leqslant 1\}$. By the corollary to Lemma 1 $\operatorname{ext} K = S$ is the set consisting of the norm 1 lattice homomorphism on K and the functional 0. Clearly any net of lattice homomorphisms which converges in the weak* topology converges to a lattice homomorphism. Let T be the weak* closure of the set S of norm 1 lattice homomorphisms and write $T \setminus S = \{x_i^*\}_{i \in I}$. For each x_i^*, $x_i^* = a_i y_i^*$ for some $0 \leqslant a_i < 1$ and $y_i^* \in S$. Thus the set of relations is $\{x_i^*, y_i^*, a_i\}_{i \in I}$. Let f be an element of $C(T, \mathbb{R})$ such that $f(x_i^*) = a_i f(y_i^*)$ for all $i \in I$. Then for any x^*, y^* in T, there is an $x_0 \in X$ such that $x^*(x_0) = f(x^*)$ and $y^*(x_0) = f(y^*)$. This is clear if $\{x^*, y^*\}$ forms a linearly dependent set since in such a case they form one of the terms of the above relations. If this is not the case, then there are x, y in X such that $a = x^*(x)\,y^*(y) - x^*(y)\,y^*(x) \neq 0$. Hence if $c = \frac{1}{a}(y^*(y) f(x^*) - x^*(y) f(y^*))$ and $d = \frac{1}{a}(y^*(x) f(x^*) - x^*(x) f(y^*))$, then $x_0 = c\,x - d\,y$ will do.

Now for any $\varepsilon > 0$, there is a weak* neighborhood $U(x^*)$ of x^* such that $z^*(x_0) > f(z^*) - \varepsilon$ for all $z^* \in U(x^*)$. Thus there are $x_1^*, \ldots, x_n^*$

and $x_1, \ldots, x_n$ such that $x_k^*(x_k) = f(x_k^*)$ and $y^*(x_k) = f(y^*)$ for $k = 1, \ldots, n$ and $T = U(x_1^*) \cup \cdots \cup U(x_n^*)$ where $z^*(x_k) > f(z^*) - \varepsilon$ for $z^* \in U(x_k^*)$, $k = 1, \ldots, n$. Let $x = x_1 \vee \cdots \vee x_n$. Then $y^*(x) = \max y^*(x_k) = f(y^*)$ and $z^*(x) = \max z^*(x_k) > f(z^*) - \varepsilon$ for all z^* in T.

That is, for any y^* in T and $\varepsilon > 0$ there is an $x \in X$ such that $y^*(x) = f(y^*)$ and $z^*(x) > f(z^*) - \varepsilon$ for all $z^* \in T$. There is a weak* neighborhood $V(y^*)$ of y^* such that $z^*(x) < f(z^*) + \varepsilon$ for all $z^* \in V(y^*)$. Thus there are $y_1^*, \ldots, y_m^*$ in T and $y_1, \ldots, y_m$ in X such that $T = \bigcup_{j=1}^{m} V(y_j^*)$ and $z^*(y_i) > f(z^*) - \varepsilon$ for any $z^* \in T$ and $z^*(y_i) < f(z^*) + \varepsilon$ for any $z^* \in V(y_j^*)$. Hence for $x = y_1 \wedge \cdots \wedge y_m$, $f(z^*) - \varepsilon < z^*(x) = \min z^*(y_i) < f(z^*) + \varepsilon$ for all $z^* \in T$.

That is, for any $\varepsilon > 0$ there is an x in X such that $|z^*(x) - f(z^*)| < \varepsilon$ for all $z^* \in T$. In particular, there is a sequence $\{x_n\}$ such that $|z^*(x_n) - f(z^*)| < \frac{1}{n}$ for all $z^* \in T$. Hence $|z^*(x_n) - z^*(x_m)| \leqslant \frac{1}{n} + \frac{1}{m}$ for all $z^* \in T$. Since $\operatorname{ext} K \cup (-\operatorname{ext} K) \setminus \{0\}$ is precisely the set of extreme points of the unit ball of X^*, for $x \in X$, $\|x\| = \sup_{z^* \in T} |z^*(x)| = |z^*(x)|$ for some extreme point z^* in K. Thus, $\|x_n - x_m\| < \frac{1}{n} + \frac{1}{m}$ and $\{x_n\}$ is a Cauchy sequence in X. Clearly $x = \lim x_n$ has the property that $f(z^*) = z^*(x)$ for all $z^* \in T$. In particular, the linear operator L defined by where $Lx(z^*) = z^*(x)$ for all $z^* \in T$ is a linear isometry and lattice isomorphism of X onto the subspace of $C(T, \mathbb{R})$ of all functions satisfying the set $\{x_i^*, y_i^*, a_i\}_{i \in I}$ of relations. ∎

The problem of determining an algebraic characterization of $C(T, \mathbb{R})$ is more complicated. The solution is presented through a series of lemmas and theorems below. We then apply it to obtain a similar characterization of $C(T, \mathbb{C})$.

A *Banach algebra* is a Banach space which is also an algebra and the norm satisfies the condition $\|xy\| \leqslant \|x\| \|y\|$ for all x, y in X. It is clear that $C(T, \mathbb{R})$ is a Banach algebra for every compact Hausdorff space T. The following few theorems are for real Banach spaces only.

Theorem 6. *Let X be a partially ordered real Banach space which is also a commutative algebra with an identity e and the property that $x, y \geqslant 0$ implies $xy \geqslant 0$ for all x, y in X. If X has order unit norm with respect to e and $x^2 \geqslant 0$ for all $x \in X$, then X is linearly isometric and algebraically isomorphic to $C(T, \mathbb{R})$ for some compact Hausdorff space T.*

Proof. Let $K = \{x^* : x^*(e) = 1 = \|x^*\|\}$. Note that for any $x^* \geqslant 0$, $[x^*(x)]^2 \leqslant x^*(x^2)$ for all $x \in X$. For, $x \in X$ and $t \in \mathbb{R}$ imply $0 \leqslant x^*[(tx + e)^2] = t^2 x^*(x^2) + 2t x^*(x) + 1$. If $x^*(x^2) = 0$, then the above inequality yields

$x^*(x)=0$ since t can be chosen arbitrarily. If $x^*(x^2)\neq 0$, $t=(-x^*(x))/x^*(x^2)$ yields the desired result.

We shall show that $T=\operatorname{ext} K$ consists precisely of the set of positive multiplicative linear functionals of norm one.

Suppose $x^*\neq 0$ is a positive multiplicative linear functional on X. Then $x^*(e)=1$ and it follows that $x^*\in K$. If $x^*=\frac{1}{2}(y^*+z^*)$ with y^*,z^* in K, then for any $x\in X$, $x^*(x^2)=[x^*(x)]^2=\frac{1}{4}[(y^*(x))^2+2y^*(x)z^*(x)+(z^*(x))^2]$. Thus

$$0=2y^*(x^2)-(y^*(x))^2+2z^*(x^2)-(z^*(x))^2-2y^*(x)z^*(x)\geqslant(y^*(x)-z^*(x))^2\geqslant 0.$$

Hence $y^*(x)=z^*(x)=x^*(x)$ and it follows that x^* is an extreme point of K.

If x^* is an extreme point of K, then for $x\in X$ with $0<x<e$ and $0<x^*(x)<1$, let y^*,z^* be defined by $y^*(y)=x^*(xy)/x^*(x)$ and $z^*(z)=x^*(z-xz)/x^*(e-x)$ for $y,z\in X$. Then y^* and z^* are in K and $x^*=x^*(x)y^*+(x^*(e-x)z^*)$. Hence $x^*=y^*=z^*$. We shall show that $x^*(xy)=x^*(x)x^*(y)$ for all $y\in X$. If $0\leqslant x$ and $x^*(x)=0$, then for $y\in X$ choose $t>0$ such that $-te\leqslant y\leqslant te$. Thus $-tx\leqslant xy\leqslant tx$ and $x^*(xy)=0$ $=x^*(x)x^*(y)$. If $0<x<e$ and $x^*(x)=1$, then $x^*(e-x)=0$ and $x^*(y-yx)=0$, that is, $x^*(y)x^*(x)=x^*(yx)$. If $0<x^*(x)<1$, then by the above $x^*(xy)=x^*(x)x^*(y)$. This clearly extends to all positive x and since the positive cone is generating, to all of x.

The set T of (nonzero) multiplicative linear functionals is a weak* closed set in K since the limit of a net multiplicative linear functional in the weak* topology is clearly a multiplicative linear functional.

The linear operator L defined by $(Lx)(x^*)=x^*(x)$ for $x^*\in T$ and $x\in X$ is an algebraic isomorphism and a linear isometry of X onto a subalgebra of $C(T,\mathbb{R})$ which contains 1 and separates the points of T. Thus by the Stone-Weierstrass theorem L is onto. ∎

For the next few lemmas and theorems X denotes a real commutative Banach algebra with unit e of norm 1. An element $x\in X$ is said to be *regular* if x^{-1} exists. Otherwise x is said to be *singular*.

Lemma 2. *For each $x\in X$ with $\|x-e\|<1$, x is regular. In particular, the set of regular elements in X is open in the norm topology on X.*

Proof. Since $\sum_{n=0}^{\infty}\|e-x\|^n<\infty$, $y=\sum_{n=0}^{\infty}(e-x)^n$ exists in X. Moreover, $xy=e$ so $y=x^{-1}$. Let T_x be defined on X by $T_x(y)=xy$ for $y\in X$. Then T_x is a norm homeomorphism of X onto X when x is regular. Hence the set of regular elements is open in the norm topology on X. ∎

Let $K=\{x^*\in X^*: x^*(e)=1=\|x\|\}$ and T denote the set of nonzero multiplicative linear functionals on X.

Lemma 3. *$T \subset K$ and T is weak* closed.*

Proof. Clearly if $x^* \in T$, then $x^*(e) \neq 0$. In fact, if $\|x-e\|<1$, then x^{-1} exists and $x^*(x) \neq 0$. In particular, the kernel of x^* is not dense in X and thus must be closed so that x^* is continuous. If $x \in X$ and $\|x\|<1$, then $x^n \to 0$ and hence $|x^*(x)|^n = |x^*(x^n)| \to 0$ and $|x^*(x)|<1$. Thus $\|x^*\| = x^*(e) = 1$. Clearly T is weak* closed in K. ∎

Theorem 7. *If $T \neq \emptyset$, then there is an algebra homomorphism of X onto a dense set in $C(T, \mathbb{R})$.*

Proof. Let A be defined by $(Ax)(x^*) = x^*(x)$ for $x \in X$ and $x^* \in T$. Then the range of A is a subalgebra of $C(T, \mathbb{R})$ containing 1 and separating the points of T. ∎

The goal is to put enough conditions on X to guarantee that the linear operator A defined in theorem 7 is a linear isometry. In such a case, by the Stone-Weierstrass theorem X is algebraically isomorphic and linearly isometric to $C(T, \mathbb{R})$.

In addition to X being a real commutative Banach algebra with unit e of norm one, it is now further assumed that $\|x^2\| = \|x\|^2$ and that $\|x^2+y^2\| \geqslant \|x^2\|$ for all x, y in X.

The positive cone C of X is taken to be the norm closure of $\{x_1^2 + \cdots + x_n^2 : x_i \in X, i=1, \ldots, n\}$.

The element x in X is said to be a *square* if $x = y^2$ for some $y \in X$.

Lemma 4. *If $\|x\| \leqslant 1$, then $e-x$ is a square. Moreover, x and $e-x$ are squares if and only if $\|x\| \leqslant 1$ and $\|e-x\| \leqslant 1$.*

Proof. Consider the power series expansion of $(1-t)^{1/2}$ about 0 for $|t| \leqslant 1$. This series converges absolutely for $|t| \leqslant 1$ and $(1-t)^{1/2} = \sum_{n=0}^{\infty} a_n t^n$. Hence for $\|x\| \leqslant 1$, we can put $(e-x)^{1/2} = \sum_{n=0}^{\infty} a_n x^n$ which exists in X. If $\|x\| \leqslant 1$ and $\|e-x\| \leqslant 1$, then by the above x and $e-x$ are both squares. If x and $e-x$ are both squares, then by the assumption made on the norm, $\|x\| \leqslant \|x+(e-x)\| = 1$ and $\|e-x\| \leqslant \|x+(e-x)\| = 1$. ∎

Lemma 5. (1) $X = C - C$.

(2) *If x and y are squares, so is $x+y$.*

(3) *If x and y are squares with $\|x\|$ and $\|y\| \leqslant 1$, then $\|x-y\| \leqslant 1$.*

Proof. (1) $x = \left(\dfrac{x+e}{2}\right)^2 - \left(\dfrac{x-e}{2}\right)^2$ for $x \in X$.

(2) Let n be such that $\left\|\frac{x}{n}\right\|, \left\|\frac{y}{n}\right\| \leqslant 1$. Since $\frac{x}{n}, \frac{y}{n}, e-\frac{x}{n}$, and $e-\frac{y}{n}$ are squares and it follows that $\left\|e-\frac{x}{n}\right\|$ and $\left\|e-\frac{y}{n}\right\| \leqslant 1$. Now $\left\|\frac{1}{2}\left(\frac{x}{n}+\frac{y}{n}\right)\right\| \leqslant 1$ and $\left\|e-\frac{1}{2}\left(\frac{x}{n}+\frac{y}{n}\right)\right\| \leqslant \frac{1}{2}\left\|e-\frac{x}{n}\right\|+\frac{1}{2}\left\|e-\frac{y}{n}\right\| \leqslant 1$. Thus by lemma 4, $\frac{1}{2}\left(\frac{x}{n}+\frac{y}{n}\right)$ is a square and so is $x+y$.

(3) $e-(x-y)^2=[(e-x)+y]\,[(e-y)+x]$ and thus is a square. Hence $\|x-y\| \leqslant 1$. ∎

It follows by induction that $x_1^2+\cdots+x_n^2$ is a square for $x_1, \ldots, x_n$ in X. Thus the condition $\|x^2+y^2\| \geqslant \|x^2\|$ clearly yields the condition $\|x_1^2+\cdots+x_n^2\| \geqslant \|x_1^2\|$ for all $x_1, \ldots, x_n$.

Let C^* denote the dual cone to C and $L=\{x^* \in C: \|x^*\| \leqslant 1\}$. Clearly L is convex and weak* compact and contains 0. It has not been shown that $C^* \neq \{0\}$. However, this follows from the lemma below and lemma 4. The proof of lemma 6 is given by a Hahn-Banach type argument which we only outline.

Lemma 6. *Let X be a real linear space and suppose C is a positive cone in X. Assume there is an $e \in C$ such that for each nonzero $x \in X$ there is an $a>0$ such that $(e-ax) \in C$ and $(e+ax) \in C$. If $Y \subset X$ is a linear subspace containing e and f is a linear functional on Y such that $f(x) \geqslant 0$ for all $x \in C \cap Y$, then there is a linear functional x^* on X such that $x^*=f$ on Y and $x^*(x) \geqslant 0$ for all $x \in C$.*

Proof. For $y \in X \setminus Y$, if a is such that $\inf\{f(z): z \in Y, (z-y) \in C\} \geqslant a \geqslant \sup\{f(z): z \in Y, (y-z) \in C\}$, then define x^* on $Y \oplus \operatorname{span}(y)$ by $x^*(z+by)=f(z)+ab$. The usual application of Zorn's lemma finishes the proof. ∎

Lemma 7. *If $x^* \in C^*$, then $\|x^*\|=x^*(e)$ and if x^* is an extreme point of L, then for any $y^* \in L$ with $y^* \leqslant x^*$, then $y^*=y^*(e)x^*$.*

Proof. Let $\|x\| \leqslant 1$. Then $(e \pm x) \in C$ and $x^*(e \pm x) \geqslant 0$. Hence $x^*(e) \geqslant |x^*(x)|$ and it follows that $x^*(e)=\|x^*\|$.

If $y^*(e)=0$ or 1, then $y^*=0$ or $y^*=x^*$ according as to whether $y^*(e)=0$ or 1. Suppose $0<y^*(e)<1$. Then $y^*/y^*(e)=z_1^*$ and $(x^*-y^*)/1-y^*(e)=z_2^*$ are in L and $x^*=y^*(e)z_1^*+(1-y^*(e))z_2^*$. Hence $y^*=y^*(e)x^*$ since x^* is an extreme point of L. ∎

Theorem 8. *The nonzero extreme points of L are multiplicative linear functionals.*

Proof. Let $x \in X$ be a square with $\|x\| \leqslant 1$ and x^* be a nonzero extreme point of L. Then for y^* defined by $y^*(y) = x^*(xy)$ for $y \in X$, $y^* \in L$. By lemma 4, $e - x$ is a square and $\|e - x\| \leqslant 1$. Thus $(x^* - y^*)(y) = x^*[(e-x)y] \geqslant 0$ for all $y \in C$ and it follows that $y^* = y^*(e)x^*$. That is, $x^*(xy) = x^*(x)x^*(y)$ for all squares x and all $y \in X$. Since $X = C - C$, it follows that x^* is multiplicative. ∎

The structure theorem characterizing $C(T, \mathbb{R})$ as a real commutative Banach algebra is now derived from the preceeding lemmas and theorems. It is due to Arens [17].

Theorem 9. *Let X be a real commutative Banach algebra with unit e of norm 1. Suppose further that $\|x^2\| = \|x\|^2$ for all $x \in X$ and $\|x^2 + y^2\| \geqslant \|x^2\|$ for all x, y in X. Then X is algebraically isomorphic and linear isometric to $C(T, \mathbb{R})$ (where T is as in lemma 3).*

Proof. Note that if $x \in X$, then $-\|x\| e \leqslant x \leqslant \|x\| e$ in the ordering induced by C. By hypothesis, $\|e + x_1^2 + \cdots + x_n^2\| \geqslant \|e\| = 1$ for all $x_1, \ldots, x_n$ in X. Thus $-e \notin C$ and it is possible to define a nonzero positive functional on the span of $\{e\}$. By lemma 6 it can be extended to a positive functional x^* on all of X so $C^* \neq \{0\}$. Thus $L \neq \{0\}$ and since L is clearly weak* closed $\operatorname{ext} L \neq \{0\}$. If $y \in C \cap (-C)$, then one can readily see from the hypothesis $\|x_1^2 + \cdots + x_n^2\| \geqslant \|x_1^2\|$ for all $x_1, \ldots, x_n$ in X that $y = 0$. Hence by lemma 6 for each $x \in C$ there is an $x^* \in L$ with $x^*(x) = \|x\|$. That is, $\{y^* \in X^* : y^*(x) = \|x\|\}$ is a supporting hyperplane for L. Hence there is an extreme point x^* in L with $x^*(x) = \|x\|$. From this it follows that $\|x\| = \sup\{|x^*(x)| : x^* \in T\}$ for all $x \in C$. Now for $x \in X$, $\|x\|^2 = \|x^2\| = \|A(x^2)\| = \|(Ax)^2\| = \|Ax\|^2$ where A is the operator defined in theorem 7, i.e., $(Ax)(x^*) = x^*(x)$ for all $x^* \in T$ and $x \in X$. Thus A is an isometry and must be onto. ∎

For the complex case there must be an operation like complex conjugation. Specifically, if X is a complex Banach algebra, an *involution* on X is a mapping from X to X with the following properties: (1) $\|\bar{x}\| = \|x\|$ for all $x \in X$; (2) $\overline{x+y} = \bar{x} + \bar{y}$ for all x, y in X; (3) $\overline{ax} = \bar{a}\bar{x}$ for all $a \in \mathbb{C}$ and $x \in X$; (4) $\bar{\bar{x}} = x$ for all $x \in X$; (5) $\overline{xy} = \bar{y}\bar{x}$ for all x, y in X; (6) if X has a unit e, we assume that $\bar{e} = e$.

Theorem 10. *Let X be a commutative complex Banach algebra with unit e of norm 1. If there is an involution on X such that $\|x\bar{x}\| = \|x\|^2$ for all $x \in X$, then X is algebraically isomorphic and linearly isometric to $C(T, \mathbb{C})$ for some compact Hausdorff space T.*

Proof. Let $X_0 = \{x \in X; x = \bar{x}\}$. Then X_0 is a real linear subalgebra of X containing e. Moreover, for any $x \in X$, $x = x_1 + i x_2$ where $x_1 = (x + \bar{x})/2$ and $x_2 = (x - \bar{x})/2i$ are in X_0. Furthermore, this decomposition is unique. Now for x, y in X_0, $\|x^2+y^2\| = \|(x+iy)(x-iy)\| = \|x+iy\|^2 = [\frac{1}{2}(\|x+iy\|+\|x-iy\|)]^2 \geqslant \frac{1}{4}\|2x\|^2 = \|x^2\|$. Thus by Theorem 9 X_0 is linearly isometric to $C(T,\mathbb{R})$ where T is the set of non-zero multiplicative linear functionals in X_0 with the weak* topology. Thus X is algebraically isomorphic and linearly isometric to $C(T,\mathbb{C})$ under the linear operator A defined by $(Ax)(x^*) = x^*(x_1) + i x^*(x_2)$ for $x^* \in T$, $x \in X$, x_1, x_2 in X_0 with $x = x_1 + i x_2$ since $\|x\|^2 = \|x\bar{x}\| = \|A(x\bar{x})\| = \|Ax\overline{Ax}\| = \|Ax\|^2$. ∎

If an algebra X does not have a unit, there is a standard method of adjoining one which goes as follows. Let $Y = X \times \mathbb{R}$ ($X \times \mathbb{C}$ in the complex case) and define the usual linear structure on Y. The multiplication on Y is given by $(x,a)(y,b) = (xy + ay + by, ab)$ for all (x,a) and (y,b) in Y. It is easy to check that Y is an algebra under this algebraic structure and $(0,1)$ is the unit of Y. If X has an involution (in the complex case), then so does Y. For, if $(x,a) \in Y$, we define $(\overline{x,a}) = (\bar{x}, \bar{a})$. It is a simple matter to check that this indeed defines an involution on Y.

Now there are exactly two multiplicative linear functionals on $\mathbb{R}$ (resp., $\mathbb{C}$). They are, in fact, the zero functional and the identity functional. Since each linear functional y^* on Y is uniquely expressible in the form $y^*(x,a) = x^*(x) + f(a)$ where x^* is a linear functional on X and f is a linear functional on $\mathbb{R}$ (resp., $\mathbb{C}$), it is easy to see that y^* is multiplicative if and only if x^* is multiplicative and f is 0 or f is multiplicative and x^* is 0.

We recall that a *locally compact Hausdorff space* is a Hausdorff topological space T such that each $t \in T$ is contained in an open set whose closure is compact. A function $f \in C(T,\mathbb{C})$ is said to *vanish at infinity* if for each $\varepsilon > 0$, $\{t : |f(t)| \geqslant \varepsilon\}$ is compact. We denote the space of all continuous real valued functions which vanish at infinity by $C_0(T,\mathbb{R})$ (the complex case is denoted by $C_0(T,\mathbb{C})$). Clearly $C_0(T,\mathbb{R})$ (resp., $C_0(T,\mathbb{C})$) is a subspace of $C(T,\mathbb{R})$ (resp., $C(T,\mathbb{C})$) and is, in fact, a Banach space under the supremum norm. Moreover $C_0(T,\mathbb{R})$ is an abstract M space and a real Banach algebra under pointwise lattice and algebraic operations respectively and $C_0(T,\mathbb{C})$ is a complex Banach algebra with involution under pointwise algebraic operations and complex conjugation.

The proofs of the following three theorems are a direct consequence of the above remarks and the preceeding theorems. We leave their proofs as exercises.

Theorem 11. *Let X be a real Banach algebra such that $\|x^2\| = \|x\|^2$ and $\|x^2 + y^2\| \geqslant \|x^2\|$ for all $x, y \in X$. Then X is linearly isometric and algebraically isomorphic to $C_0(T,\mathbb{R})$ for some locally compact Hausdorff space T.* ∎

Theorem 12. *Let X be a complex Banach algebra with involution and suppose that $\|x\bar{x}\| = \|x\|^2$ for all $x \in X$. Then X is linearly isometric and algebraically isomorphic to $C_0(T, \mathbb{C})$ for some locally compact Hausdorff space T.* ∎

The only case necessary to prove in the above theorems is where X does not have a unit (of norm one). In such a case we adjoin a unit to X in the above described manner to obtain an algebra Y with unit such that X is naturally isomorphic to a hyperplane in Y, namely, $\{(x,0): x \in X\}$. We must norm Y in such a manner that Y becomes a Banach algebra, $\|(0,1)\| = 1$, and $\|x\| = \|(x,0)\|$ for all $x \in X$. In the real case, Y must also satisfy the conditions of theorem 9 and in the complex case it must satisfy the conditions of theorem 10. We norm Y as follows: $\|(x,a)\| = \sup\{\|xy + ay\| : y \in X \text{ and } \|y\| \leqslant 1\}$. Then one checks that with this norm Y has the desired properties.

We can also characterize $C_0(T, \mathbb{R})$ as an abstract M space. We shall use the following notation in the theorem below. Let X be an abstract M space and S be the weak* closure of the set of norm one lattice preserving linear functions on X. Then by theorem 5 there is a set $\{s_i, s_i', a_i\}$ of relations on S with $s_i, s_i' \in S$ and $0 \leqslant a_i < 1$ $(i \in I)$ such that X is linearly isometric and lattice isomorphic to $\{f : f(s_i) = a_i f(s_i')$ for all $i \in I\}$.

Theorem 13. *Let X be an abstract M space. Then the following are equivalent:*

(1) *X is linearly isometric and order isomorphic to $C_0(T, R)$ for some locally compact Hausdorff space T,*

(2) *S is the set of norm one lattice preserving linear functionals on X together with at most the zero functional,*

(3) *if f satisfies the relations determined by X, then so does f^2.* ∎

Exercises. 1. Let S and T be compact Hausdorff spaces and suppose $h: S \to T$ is a continuous map. Let $C(T, \mathbb{C}) \to C(S, \mathbb{C})$ be defined by $L(f) = f \circ h$ for all $f \in C(T, \mathbb{C})$. Show that L is an isometry if and only if h is onto. Show that L is onto if and only if h is one to one.

2. Suppose $L: C(T, \mathbb{C}) \to C(S, \mathbb{C})$ is a linear operator such that $L(1) = 1$ and $L(fg) = L(f)L(g)$ for all $f, g \in C(T, \mathbb{C})$. Show that there is a continuous map $h: S \to T$ such that $L(f) = f \circ h$ for all $f \in C(T, \mathbb{C})$.

3. Suppose $L: C(T, \mathbb{R}) \to C(S, \mathbb{R})$ is a linear operator such that $L(1) = 1$ and $L(f \vee g) = L(f) \vee L(g)$ for all $f, g \in C(T, \mathbb{R})$. Show that there is a continuous map $h: S \to T$ such that $L(f) = f \circ h$ for all $f \in C(T, \mathbb{C})$.

4. Let X be a normed linear lattice and suppose that S is a separable subset of X. Show that S is contained in a separable (closed) sublattice of X.

5. Let T be a compact Hausdorff space. Show that T is metrizable if and only if $C(T,\mathbb{C})$ (resp., $C(T,\mathbb{R})$) is separable.

6. Let T be a compact Hausdorff space and suppose that S is a separable subset of $C(T,\mathbb{C})$. Show that S is contained in a separable subspace Y of $C(T,\mathbb{C})$ which is linearly isometric to $C(T',\mathbb{C})$ for some compact metrizable space T'.

7. Give an example of a compact Hausdorff space T and a closed linear subspace X of $C(T,\mathbb{R})$ which is linearly isometric to $C(S,\mathbb{R})$ for some compact Hausdorff space S and X is not a sublattice of $C(T,\mathbb{R})$.

8. Let T be a compact Hausdorff space. Prove that T is totally disconnected if and only if $C(T,\mathbb{R})$ is the smallest closed sublattice containing all finite dimensional sublattices of $C(T,\mathbb{R})$ containing 1.

9. Give an abstract M space proof of the following topological result. If S and T are compact Hausdorff spaces with S metrizable and T a continuous image of S, then T is metrizable.

10. Give an abstract M space proof of the following topological result. If T is an infinite extremally disconnected compact Hausdorff space, then $\beta\mathbb{N}$ is a continuous image of T.

11. Give an abstract M space proof of the following topological result. If T is an infinite compact metric space, then there is a bounded noncontinuous function of first Baire class (i.e., a pointwise limit of continuous functions) on T. Use this to prove that $C(T,\mathbb{R})$ is not reflexive.

12. Let T be a compact Hausdorff space and suppose $X\subset C(T,\mathbb{R})$ is a closed linear subspace. Show that X is a sublattice of $C(T,\mathbb{R})$ if and only if there is a family of relations $\{t_i,t_i',a_i\}$ for i in some index set I where $0\leqslant a_i\leqslant 1$ and $X=\{f\in C(T,\mathbb{R}): f(t_i)=a_i f(t_i') \text{ for all } i\in I\}$. Moreover, show that X is a subalgebra of $C(T,\mathbb{R})$ if and only if the relations can be chosen so that $a_i\in\{0,1\}$ for all $i\in I$.

13. Let X,T be as in 12. Show that if X is a sublattice of $C(T,\mathbb{R})$ then either there is a $t\in T$ with $f(t)=0$ for all $f\in X$ or there is an $f\in X$ with $f(t)>0$ for all $t\in T$.

14. Let X,T be as in 12 and suppose X is a sublattice of $C(T,\mathbb{R})$. Show that X is a subalgebra of $C(T,\mathbb{R})$ if and only if $f\wedge 1$ is in X when $f\in X$.

15. Let X be an abstract M space. Then the following are equivalent.

(1) M is linearly isometric and lattice isomorphic to $C_0(T,\mathbb{R})$ for some locally compact Hausdorff space T,

(2) If $J: X\to X^{**}$ is the natural map, then $JX+\mathbb{R}\cdot 1$ is a vector lattice of X^{**},

(3) If $0\leqslant x\in X$, then $\sup\{y\in X: \|y\|\leqslant 1, y\leqslant x\}$ exists.

§ 10. Banach Space Characterizations of Spaces of Continuous Functions

In this section we shall not assume a priori a lattice or algebraic structure on a Banach space in order to characterize $C(T,\mathbb{R})$, but shall make assumptions on the geometry of the space. We present some characterizations of $C(T,\mathbb{R})$ due to Clarkson, Arens and Kelley, and Jerison.

We first discuss a result due to Clarkson [67]. Let X be a real Banach space and for each $x,y\in X$ let $[x,y]=\{ax+(1-a)y:0\leqslant a\leqslant 1\}$ (i.e., $[x,y]$ is the line segment joining x and y). We say that X satisfies *Clarkson's condition* if there is an $e\in X$ of norm one such that for each $x\in X$ of norm one, either all elements of $[x,e]$ or all elements of $[x,-e]$ have norm one. Note that clearly if $X=C(T,\mathbb{R})$ for some compact Hausdorff space and e is the constant function 1 on T, then for all $f\in X$ with $\|f\|=1$, $\|af+(1-a)e\|=1$ if there is a $t\in T$ with $1=f(t)$ and $\|af+(1-a)(-e)\|=1$ if there is a $t\in T$ with $1=-f(t)$.

Lemma 1. *Suppose X is a two dimensional real Banach space. If X satisfies Clarkson's condition, then X is linearly isometric to $l_\infty(2,\mathbb{R})$.*

Proof. Let e be such that X satisfies Clarkson's condition with respect to e. Let $x\in X$ with $\|x\|=1$ and $x\neq\pm e$ and suppose (without loss of generality) that $\|ax+(1-a)e\|=1$ for all $a\in[0,1]$. Put $a_0=\sup\{a>0:\|ax+(1-a)e\|=1\}$. Then clearly $\|a_0x+(1-a_0)e\|=1$ and for $x_0=a_0x+(1-a_0)e$, each element of $[x_0,e]$ has norm one. Moreover, $\{e,x_0\}$ is a basis for X. We define $L:X\to l_\infty(2,\mathbb{R})$ by $L(ax_0+be)=a(1,1)+b(1,-1)$. Then $\|L(ax_0+be)\|=\max(|a+b|,|a-b|)$ $=\|a(1,1)+b(1,-1)\|$ and L is a linear isometry. ∎

Lemma 2. *Let X be a real Banach space satisfying Clarkson's condition. Then X can be partially ordered so that it is a partially ordered Banach space with a strong order unit and strong order unit norm.*

Proof. Clearly we can assume that the Hamel dimension of X is greater than one since if it is one, $X=\mathbb{R}$ and we are done. Let $e\in X$ be such that X satisfies Clarkson's condition with respect to e and let $C=\{x\in X:x=a(e+y)$ where $a\geqslant 0$ and $\|y\|\leqslant 1\}$. Then it is easy to see that $C+C=C$ and $aC\subset C$ for all $a\geqslant 0$. Suppose $x\in X$ and let Y be a two dimensional subspace of X containing x and e. By lemma 1 there is a linear isometry L of Y onto $l_\infty(2,\mathbb{R})$ such that $L(e)=(1,1)$. In particular, if $y\in Y\cap C$, then $L(y)\geqslant 0$. Hence it follows that if $x\in C\cap(-C)$, then $x=0$ and C is a positive cone.

Suppose $x\in X$ and $\|x\|\leqslant 1$. Then $e-x=e+(-x)$ and it follows that $e\geqslant x$. Similarly, $e+x\in C$ and $e\geqslant -x$. Now suppose $-e\leqslant x\leqslant e$. Let Y and L be as above. Then $(-1,1)\leqslant L(x)\leqslant(1,1)$ and it follows that $\|x\|=\|L(x)\|\leqslant 1$. Hence e is a strong order unit and X has strong order unit norm.

Finally, note that C is closed. For, if $x_n=a_n(e+y_n)$ with $a_n\geqslant 0$ and $\|y_n\|\leqslant 1$, and $0\neq x=\lim x_n$, then since $\|e+y_n\|\leqslant 2$ for all n, $\{a_n\}$ is bounded. Thus we may suppose that $\{a_n\}$ converges to some $a\neq 0$ and $\{y_n\}$ converges to some y, $\|y\|\leqslant 1$, and $x=a(e+y)$. Hence it follows that $x\in C$ and C is closed. ∎

Applying the above and theorem 2 of section 1 we get the following characterization of $C(T,\mathbb{R})$.

Theorem 1. *Let X be a real Banach space satisfying Clarkson's condition and let C be the cone defined in lemma 2. Then X is linearly isometric to $C(T,\mathbb{R})$ for some compact Hausdorff space T if the intersection of two translates of C is again a translate of C. Conversely, the natural cone in $C(T,\mathbb{R})$ has the property that the intersection of two translates of it is again a translate of it and $C(T,\mathbb{R})$ satisfies Clarkson's condition.*

Proof. By the above, X is a partially ordered Banach space with strong order unit norm and a vector lattice so that it is an abstract M space with a strong order unit (clearly $-ae\leqslant x\leqslant ae$ if and only if $-ae\leqslant|x|\leqslant ae$ so that $\|x\|=\||x|\|$ and $0\leqslant x\leqslant y$ implies that $\|y\|\leqslant\|x\|$). Hence by theorem 4 of section 9 X is linearly isometric to $C(T,\mathbb{R})$ for some compact Hausdorff space T. ∎

We now present another approximation theorem in order to obtain a characterization of $C(T,\mathbb{R})$ due to Arens and Kelley [18]. Suppose T is a nonempty set and $B(T,\mathbb{R})$ is the Banach space of all bounded real valued functions on T with the supremum norm and pointwise algebraic operations. Let X be a closed linear subspace of $B(T,\mathbb{R})$. A subspace Y of X is said to be *normal* in X if for each $f\in X$ there is a $g\in Y$ such that $\|g\|=1$, $g(s)=1$ if $f(s)\geqslant\frac{1}{3}\|f\|$ and $g(s)=-1$ if $f(s)\leqslant-\frac{1}{3}\|f\|$. The following approximation theorem is motivated by the usual proof of the Tietze extension theorem (see [91]).

Theorem 2. *If Y is normal in X, then it is dense in X.*

Proof. Suppose $f_0\in X$ and $\|f_0\|=1$. By the normality of Y there is a $g_0\in Y$ such that $\|g_0\|=\frac{1}{3}$ and $g_0(s)=\frac{1}{3}$ if $f(s)\geqslant\frac{1}{3}$ and $g_0(s)=-\frac{1}{3}$ if $f_0(s)\leqslant-\frac{1}{3}$. Let $f_1=f_0-g_0$. Then $f_1\in X$ and $\|f_1\|\leqslant\frac{2}{3}$. Thus there is a $g_1\in Y$ such that $\|g_1\|=\frac{1}{3}(\frac{2}{3})$, $g_1(s)=\frac{1}{3}(\frac{2}{3})$ if $f_1(s)\geqslant\frac{1}{3}(\frac{2}{3})$ and $g_1(s)=-\frac{1}{3}(\frac{2}{3})$ if $f_1(s)\leqslant-\frac{1}{3}(\frac{2}{3})$. If $f_2=f_1-g_1$, then $\|f_2\|\leqslant(\frac{2}{3})^2$. By induction we obtain a sequence $\{g_n\}$ in Y such that $\|g_n\|\leqslant\frac{1}{3}(\frac{2}{3})^n$ and $\left\|f_0-\sum_{i=0}^{n} g_i\right\|\leqslant(\frac{2}{3})^{n+1}$. Hence Y is dense in X. ∎

Corollary. *Let T be a compact Hausdorff space and T_0 be a dense subset of T. If X is a linear subspace of $C(T, \mathbb{R})$ such that for each pair of sets A and B contained in T_0 with disjoint closures (in T) there is an $f \in X$ with $\|f\|=1$ and $f=1$ on A and $f=-1$ on B, then X is dense in $C(T, \mathbb{R})$.* ∎

Let X be a Banach space and $V^*=b(X^*)$ (the closed unit ball of X^*). Recall that a weak* closed *hyperplane of support* for V^* is a set $H=\{x^* \in X^*: x^*(x)=a\}$ for some $a>0$ and some $x \in X$ such that $H \cap V^* \neq \emptyset$ and $y^*(x) \leqslant a$ for all $y^* \in V^*$. Clearly such an H is weak* closed and if $y^* \in H \cap V^*$, then $\|y^*\|=1$.

Theorem 3 (Arens-Kelley). *Let X be a real Banach space. Then X is linearly isometric to $C(T, \mathbb{R})$ for some compact Hausdorff space T if and only if V^* satisfies the following two conditions:*

(1) *There is a weak* closed hyperplane H of support for V^* such that* $\operatorname{ext} V^* \subset H \cup(-H)$.

(2) *For each $A \subset \operatorname{ext} V^*$ such that $\bar{A} \cap(-\bar{A})=\emptyset$ there is a weak* closed hyperplane of support for V^* containing A.*

Proof. For the necessity put $H=\{x^*: x^*(1)=1\}$. The result is then immediate from theorem 6 of section 2 and Urysohn's lemma.

Suppose conditions (1) and (2) hold and H is a hyperplane of support for V^* with $\operatorname{ext} V^* \subset H \cup(-H)$. Let T be the weak* closure of $H \cap \operatorname{ext} V^*$. Then the mapping $L: X \rightarrow C(T, \mathbb{R})$ defined by $(L x)(x^*)=x^*(x)$ for $x \in X$ and $x^* \in T$ is a linear isometry since $\|x\|=\sup\{|x^*(x)|: x^* \in V^*\}$ $=\sup\{|x^*(x)|: x^* \in \operatorname{ext} V^*\}=\sup\{|x^*(x)|: x^* \in H \cap \operatorname{ext} V^*\}$ by the Krein-Milman theorem and the fact that $\operatorname{ext} V^* \subset H \cup(-H)$. Thus it remains only to show that $L(X)$ is dense in $C(T, \mathbb{R})$. Suppose $A, B \subset H \cap(\operatorname{ext} V^*)$ and $\bar{A} \cap \bar{B}=\emptyset$. Then $\overline{A \cup(-B)} \cap \overline{(-A) \cup B}=\emptyset$ and by condition (2) there is a weak* closed hyperplane of support H' for V^* which contains $A \cup(-B)$. Now $H'=\{x^*: x^*(x)=a\}$ for some $a>0$ and some $x \in X$. Since $H' \cap V^* \neq \emptyset$ and $y^*(x) \leqslant a$ for all $y^* \in V^*$, it follows that $\|x\|=\sup\{y^*(x): y^* \in V^*\}=a$. Moreover, $L(x)(x^*)=a$ for $x^* \in A$ and $L(x)(x^*)=-a$ for $x^* \in B$ so that $L(X)$ is a normal subspace of $C(T, \mathbb{R})$ and is dense in $C(T, \mathbb{R})$. Since L is a linear isometry, it is onto $C(T, \mathbb{R})$. ∎

We now discuss a characterization of $C(T, \mathbb{R})$ due to Arens and Kelley [18] and refined by Jerison [143]. The presentation here follows [165] and [193].

Let T be a compact Hausdorff space and $\sigma: T \rightarrow T$ be a continuous function such that $\sigma \circ \sigma$ is the identity (that is, σ is an involuntary homeomorphism). We set $C_\sigma(T, \mathbb{R})=\{f \in C(T, \mathbb{R}): f=-f \circ \sigma\}$. Then $C_\sigma(T, \mathbb{R})$ is a closed linear subspace of $C(T, \mathbb{R})$. If σ is fixed point free, we denote it by Σ and the corresponding space by $C_\Sigma(T, \mathbb{R})$.

Suppose $\sigma(t)=t$. Then for each $f\in C_\sigma(T,\mathbb{R})$, $f(t)=0$. On the other hand, if $\sigma(t)\neq t$, then there is an $f\in C(T,\mathbb{R})$ such that $\|f\|=1$, $f(\sigma(t))=-1$ and $f(t)=1$. Thus for $g=\frac{1}{2}(f-f\circ\sigma)$, $g\in C_\sigma(T,\mathbb{R})$, $\|g\|=1$, and $g(t)=1$. That is, if t_σ denotes the continuous linear functional on $C_\sigma(T,\mathbb{R})$ given by $t_\sigma(f)=f(t)$ for $f\in C_\sigma(T,\mathbb{R})$ (i.e., $t_\sigma=\varepsilon_t|C_\sigma(T,\mathbb{R})$), then $t_\sigma=0$ when $\sigma(t)=t$ and $\|t_\sigma\|=1$ when $\sigma(t)\neq t$. We note further that the mapping $P:C(T,\mathbb{R})\to C_\sigma(T,\mathbb{R})$ defined by $P(f)=\frac{1}{2}(f-f\circ\sigma)$ is a contractive projection of $C(T,\mathbb{R})$ onto $C_\sigma(T,\mathbb{R})$.

We wish to show that the set of extreme points of the closed unit ball V^* of $C_\sigma(T,\mathbb{R})^*$ consists precisely of those functionals t_σ such that $\sigma(t)\neq t$. Clearly any extreme point of V^* is of the form $\pm t_\sigma$ where $\sigma(t)\neq t$ and since $-t_\sigma=(\sigma(t))_\sigma$, every extreme point of V^* is of the form t_σ where $\sigma(t)\neq t$ (see exercise 20 at the end of chapter 2).

Let us note that for every $x^*\in C_\sigma(T,\mathbb{R})^*$ with $\|x^*\|\leqslant 1$ there is a positive normalized regular Borel measure μ on T such that $x^*(f)=\int f\,d\mu$ for all $f\in C_\sigma(T,\mathbb{R})$. For, by the Krein-Milman theorem there is a net $\{x_d^*\}$ of convex combinations of extreme points of V^* which converges to x^* in the weak* topology. That is, $x_d^*=\sum_{i=1}^{n_d} a_{i,d}(t_\sigma)_{i,d}$ where $a_{i,d}\geqslant 0$ and $\sum_{i=1}^{n_d} a_{i,d}=1$. Let $\mu_d=\sum_{i=1}^{n_d} a_{i,d}\varepsilon_{t_{i,d}}$ (where $(t_{i,d})_\sigma=(t_\sigma)_{i,d}$). Then $\{\mu_d\}$ is a net of positive normalized regular Borel measures and by passing to a subnet if necessary, we may assume that $\{\mu_d\}$ converges to μ in the weak* topology (relative to $C(T,\mathbb{R})$). Clearly we have that $x^*(f)=\lim x_d^*(f)=\lim\int f\,d\mu_d=\int f\,d\mu$ for all $f\in C_\sigma(T,\mathbb{R})$.

We preserve the above notation in the following lemma.

Lemma 3. *If $\sigma(t)\neq t$, then t_σ is an extreme point of V^*.*

Proof. Suppose μ is a positive normalized regular Borel measure on T and $f(t)=\int f\,d\mu$ for all $f\in C_\sigma(T,\mathbb{R})$. We shall show that $\mu=\varepsilon_t$. Let V be an open set containing t and let W be an open subset of V containing t and such that $W\cap\sigma(W)=\emptyset$. Choose $f\in C(T,\mathbb{R})$ such that $0\leqslant f\leqslant 1$ and $f(t)=1$ and $f=0$ on $T\setminus W$. Then $g=f-f\circ\sigma$ is in $C(T,\mathbb{R})$, $g(t)=1$, $g=f$ on W, and $g=-f\circ\sigma$ on $T\setminus V$. Now $1=g(t)=\int g\,d\mu=\int_W g\,d\mu+\int_{T\setminus W} g\,d\mu=\int_V f\,d\mu+\int_{T\setminus V}-f\circ\sigma\,d\mu$. Since f is positive and $0\leqslant f\leqslant 1$, it follows that $\int_{T\setminus V}-f\circ\sigma\,d\mu=0$ and $\int_W f\,d\mu=1$. Thus $\mu(V)\geqslant\mu(W)\geqslant\int_V f\,d\mu=1$ and it must be that $\mu(V)=1$. By the regularity of μ we obtain that $\mu(\{t\})=1$ and $\mu=\varepsilon_t$.

Now, if t_σ is not an extreme point of V^*, then there are $x^*,y^*\in V^*$ such that $x^*\neq t_\sigma\neq y^*$ and $t_\sigma=\frac{1}{2}(x^*+y^*)$. In particular, since $\|t_\sigma\|=1$, we must have that $\|x^*\|=\|y^*\|=1$. By the above remarks there are positive normalized regular Borel measures μ and ν on T such that

$x^*(f)=\int f\,d\mu$ and $y^*(f)=\int f\,dv$ for all $f\in C_\sigma(T,R)$. Now clearly $\mu\neq\varepsilon_t\neq v$ since $x^*\neq t_\sigma\neq y^*$. But for $f\in C_\sigma(T,R)$, $f(t)=\frac{1}{2}(x^*(f)+y^*(f))$ $=\frac{1}{2}(\int f\,d\mu+\int f\,dv)=\int f\,d[\frac{1}{2}(\mu+v)]$. By the first part it follows that $\frac{1}{2}(\mu+v)=\varepsilon_t$ which is a contradiction since ε_t is an extreme point of the closed unit ball of $C(T,\mathbb{R})^*$ (theorem 11 of section 8). ∎

We can use the above to give a natural representation of $C_\sigma(T,\mathbb{R})$. Suppose $X=C_\sigma(T,\mathbb{R})$ and T^* is the weak* closure of the extreme points of the closed unit ball V^* of X^*. By the above lemma, $T^*=\{t_\sigma : t\in T\}$. For, suppose $\{x_d^*\}$ is a net of extreme points of V^* and $\lim x_d^*=x^*$. Now each x_d^* is of the form t_σ for some $t\in T$. Thus we obtain a net $\{t_d\}$ in T such that $(t_d)_\sigma=x_d^*$. By compactness some subnet $\{t_{d'}\}$ of $\{t_d\}$ converges to some $t\in T$. Clearly by the definition of the weak* topology, $\{x_{d'}^*\}$ converges to t_σ. Hence $x^*=t_\sigma$ and the result follows. Now T^* has a natural involutory homeomorphism given by $\sigma^*(x^*)=-x^*$ for all $x^*\in T^*$. Moreover, $\sigma^*(t_\sigma)=-t_\sigma=(\sigma(t))_\sigma$ by the definition of $C_\sigma(T,\mathbb{R})$. We preserve the above notation in the following theorem.

Theorem 4. *The Banach space* $C_\sigma(T,\mathbb{R})$ *is linearly isometric to* $C_{\sigma^*}(T^*,\mathbb{R})$.

Proof. The mapping $L: C_{\sigma^*}(T^*,\mathbb{R})\to C_\sigma(T,\mathbb{R})$ given by $L(f)(t)=f(t_\sigma)$ for $f\in C_{\sigma^*}(T^*,\mathbb{R})$ and $t\in T$ is a linear isometry of $C_{\sigma^*}(T^*,\mathbb{R})$ onto $C_\sigma(T,\mathbb{R})$. ∎

We showed above that $C_\sigma(T,\mathbb{R})$ is the range of a contractive projection on $C(T,\mathbb{R})$. We now wish to show the converse is true. That is, the range of a contractive projection on a space $C(T,\mathbb{R})$ is linearly isometric to $C_\sigma(S,\mathbb{R})$ for some compact Hausdorff space S and some involutory homeomorphism σ. We first note that any $C(T,\mathbb{R})$ can be represented as a space $C_\sigma(S,\mathbb{R})$. For, let $S=T\times\{0,1\}$ and define $\sigma: S\to S$ by $\sigma(t,0)=(t,1)$ and $\sigma(t,1)=(t,0)$ for all $t\in T$. Then clearly $\sigma\circ\sigma$ is the identity. Let $L: C(T,\mathbb{R})\to C_\sigma(S,\mathbb{R})$ be defined by $L(f)(t,0)$ $=f(t)$ and $L(f)(t,1)=-f(t)$ for all $t\in T$. Then clearly L is a linear isometry which maps $C(T,\mathbb{R})$ onto $C_\sigma(S,\mathbb{R})$.

Now suppose $Y\subset C(T,\mathbb{R})$ is the range of a contractive projection P. Then Y is linearly isometric to a subspace Z of a space $C(S,\mathbb{R})$ such that Z separates the points of S and Z is the range of a contractive projection on $C(S,\mathbb{R})$. For, let S be the closed unit ball of Y^* with the weak* topology and let $\phi: Y\to C(S,\mathbb{R})$ be defined by $\phi(y)(y^*)=y^*(y)$ for all $y\in Y$ and $y^*\in S$. Then clearly ϕ is a linear isometry. Let $\psi: C(S,\mathbb{R})\to C(T,\mathbb{R})$ be defined by $\psi(f)(t)=f(\varepsilon_t|Y)$ for all $f\in C(S,\mathbb{R})$ and $t\in T$. Then ψ is a linear operator and since $\|\psi(1)\|=1$ and $\|\psi(f)\|\leqslant\|f\|$ for all f, we obtain that $\|\psi\|=1$. Let $Q: C(S,\mathbb{R})\to Z$ (where $Z=\phi(Y)$) be defined by $Q=\phi\circ P\circ\psi$. Then Q is a contractive projection of $C(S,\mathbb{R})$ onto Z since $\psi\circ\phi$ is the identity on Y and P is a contractive projection.

Theorem 5. *Let T be a compact Hausdorff space and suppose that Y is the range of a contractive projection on $C(T,\mathbb{R})$. Then Y is linearly isometric to $C_\sigma(S,\mathbb{R})$ for some compact Hausdorff space S and some involutory homeomorphism σ on S.*

Proof. By the above remarks we may assume that $Y\subset C_{\sigma^*}(V^*,\mathbb{R})$ where V^* is the closed unit sphere on Y^* and $\sigma^*(x^*)=-x^*$ for all $x^*\in V^*$ and Y is the range of a contractive projection on $C_{\sigma^*}(V^*,\mathbb{R})$. Let S be the weak* closure of $\operatorname{ext} V^*$ and $\sigma=\sigma^*|S$. Then clearly σ is an involutory homeomorphism on S and the mapping $L: Y\to C_\sigma(S,\mathbb{R})$ defined by $L(y)(y^*)=y^*(y)$ for $y\in Y$ and $y^*\in S$ is a linear isometry. Hence we need only show that L is onto. Let P be a contractive projection of $C_{\sigma^*}(V^*,\mathbb{R})$ onto Y and suppose $g\in C_\sigma(S,R)$. Let $f\in C(V^*,\mathbb{R})$ be such that $f|S=g$ and $\|f\|=\|g\|$ (such an f exists by the Tietze extension theorem). Then $h=\frac{1}{2}(f-f\circ\sigma)$ is in $C_{\sigma^*}(V^*,\mathbb{R})$ and $Ph\in Y$.

Let y^* be an extreme point of V^*. Then there is only one norm preserving extension of y^* to $C_{\sigma^*}(V^*,\mathbb{R})$. For, the set K of all such norm preserving extensions of y^* is clearly a weak* compact convex set in $C_{\sigma^*}(V^*,\mathbb{R})^*$ and thus if K consists of more than one point, then it must have more than one extreme point. But, each extreme point of K is an extreme point of the closed unit ball of $C_{\sigma^*}(V^*,\mathbb{R})^*$ since y^* is an extreme point of V^*. Thus each extreme point of K can be thought of as an element of V^*. But Y separates the points of V^*. Let P^* be the adjoint of P (i.e., $P^*: Y^*\to C_{\sigma^*}(V^*,\mathbb{R})^*$). Since $\varepsilon_{y^*}|C_{\sigma^*}(V^*,\mathbb{R})$ and $P^*(y^*)$ are both norm preserving extensions of y^*, it follows that $P^*(y^*)=\varepsilon_{y^*}|C_{\sigma^*}(V^*,R)$. Now $(Ph)(y^*)=P^*(y^*)(h)=h(y^*)=g(y^*)$ and it follows that $L(Ph)=g$. Hence L is onto. ∎

We wish to present a characterization of $C(T,\mathbb{R})$ which was given by Arens and Kelley [18] (with an additional assumption) and refined by Jerison in [143]. As mentioned above, the techniques and terminology of these papers differs from this presentation. The characterization is that a $C_\sigma(T,\mathbb{R})$ space is linearly isometric to $C(S,\mathbb{R})$ for some compact Hausdorff space S if and only if the closed unit ball of $C_\sigma(T,\mathbb{R})$ contains an extreme point.

We first prove that if the closed unit ball of $C_\sigma(T,\mathbb{R})$ contains an extreme point, then $C_\sigma(T,\mathbb{R})$ can be partially ordered so that it becomes a partially ordered Banach space with a strong order unit and strong order unit norm. In particular, the extreme points of the closed unit ball V^* of $C_\sigma(T,\mathbb{R})^*$ must be contained in $K\cup(-K)$ where $K=\{x^*\in C_\sigma(T,\mathbb{R})^*: \|x^*\|=1, x^*\geqslant 0\}$ (see theorem 6 of section 2). Moreover, since K is weak* compact, it follows that $\operatorname{ext} V^*$ is weak* compact since the closure of $\operatorname{ext} V^*$ adds at most the functional 0 (see lemma 3). In particular, it can be considered that σ is fixed point free since, as above, we can take for T the weak* closure of $\operatorname{ext} V^*$ and for σ we can take negation.

In the proof of the theorem we will need the fact that if $f_i \in C(T, \mathbb{R})$, $r_i > 0$, $S(f_i, r_i) = \{f : \|f - f_i\| \leqslant r_i\}$ and the $S(f_i, r_i)$'s intersect in pairs, then $\bigcap_{i=1}^{n} S(f_i, r_i) \neq \emptyset$. This follows since in such a case $\sup(-f_i - r_i) \leqslant \inf(f_i + r_i)$ and both of these functions are in each $S(f_i, r_i)$.

We shall also use the following theorem whose proof we leave as an exercise.

Theorem 6. *Let T be a compact Hausdorff space. Then $f \in C(T, \mathbb{R})$ (resp., $C(T, \mathbb{C})$) is an extreme point of the closed unit ball of $C(T, \mathbb{R})$ (resp., $C(T, \mathbb{C})$) if and only if $|f| = 1$.* ∎

Theorem 7. *Let T be a compact Hausdorff space and σ be an involutory homeomorphism on T. If the closed unit ball of $C_\sigma(T, \mathbb{R})$ has an extreme point f, then there is a positive cone C on $C_\sigma(T, \mathbb{R})$ such that f is a strong order unit for C, C is closed, and the norm is a strong order unit norm with respect to f and C.*

Proof. Let $X = C_\sigma(T, \mathbb{R})$. We define C as we did in lemma 2. That is, $C = \{g \in X : g = a(f + h), a \geqslant 0, \|h\| \leqslant 1\}$. The proof that C is a closed positive cone is the same as that in lemma 2. Hence it suffices to show that $\|g\| \leqslant 1$ if and only if $-f \leqslant g \leqslant f$. Clearly if $\|g\| \leqslant 1$, then by the definition of C, $-f \leqslant g \leqslant f$. Suppose $-f \leqslant g \leqslant f$ and $g \neq 0$. Then $g = -a(f + h) + f = a(f + k) - f$ for some $a > 2$ and $\|h\|, \|k\| \leqslant 1$. Let $S_1 = S(g + (a-1)f, a-1)$, $S_2 = S(g - (a-1)f, a-1)$, and $S_3 = S\left(-\frac{a-2}{\|g\|} g, a-1\right)$. Then $\{g\} = S_1 \cap S_2$, $-h \in S_1 \cap S_3$, and $k \in S_2 \cap S_3$. Thus by the above remarks there is an $f_0 \in C(T, R)$ such that $f_0 \in S_1 \cap S_2 \cap S_3$. Now $f_1 = \frac{1}{2}(f_0 - f_0 \circ \sigma)$ is in $C_\sigma(T, \mathbb{R})$ and $f_1 \in S_1 \cap S_2 \cap S_3$. Thus $f_1 = g$ and it follows that $\|g\| \leqslant 1$ since $g \in S_3$. ∎

Theorem 8 (Arens and Kelley; Jerison). *Let T be a compact Hausdorff space and σ be an involutory homeomorphism on T. Then $C_\sigma(T, \mathbb{R})$ is linearly isometric to $C(S, \mathbb{R})$ (for some compact Hausdorff space S) if and only if the closed unit ball of $C_\sigma(T, \mathbb{R})$ contains an extreme point.*

Proof. The necessity is clear.

Suppose f is an extreme point of the closed unit ball of $C_\sigma(T, \mathbb{R})$. Then by theorem 7 we may assume that $\sigma = \Sigma$ is fixed point free. We shall show that f is an extreme point of the closed unit ball of $C(T, \mathbb{R})$, i.e., $|f| = 1$. Suppose $g \in C_\Sigma(T, \mathbb{R})$, $\|g\| = 1$, and $|g| \neq 1$. Then we can suppose that there is a $t \in T$ with $0 \leqslant g(t) < 1$. Let U be an open set in T with $U \cap \Sigma(U) = \emptyset$ and $t \in U$. Choose an open set V with $t \in V \subset U$ and $|g(s)| < 1$ for all $s \in V$. Let $g' \in C(T, \mathbb{R})$, $g'(t) = 1 - g(t)$ and $|g'(s)| \leqslant 1 - |g(s)|$ for all $s \in T$ and $g'(s) = 0$ for $s \notin U$ and put $h = g' - g' \circ \Sigma$. Then $h \in C_\Sigma(T, \mathbb{R})$ and $g \pm h$ are in $C_\Sigma(T, \mathbb{R})$ with $\|g \pm h\| \leqslant 1$ and $g = \frac{1}{2}(g+h) + \frac{1}{2}(g-h)$. Thus g is not an extreme point of the unit sphere of $C_\Sigma(T, \mathbb{R})$. We put $U = \{s \in T : f(s) = 1\}$. Then $\Sigma(U) = \{s \in T : f(s) = -1\}$ and U is open. Moreover, $U \cap \Sigma(U) = \emptyset$, and $T = U \cup \Sigma(U)$. Clearly $C_\Sigma(T, \mathbb{R})$ is linearly isometric to $C(U, \mathbb{R})$. ∎

It turns out that all $C_\Sigma(T, \mathbb{R})$ spaces are linearly isometric to some $C(S, \mathbb{R})$ for special conditions on T. We shall shortly see that this is true for T either extremally disconnected or dispersed.

Theorem 9. *Let T be an extremally disconnected compact Hausdorff space and Σ be a fixed point free involutory homeomorphism on T. Then $C_\Sigma(T, \mathbb{R})$ is linearly isometric to some $C(S, \mathbb{R})$.*

Proof. First note that for any T, $T = U \cup \Sigma(U) \cup B$ where U is open and B is closed and nowhere dense and $U, \Sigma(U), B$ are pairwise disjoint. For, by Zorn's lemma it is possible to choose a maximal open set U such that $U \cap \Sigma(U) = \emptyset$. Thus we let $B = T \setminus (U \cup \Sigma(U))$. Then B is closed and nowhere dense and $\Sigma(B) = B$. Now suppose T is extremally disconnected. Then $\bar{U}$ is open and $\bar{U} \cap \Sigma(\bar{U}) = \emptyset$. Hence $\bar{U} = U$ and $\Sigma(\bar{U}) = \Sigma(U)$ so that $B = \emptyset$. Clearly $C_\Sigma(T, \mathbb{R})$ is linearly isometric to $C(U, \mathbb{R})$. ∎

In order to prove the result for dispersed spaces we need the following technical lemma and theorem. For the definition of the derivative $A^{(\eta)}$ of a set in a topological space refer to section 5.

Lemma 4. *Let T be a topological space, A and B subsets of K, and η an ordinal. Then $(A \cup B)^{(\eta)} = A^{(\eta)} \cup B^{(\eta)}$.* ∎

The proof of the above lemma is a straightforward application of transfinite induction.

Theorem 10. *Let T be a compact Hausdorff space and Σ a fixed point free involutary homeomorphism on T. If for some ordinal α we have $T^{(\alpha)} = A \cup \Sigma(A)$ where $A \cap \Sigma(A) = \emptyset$ and A is closed, then $C_\Sigma(T, \mathbb{R})$ is linearly isometric to some $C(S, \mathbb{R})$.*

Proof. Case 1: α is not a limit ordinal, i.e., $\alpha = \beta + 1$.

Since A and $\Sigma(A)$ are disjoint compact subsets in $T^{(\beta)}$, there are disjoint relatively open sets U_1, V_1 in $T^{(\beta)}$ such that $A \subset U_1$ and $\Sigma(A) \subset V_1$. Since Σ is continuous there is a relatively open set W in $T^{(\beta)}$ such that $A \subset W \subset U_1$ and $\Sigma(W) \subset V_1$. Thus $W \cap \Sigma(W) = \emptyset$. By

Zorn's lemma there is a maximal relatively open set U such that $A \subset U$ and $\Sigma(U) \cap U = \emptyset$. Now suppose $t \in T$ is a limit point of U and $\{t_\gamma\}$ is a net in $U \setminus \{t\}$ which converges to t. Then $t \in T^{(\alpha)} = A \cup \Sigma(A)$. If $t \in \Sigma(A)$, then since $\Sigma(U) \supset A$, for some γ_0, $t_\gamma \in \Sigma(U)$ for $\gamma \geqslant \gamma_0$. Hence $U \cap \Sigma(U) = \emptyset$ which is a contradiction. Thus $t \in A$ and U is closed. Since $\bar{U} \cap \Sigma(\bar{U}) \neq \emptyset$ and U is maximal, $T^{(\beta)} \setminus (U \cup \Sigma(U)) = \emptyset$. Hence $T^{(\beta)} = U \cup \Sigma(U)$ where U is closed and open in $T^{(\beta)}$.

Case 2: α is a limit ordinal.

By definition $T^{(\alpha)} = \bigcap T^{(\beta)}$ and by hypothesis $T^{(\alpha)} = A \cup \Sigma(A)$ with A closed and $A \cap \Sigma(A) = \emptyset$.

Let U be an open set such that $A \subset U$, $U \cap \Sigma(U) = \emptyset$ and $B = T \setminus (U \cup \Sigma(U))$. Choose an open set V such that $A \subset V \subset \bar{V} \subset U$. Then $T = \bar{V} \cup (\bar{U} \setminus V) \cup \Sigma(\bar{V}) \cup \Sigma(\bar{U} \setminus V) \cup B$. Since $A \subset V$, $(\bar{U} \setminus V) \cap A = \emptyset$. Also, since $\bar{U} \subset T \setminus \Sigma(V)$ and $\Sigma(A) \subset \Sigma(V)$, $(\bar{U} \setminus V) \cap \Sigma(A) = \emptyset$. Moreover, $\Sigma(\bar{U} \setminus V) \cap A = \emptyset$ and $\Sigma(\bar{U} \setminus V) \cap \Sigma(A) = \emptyset$. Let $C = (\bar{U} \setminus V) \cup \Sigma(\bar{U} \setminus V) \cup B$. Clearly C is closed and disjoint from $A \cup \Sigma(A)$ and $T = \bar{V} \cup \Sigma(\bar{V}) \cup C$. By lemma 3, $T^{(\alpha)} = V^{(\alpha)} \cup (\Sigma(\bar{V}))^{(\alpha)} \cup C^{(\alpha)}$. Since $C^{(\alpha)} \subset C$ and $C \cap (A \cup \Sigma(A)) = \emptyset$, $C^{(\alpha)} = \emptyset$. Thus $C^{(\beta)} = \emptyset$ for some $\beta < \alpha$. Hence $T^{(\alpha)} = \bar{V}^{(\beta)} \cup (\Sigma(\bar{V})^{(\beta)})$ where $\bar{V}^{(\beta)}$ is closed and $\bar{V}^{(\beta)} \cap (\bar{V}^{(\beta)}) = \emptyset$.

Hence it has been shown that if $\alpha \geqslant 1$ and $T^{(\alpha)} = A \cup \Sigma(A)$ with A closed and $A \cap \Sigma(A) = \emptyset$, then for some $\beta < \alpha$, $T^{(\beta)} = D \cup \Sigma(D)$ with $D \cap \Sigma(D) = \emptyset$ and D closed.

If $\beta = 0$, then $C_\Sigma(T, \mathbb{R})$ is linearly isometric to $C(U, \mathbb{R})$ where U is chosen in case 1.

If $\beta > 0$, then by choosing the first ordinal γ such that $T^{(\gamma)} = D \cup \Sigma(D)$ with D closed and $D \cap \Sigma(D) = \emptyset$, by the above it follows that $\gamma = 1$. Hence by case 1, $T = D \cup \Sigma(D)$ with D closed and $D \cap \Sigma(D) = \emptyset$. Thus $C_\Sigma(T, \mathbb{R})$ is linearly isometric to $C(D, \mathbb{R})$. ∎

Corollary. *If T is a dispersed compact Hausdorff space and Σ is a fixed point free involutary homeomorphism on K, then $C_\Sigma(T, \mathbb{R})$ is linearly isometric to some $C(S, \mathbb{R})$.*

Proof. Since T is dispersed there is a smallest (non limit) ordinal α such that $T^{(\alpha)} = \emptyset$. ∎

We briefly connect the notions above with those developed by Arens and Kelley [18] and Jerison [143] to prove theorem 8.

Let X be a real Banach space. We shall call a maximal convex set F of norm one elements of X a *face* of $b(X)$. Every face of $b(X)$ is of the form $H \cap b(X)$ where H is a closed supporting hyperplane of $b(X)$. For, clearly F is closed by the maximality of F and $b_0(X)$ is open and F and $b_0(X)$ are disjoint. Thus by the first basic separation theorem there is a closed hyperplane H separating F and $b_0(X)$. That is, there is an $x^*\in X^*$ and an $a>0$ such that $x^*(x)\leqslant a$ if $\|x\|<1$ and $x^*(y)\geqslant a$ if $x\in F$. Clearly we must have $x^*(x)<a$ for all $x\in b_0(X)$ and $x^*(y)=a$ for $x\in F$ so that $F\subset H\cap b(X)$ and H is a hyperplane of support for $b(X)$. By the maximality of F, $F=H\cap b(X)$.

The following theorem is due to Eilenberg [99].

Theorem 11. *Let T be a compact Hausdorff space and $U=b(C(T,R))$. Then F is a face of U if and only if $F=F_t$ or $F=-F_t$ where $F_t=\{f\in U: f(t)=1\}$.*

Proof. Suppose $t\in T$ and $f\in U$ with $f(t)<1$. Let V be an open set and $\varepsilon>0$ be such that $f(s)<1-\varepsilon$ if $s\in V$. There is a $g\in U$ such that $g(t)=1$ and $\overline{\{s: g(s)\neq 0\}}\subset V$. Clearly $\frac{1}{2}(f+g)\notin F_t$. Hence each F_t (and $-F_t$) is a face of U.

Let F be a face of U. If $F\neq\pm F_t$ for all $t\in T$, then $\bigcap\{t: f(t)=1, f\in F\}=\emptyset=\bigcap\{t: f(t)=-1, f\in F\}$. Since for each f, $\{t: f(t)=1\}$ and $\{t: f(t)=-1\}$ are closed, it follows that there are $f_1,\dots,f_n$ in F such that $\bigcap_{i=1}^{n}\{t: f_i(t)=1\}=\emptyset=\bigcap_{i=1}^{n}\{t: f_i(t)=-1\}$. In particular, $\left\|\frac{1}{n}\sum_{i=1}^{n} f_i\right\|<1$ which is impossible since F is convex. Thus there is a t such that either $F\subset F_t$ or $F\subset -F_t$. By the maximality of F, either $F=F_t$ or $F=-F_t$. ∎

We shall call the cone T generated by a face of F of $b(X)$ a *T-set*, i.e., $T=\{ax: a\geqslant 0, x\in F\}$. Note that if $x,y\in F$ and $a,b\geqslant 0$, then $\|ax+by\|=a+b$. That is, the norm is additive on T. Moreover, since F is maximal, T is a maximal set on which the norm is additive. Conversely, if a given set T is a maximal set on which the norm is additive, then it is easy to see that T is a positive cone and the set F of elements of norm one in T is a face in $b(X)$.

We associate with each T-set a function F_T defined as follows: $F_T(x)=\inf\{\|x+y\|-\|y\|: y\in T\}$. Then F_T has the following easily verified properties.

(1) $F_T(0)=0$.
(2) $|F_T(x)|\leqslant\|x\|$ *for all* $x\in X$.
(3) $F_T(x)=\|x\|$ *if and only if* $x\in T$.
(4) $F_T(x)=-\|x\|$ *if and only if* $x\in(-T)$.
(5) $F_T(ax)=aF_T(x)$ *for all* $a\geqslant 0$ *and all* $x\in X$.
(6) $F_T(x+y)\leqslant F_T(x)+F_T(y)$ *for all* $x,y\in X$.
(7) F_T *is continuous.*
(8) F_T *is linear on* $T-T$.
(9) *if* $F_{T_1}=F_{T_2}$, *then* $T_1=T_2$.
(10) F_T *is linear on* X *if and only if* $F_T(-x)=-F_T(x)$ *for all* $x\in X$.

Let X be a real Banach space. We say that X has the *Arens-Kelley property* if for each family $\mathscr{F}$ of faces of $b(X)$ with $\bigcap \mathscr{F}=\emptyset$ there are two nets $\{F_d\}$ and $\{F'_d\}$ (indexed by the same set) of elements of $\mathscr{F}$ such that for each $x \in b(X)$, $\lim(d(x,F_d)+d(x,F'_d))=2$.

Lemma 5. *Let X be a real Banach space with the Arens-Kelley property. Then for each collection $\mathscr{I}$ of T-sets in X such that $\bigcap \mathscr{I}=\{0\}$ there are nets $\{T_d\}$ and $\{T'_d\}$ of elements of $\mathscr{I}$ such that $\lim(F_{T_d}(x)+F_{T'_d}(x))=0$ for all $x\in X$.*

Proof. Let T be a T-set and $F=\{x\in T: \|x\|=1\}$. Then for any $x\in X$, $y,z\in T$ we have that $\|x+y\|-\|y\| \geqslant \|z+y\|-\|x-z\|-\|y\|=\|z\|-\|x-z\|$. Hence $F_T(x)\geqslant \|z\|-\|x-z\|$. If $z\in F$, then $F_T(x)\geqslant 1-d(x,F)$ and $F_T(x)\leqslant \inf\{\|x+z\|-\|z\|: z\in F\}=d(-x,F)-1$.

Now, given a collection $\mathscr{T}$ of T-sets with corresponding collection $\mathscr{F}$ of faces, if $\bigcap\mathscr{T}=\{0\}$, then $\bigcap\mathscr{F}=\emptyset$. Hence by the Arens-Kelley property there are two nets $\{F_d\}$ and $\{F'_d\}$ in $\mathscr{F}$ such that if $x\in b(X)$ and $\varepsilon>0$ is given, there is a d_0 such that $d(x,F_d)+d(x,F'_d)<2+\varepsilon$ and $d(-x,F_d)+d(-x,F'_d)<2+\varepsilon$ for all $d\geqslant d_0$. Let T_d, T'_d be the corresponding cones to F_d and F'_d respectively. Then $F_{T_d}(x)+F_{T'_d}(x) \geqslant 1-d(x,F_d)+1-d(x,F'_d)>-\varepsilon$ and $F_{T_d}(x)+F_{T'_d}(x)\leqslant d(-x,F_d)-1 + d(-x,F'_d)-1<\varepsilon$ for all $d\geqslant d_0$. ∎

Corollary. *If X has the Arens-Kelley property, then for each T-set T in X, F_T is linear on X.*

Proof. Let T be a T-set in X. Then $-T$ is a T-set and $T\cap(-T)=\{0\}$. Thus by the lemma $F_T(x)=-F_{-T}(-x)$ for all $x\in X$. ∎

The following theorem is due to Jerison [143].

Theorem 12. *Let X be a real Banach space. Then X is linearly isometric to $C_\sigma(S,\mathbb{R})$ for some compact Hausdorff space S and some involutory homeomorphism σ if and only if X has the Arens-Kelley property.*

Proof. Suppose $X=C_\sigma(S,\mathbb{R})$. Let $\mathscr{F}$ be a collection of faces of $b(X)$ such that $\bigcap\mathscr{F}=\emptyset$. By theorem 11, each $F\in\mathscr{F}$ is of the form $F=F_s\cap b(X)$ or $F=-F_s\cap b(X)$ for some $s\in S$. Since $f=-f\circ\sigma$ for all $f\in X$, we can assume that $F=F_s\cap b(X)$. Let $A=\{s\in S: F=F_s\cap b(X), F\in\mathscr{F}\}$. If $\bar{A}\cap\sigma(\bar{A})=\emptyset$, then there is an open set $V\supset A$ with $V\cap\sigma(V)=\emptyset$. Thus there is an $f\in C(S,\mathbb{R})$ with $0\leqslant f\leqslant 1$, $f=1$ on A and $f=0$ on $S\setminus V$. Then $g=f-f\circ\sigma$ is in $\bigcap\mathscr{F}$ which is impossible. Let $s\in\bar{A}\cap\sigma(\bar{A})$ and choose nets $\{s_d\}$ in A and $\{\sigma(t_d)\}$ in $\sigma(A)$ such that $s=\lim s_d=\lim\sigma(t_d)$. Let $\{F_d\}, \{F'_d\}$ be the faces corresponding to $\{s_d\}$ and $\{t_d\}$ respectively. Then for all $f\in b(X)$, $d(f,F_d)+d(f,F'_d)=1-f(s_d)+1-f(\sigma(t_d))$ so that $\lim(d(f,F_d)+d(f,F'_d))=1-f(s)+1-f(\sigma(s))=2$. Hence X has the Arens-Kelley property.

Suppose X has the Arens-Kelley property. Then for each T-set T of X, F_T is linear and, in fact, $\|F_T\| \leqslant 1$. Let S be the weak* closure of the collection S_0 of all F_T's such that T is a T-set of X and let $L: X \to C(S, \mathbb{R})$ be defined by $L(x)(x^*) = x^*(x)$ for all $x \in X$ and $x^* \in S$. Clearly L is linear and $\|L(x)\| \leqslant \|x\|$ for all $x \in X$. Since each x is contained in some T-set, we obtain $\|L(x)\| = F_T(x) = \|x\|$. Let σ be defined on S by $\sigma(x^*) = -x^*$ for all $x^* \in S$. Then σ is an involutory homeomorphism on S since $F_T = -F_{-T}$ for all T-sets T in X. In particular, the range of L is contained in $C_\sigma(S, \mathbb{R})$. To show that L is onto it suffices to show that $L(X)$ is normal in $C_\sigma(S, \mathbb{R})$ (see theorem 2). Suppose $A \subset S$ and $\bar{A} \cap \sigma(\bar{A}) = \emptyset$. If $\bigcap \{T: F_T \in A\}$ contains a nonzero element, then it contains some x of norm one. That is, $F_T(x) = \|x\| = 1$ for all $F_T \in A$. Hence $L(x) = 1$ on A (and $L(x) = -1$ on $\sigma(A)$). If $\bigcap \{T: F_T \in A\} = \{0\}$, then choose two nets $\{T_d\}$, $\{T'_d\}$ such that $\lim(F_{T_d}(x) + F_{T'_d}(x)) = 0$ for all x. Since S is compact, $\{F_{T_d}\}$ has a cluster point $x^* \in S$ and $\sigma(x^*)$ is a cluster point of $\{F_{T'_d}\}$. Thus x^*, $\sigma(x^*) \in A$ which is a contradiction. The above shows that the range of L is normal in $C_\sigma(S, \mathbb{R})$. ∎

Thus we can state theorem 8 as follows.

Theorem 13 (Arens-Kelley; Jerison). *Let X be a real Banach space. Then X is linearly isometric to $C(S, \mathbb{R})$ for some compact Hausdorff space S if and only if X has the Arens-Kelley property and the closed unit ball of X has an extreme point.* ∎

Arens and Kelley proved theorem 13 under the additional assumption that the closed unit ball of X is the convex hull of $F \cup (-F)$ for each face F [18] and Jerison removed this condition in [143].

We now present a characterization of $C_0(T, \mathbb{R})$ for T a locally compact Hausdorff space.

Theorem 14. *Let S be a compact Hausdorff space and σ be an involutory homeomorphism on S. Then $C_\sigma(S, \mathbb{R})$ is linearly isometric to $C_0(T, \mathbb{R})$ for some locally compact Hausdorff space T if and only if $T = U \cup \sigma(U) \cup B$ where U is open, $U, \sigma(U), B$ are pairwise disjoint, and $\bar{U} \cap \sigma(\bar{U}) \subset B = \{s: \sigma(s) = s\}$.*

Proof. Suppose $C_\sigma(S, \mathbb{R}) = C_0(T, \mathbb{R})$. Since $C_0(T, \mathbb{R})$ is an abstract M space, the positive extreme points of the closed unit ball of $C_0(T, \mathbb{R})^*$ are the lattice preserving linear functionals of norm one. In particular, the set of extreme points is of the form $A \cup (-A)$ where $A = \{\varepsilon_t | C_0(T, \mathbb{R}): t \in T\}$. Moreover, clearly $\bar{A} = A \cup \{0\}$. Let $U = \{s: \varepsilon_s | C_0(T, \mathbb{R}) \in A\}$. Then U is open since A is open in $\overline{A \cup (-A)}$ and $s \to \varepsilon_s | C_0(T, \mathbb{R})$ is continuous with respect to the weak* topology. Clearly $B = \{s: \varepsilon_s | C_0(T, \mathbb{R}) = 0\} = \{s: \sigma(s) = s\}$ and $\bar{U} \cap \sigma(\bar{U}) = B$.

Suppose $T = U \cup \sigma(U) \cup B$ where U is open, $U, \sigma(U), B$ are pairwise disjoint and $\bar{U} \cap \sigma(\bar{U}) \subset B = \{s : \sigma(s) = s\}$. Let $L: C_\sigma(S, \mathbb{R}) \to C_0(U, \mathbb{R})$ be defined by $L(f) = f|U$. Then since $f|\sigma(U) = -f|U$ and $f|B = 0$, $\|L(f)\| = \|f\|$. Since $f|\bar{U} \setminus U = 0$, it follows that $L(f) \in C_0(U, \mathbb{R})$. If $g \in C_0(U, \mathbb{R})$ and f is defined by $f(s) = g(s)$ for $s \in U$, $f(s) = -g(\sigma(s))$ for $s \in \sigma(U)$, and $f|B = 0$, then $f \in C_\sigma(S, \mathbb{R})$ and $L(f) = g$. ∎

Corollary (Jerison). *Let X be a real Banach space. Then X is linearly isometric to $C_0(T, \mathbb{R})$ for some locally compact Hausdorff space T if and only if X has the Arens-Kelley property and* $\operatorname{ext} b(X^*) = A \cup (-A)$ *where $\bar{A} \cap (-A) = \emptyset$ ($\bar{A}$ is the closure of A in the weak* topology).* ∎

For spaces of ordinals we have the following theorem.

Theorem 15. *Let ξ be an ordinal number and $T = \Gamma(\xi)$. Then for any involutory homeomorphism σ on T, $C_\sigma(T, \mathbb{R})$ is linearly isometric to $C_0(S, \mathbb{R})$ for some locally compact Hausdorff space S.*

Proof. Let $U = \{t : \sigma(t) < t\}$ and $B = \{t : \sigma(t) = t\}$. Then $\bar{U} \subset U \cup B$ and $\sigma(\bar{U}) \subset \sigma(U) \cup B$. Moreover, $\sigma(U) = \{t : \sigma(t) > t\}$ and $U, \sigma(U), B$ are pairwise disjoint and $\bar{U} \cap \sigma(\bar{U}) \subset B$. ∎

The above theorem says, in particular, that if T is a countable dispersed space, then $C_\sigma(T, \mathbb{R}) = C_0(S, \mathbb{R})$ for each involutory homeomorphism σ on T. We present an example to show that this does not hold for all dispersed spaces.

Let Ω be the first uncountable ordinal and let $\hat{K}$ be the quotient space of $\Gamma(\Omega \cdot 2)$ which is obtained by identifying Ω and $\Omega \cdot 2$. Then $\hat{K}$ is dispersed and, hence so is $K = \hat{K} \times \hat{K}$. The points of K will be represented by $(\alpha, 0)$ if $1 \leqslant \alpha \leqslant \Omega$ and by $(\alpha, 1)$ if $\Omega < \Omega + a \leqslant \Omega \cdot 2$. Let Ω equal the equivalence class $\{(\Omega, 0), (\Omega \cdot 2, 1)\}$. Let $\sigma : K \to K$ be defined by

$$\sigma((\alpha, i), (\beta, j)) = ((\alpha, i \oplus 1), (\beta, j \oplus 1))$$

where $\oplus$ is addition modulo 2. It is obvious that σ is one-to-one onto, and $\sigma \circ \sigma$ is the identity. Thus if σ is continuous, it is an involutory homeomorphism. Suppose $((\alpha_\tau, i_\tau), (\beta_\tau, j_\tau)) \to ((\alpha, i), (\beta, j))$. If $\{i_\tau\}$ is not eventually constant, then $\alpha_\tau \to \Omega$. A similar statement holds for $\{j_\tau\}$. Since the equivalence class $[(\Omega, i)]$ is equal to the equivalence class $[(\Omega, i \oplus 1)]$ which is equal to $\hat{\Omega}$, $\Pi_1((\alpha_\tau, i_\tau), (\beta_\tau, j_\tau)) = (\alpha_\tau, i_\tau \oplus 1) \to [(\Omega, i)] = \hat{\Omega} = \Pi_1(\hat{\Omega}, (\beta, j))$. Similarly $\Pi_2((\alpha_\tau, i_\tau), (\beta_\tau, j_\tau)) = (\beta_\tau, j_\tau \oplus 1) \to [(\Omega, j)] = \hat{\Omega} = \Pi_2((\alpha, i), \hat{\Omega})$. If $\{i_\tau\}$ is eventually constant, i.e., $i_\tau = i$ for all $\tau \geqslant$ some τ_0, then $\Pi_1((\alpha_\tau, i_\tau), (\beta_\tau, j_\tau)) = (\alpha_\tau, i_\tau \oplus 1) \to (\alpha, i \oplus 1) = \Pi_1((\alpha, i), (\beta, j))$. A similar statement holds for Π_2. Hence Π_1 and Π_2 are continuous, and consequently σ is continuous. Therefore σ is an involutory homeomorphism of K onto K.

Let $\Omega^* = (\hat{\Omega}, \hat{\Omega})$. Clearly Ω^* is the one and only fixed point for σ.

In order to simplify the notation, geometric language will be used. By horizontal line through (β, j) and vertical line through (α, i) the sets of the form $\{((\alpha, i), (\beta, j)) : (\beta, j) \text{ fixed}\}$ and $\{((\alpha, i), (\beta, j)) : (\alpha, i) \text{ fixed}\}$, respectively, will be meant. The horizontal and vertical lines through $\hat{\Omega}$ will be called the co-ordinate axes. By the upper half plane, lower half plane, right half plane, and left half plane will be meant the sets $\{((\alpha, i), (\beta, j)) : j = 0\}$, $\{((\alpha, i), \beta, j)) : j = 1\}$, $\{((\alpha, i), (\beta, j)) : i = 0\}$, and $\{((\alpha, i), (\beta, j)) : i = 1\}$, respectively.

Suppose $K = U \cup \sigma(U) \cup \{\Omega^*\}$, where $\bar{U}, \sigma(U), \{\Omega^*\}$ are pairwise disjoint and $\bar{U} \cap (\bar{U}) \subset \{\Omega^*\}$. If the horizontal line through $(\beta, j) \neq \hat{\Omega}$ or if the vertical line through $(\alpha, i) \neq \hat{\Omega}$ contained uncountable many points from both U and $\sigma(U)$, then $(\Omega, (\beta, j))$ or $((\alpha, i), \Omega)$ would be a limit point of both U and $\sigma(U)$. Thus it may be assumed that no horizontal or vertical line other than the co-ordinate axes contains uncountably many points in U and $\sigma(U)$.

Now suppose the upper half plane contains uncountably many horizontal lines which contain uncountably many points of U and it contains uncountably many lines which contain uncountably many points from $\sigma(U)$. Let $A = \{(\hat{\Omega}, (\beta, 0)) : (\hat{\Omega}, (\beta, 0)) \in U\}$ and $B = \{(\hat{\Omega}, (\beta, 0)) : (\hat{\Omega}, (\beta, 0)) \in \sigma(U)\}$. Clearly if the horizontal line through $(\beta, 0)$ contains uncountably many points of U, then $(\hat{\Omega}, (\beta, 0)) \in U$ or $(\Omega, (\beta, 0)) \in (\bar{U} \cup \sigma(\bar{U})) \setminus \{\Omega^*\}$. If the second statement held, then the problem would be finished. Thus it may be assumed that A is uncountable. Similarly the problem would be finished if the horizontal line through $(\hat{\Omega}, (\beta, 0))$ contained uncountably many points of $\sigma(U)$. Thus it may be assumed that B is uncountable. Now $\bar{A}$ and $\bar{B}$ are closed subsets of a ray of ordinals and thus are heomeomorphic to a ray of ordinals (corollary 1, theorem 2, section 5). Clearly $\bar{A}$ and $\bar{B}$ must be homeomorphic to $\Gamma(\Omega)$ since they are uncountable and each proper section is countable. If $\bar{A} \cap \bar{B} = \{\Omega^*\}$, then $\Gamma(\Omega)$ would be homeomorphic to $\bar{A} \cup \bar{B}$ such that $\bar{A} \cap \bar{B} = \{\Omega^*\}$. Clearly $\bar{A} \cup \bar{B}$ is homeomorphic to K. This would imply $\Gamma(\Omega)$ is homeomorphic to K. But one can easily show from theorem 2 of section 5 that this is false. Therefore the upper half plane contains only countably many horizontal lines with uncountably many points of precisely one of U or $\sigma(U)$. It may be assumed that it is U. Then the lower half plane would contain only countably many horizontal lines which contain uncountably many points of $\sigma(U)$. Suppose there exists $\alpha_0 < \Omega$ such that each vertical line through $(\alpha, 0)$ where $\Omega > \alpha > \alpha_0$ contains no points of U in the second quadrant (i.e., intersection of upper half plane and left half plane) and no points of $\sigma(U)$ in the third quadrant. But this is clearly impossible because there would exist a vertical line other than the vertical co-ordinate axis containing uncountably many points of both U and $\sigma(U)$. Hence it may be assumed that for each ordinal $\alpha<\Omega$, there exists a horizontal line in the upper half plane such that there is a point $((\hat{\alpha},0),(\beta,0))\in U$ such that $\Omega>\hat{\alpha}>\alpha$. For $\beta=1$ (i.e., the first line in the upper half plane), let $\hat{\alpha}_1$ be the first ordinal such that there exists only finitely many points $((\alpha,0),(1,0))\in U$ with $\alpha>\hat{\alpha}_1$. Let $\delta_1=\hat{\alpha}_1+\gamma_1$ where γ_1, $0\leqslant\gamma_1<\omega$, is the first ordinal such that $(((\hat{\alpha}_1+\gamma_1),0),(1,0))\in\sigma(U)$. Let $t>1$ be an ordinal, and suppose that for each $s>t$, $\delta_s, \hat{\alpha}_s$, and γ_s have been chosen such that $\delta_s=\hat{\alpha}_s+\gamma_s$ and $\hat{\alpha}_s>\gamma_u\geqslant\hat{\alpha}_u$ for all $u<s$ and $0\leqslant\gamma_s<\omega$ is the first ordinal such that $(((\hat{\alpha}_s+\gamma_s),0),(s,0))\in\sigma(U)$. Let $\hat{\alpha}_t$ be the first ordinal such that $\hat{\alpha}_t>\delta_s\geqslant\hat{\alpha}_s$ and there is only a finite number of points $((\alpha,0),(t,0))\in U$ with $\alpha>\hat{\alpha}_t$. Let $\delta_t=\hat{\alpha}_t+\gamma_t$ where $0\leqslant\gamma_t<\omega$ is the first ordinal such that $((\hat{\alpha}_t+\gamma_t,0),(t,0))\in\sigma(U)$. Now if $\delta_t=\hat{\alpha}_t$, the problem is finished since $((\hat{\alpha}_t,0),(t,0))\in U$ and $((\alpha_t,0),(t,0))=((\hat{\alpha}_t+\gamma_t,0),(t,0))\in\sigma(U)$. Thus it may be assumed $\delta_t\neq\hat{\alpha}_t$. By transfinite induction the transfinite sequences, $\{\hat{\alpha}_t:t\in\Gamma(\Omega)\}$, $\{\delta_t:t\in\Gamma(\Omega)\}$ have been defined. There are uncountably many points in $\{\hat{\alpha}_t:t\in\Gamma(\Omega)\}$ which are limit points of U and hence in U. (If any one other than Ω^* were in $\sigma(U)$, the problem would be finished.) Let $D=\{t:((\hat{\alpha}_t,0),(t,0))\in U\}$. Then $\bar{D}$ has a limit point other than Ω since $\bar{D}$ homeomorphic to $\Gamma(\Omega)$. Let t_1 be a limit point of D other than Ω. Since $\{\hat{\alpha}_t:t<t_1\}$ is an increasing sequence of ordinals bounded above by $\hat{\alpha}_{t_1+1}$, it converges to some ordinal less than or equal to $\hat{\alpha}_{t_1+1}$. Since $\delta_t<\hat{\alpha}_{t+1}<\delta_{t+1}$, $\{\delta_t:t<t_1\}$ also converges to α^*. Thus

$$((\alpha^*,0),(t_1,0))\in(\bar{U}\cap\sigma(\bar{U}))\setminus\{\Omega^*\}.$$

This contradiction shows that the assumption that $K=U\cup\sigma(U)\cup\{\Omega^*\}$ where U is open, $U, \sigma(U)$, and $\{\Omega^*\}$ are pairwise disjoint, and $\bar{U}\cap\sigma(\bar{U})\subset\{\Omega^*\}$ is impossible. Thus theorem 14 implies that $C_\sigma(K,\mathbb{R})$ is not linearly isometric to $C_0(T,\mathbb{R})$ for all locally compact Hausdorff spaces T.

We close this section by stating some results in the isomorphic classification of Banach spaces of continuous functions.

The first well known and trivial result is that $c_0(\mathbb{N},\mathbb{R})$ is linearly isomorphic to $C(\Gamma(\omega),\mathbb{R})$, the mapping $L:C(\Gamma(\omega),R)\to c_0(\mathbb{N},\mathbb{R})$ defined by $(Lf)(1)=f(\omega)$, $(Lf)(n)=f(n-1)-f(\omega)$ for $n\geqslant 2$ is such an isomorphism.

The complete isomorphic classification of the spaces $C(\Gamma(\alpha),\mathbb{R})$, where α is a countable ordinal, was given by Bessaga and Pełczyński in [39].

Theorem (Bessaga, Pełczyński). *Let $\omega\leqslant\alpha\leqslant\beta<\Omega$. Then $C(\Gamma(\alpha),\mathbb{R})$ is linearly isomorphic to $C(\Gamma(\beta),\mathbb{R})$ if and only if $\beta<\alpha^\omega$.* ∎

From this one easily sees that there are uncountably many isomorphic types among these spaces.

We can use the above theorem and the results of this section to prove the following isomorphic classification theorem for $C_\sigma(T,\mathbb{R})$ spaces due to Samuel [302].

Theorem (Samuel). *Let T be a compact Hausdorff space and suppose X is the range of a contractive projection on $C(T,\mathbb{R})$. If the extreme points of the closed unit ball in X^* are countable, then X is isomorphic to $C(\Gamma(\alpha),\mathbb{R})$ for some countable ordinal α.*

Proof. Let $X=C_\sigma(S,\mathbb{R})$ where S is the weak* closure of the set of extreme points of the closed unit ball of X^* and σ is negation. Clearly S is either the extreme points or the extreme points plus the zero functional. Hence S is countable and by theorem 3 of section 5, S is homeomorphic to $\Gamma(\alpha)$ for some countable ordinal α. It follows from theorem 15 that $C_\sigma(S,\mathbb{R})$ is linearly isometric to $C_0(V,\mathbb{R})$ where $V=\{s\in S:\sigma(s)>s\}$ (in the order inherited from $\Gamma(\alpha)$). If V is compact, we are done since then V is homeomorphic to $\Gamma(\beta)$ for some $\beta\leqslant\alpha$ (by theorem 3 of section 5). If V is not compact, then $\bar{V}$ adds only the one fixed point of σ and $\bar{V}$ is the Alexandroff one point compactification of V. As above, $\bar{V}$ is homeomorphic to $\Gamma(\beta)$ for some ordinal $\beta\leqslant\alpha$. Let γ be the (necessarily limit) ordinal corresponding to the point in $\bar{V}\setminus V$ so that $C_0(V,\mathbb{R}) =\{f\in C(\Gamma(\beta),\mathbb{R}):f(\gamma)=0\}$. Let $L:c_0(\mathbb{N},\mathbb{R})\to C_0(V,\mathbb{R})$ be defined by $(Lf)(\delta)=f(\delta)$ if $\delta<\omega$ and $(Lf)(\delta)=0$ if $\delta\geqslant\omega$. Then clearly L is a linear isometry of $c_0(\mathbb{N},\mathbb{R})$ into $C_0(V,\mathbb{R})$ and the mapping $P:C_0(V,\mathbb{R}) \to C_0(V,\mathbb{R})$ defined by $(Pf)(\delta)=0$ if $\delta>\omega$ and $(Pf)(\delta)=f(\delta)-f(\omega)$ if $\delta\leqslant\omega$ is a projection of $C_0(V,\mathbb{R})$ onto the range of L (moreover, $\|P\|\leqslant 2$). In particular, $C_0(V,\mathbb{R})$ is linearly isomorphic to $Z\times c_0(\mathbb{N},\mathbb{R})$ where Z is the kernel of P ($Z\times c_0(\mathbb{N},\mathbb{R})$ can have the norm $\|(f,g)\| =\max(\|f\|,\|g\|)$ for example). Since clearly $c_0(\mathbb{N},\mathbb{R})$ is linearly isomorphic to $c_0(\mathbb{N},\mathbb{R})\times\mathbb{R}$, we get that $C_0(V,\mathbb{R})$ is linearly isomorphic to $C_0(V,\mathbb{R})\times\mathbb{R}$, which is linearly isomorphic to $C(\bar{V},\mathbb{R})$. ∎

In the above we used a special case of the fact that if T is a locally compact Hausdorff space and $C_0(T,\mathbb{R})$ contains a complemented subspace linearly isomorphic to $c_0(\mathbb{N},\mathbb{R})$, then $C_0(T,\mathbb{R})$ is linearly isomorphic to $C_0(T,\mathbb{R})\times\mathbb{R}$ which, in turn, is linearly isomorphic to $C(\bar{T},\mathbb{R})$, where $\bar{T}$ is the Alexandroff one point compactification of T. The following general question remains open.

Question. *Let T be a locally compact Hausdorff space. Is $C_0(T,\mathbb{R})$ linearly isomorphic to $C_0(T,\mathbb{R})\times\mathbb{R}$?*

The isomorphic classification of $C(T,\mathbb{R})$ when T is an uncountable compact metric space was settled by Milutin in [208]. The reader may see [21], [87], [228] and [256] for expositions of it. Reference [228] also contains partial results in the nonmetrizable situation.

Theorem (Milutin). *Let T be an uncountable compact metric space. Then $C(T,\mathbb{R})$ is linearly isomorphic to $C([0,1],\mathbb{R})$.* ∎

Exercises. 1. (Banach-Stone). If S and T are compact Hausdorff spaces and $C(S,\mathbb{C})$ (resp. $C(S,\mathbb{R})$) is linearly isometric to $C(T,\mathbb{C})$ (resp. $C(T,\mathbb{R})$), then S is homeomorphic to T.

2. Let X be a real Banach space and suppose V is the closed unit ball of X. Show that every convex set of elements of norm one in V is contained in a face of V. Give an example where $X=Y^*$ for some Y and V has a face which is not weak* closed.

3. Let T be a compact Hausdorff space and suppose F is a face of the closed unit ball V of $C(T,\mathbb{R})$. Show that $V=\mathrm{co}(F\cup(-F))$. Show that if $f\in\mathrm{ext}\,V$, then $f\in F\cup(-F)$ for each face F.

4. Let F be as in exercise 3. Show that the affine span of F is a hyperplane in $C(T,\mathbb{R})$.

5. Let X be a real Banach space and suppose F is a convex subset of elements of norm one in X. Show that if the closed unit ball V of X is the convex hull of $F\cup(-F)$, then there is a unique $x^*\in X^*$ such that $x^*(x)=1-d(x,F)$ for all $x\in X$, where $d(x,F)$ is the distance from x to F.

6. (Arens-Kelley). Let T be a compact Hausdorff space and suppose $f\in\mathrm{ext}\,V$, where V is the closed unit ball of $C(T,\mathbb{R})$. Let S be the collection of all faces containing f and for each $F\in S$ let x_F^* be the element defined in exercise 5. Show that $\{x_F^*: F\in S\}$ with the weak* topology is homeomorphic to T.

7. Give a direct proof of the Arens-Kelley theorem (theorem 13) under the assumption that $V=\mathrm{co}(F\cup(-F))$ for every face F of V (where V is the closed unit ball of X).

8. (Jerison). Let T_1 and T_2 be compact Hausdorff spaces and suppose Σ_1 and Σ_2 are fixed point free involutory homeomorphisms on T_1 and T_2 respectively. Show that $C_{\Sigma_1}(T_1,\mathbb{R})$ is linearly isometric to $C_{\Sigma_2}(T_2,\mathbb{R})$ if and only if there is a homeomorphism ϕ of T_1 onto T_2 such that $\Sigma_1=\phi^{-1}\Sigma_2\phi$ (the same is true if Σ_1 and Σ_2 both have exactly one fixed point).

9. Let T be a compact Hausdorff space and $V=b(C(T,R))$. Prove that V is the norm closed convex hull of its extreme points if and only if T is totally disconnected.

10. Let T be as in 9 and $V=b(C(T,\mathbb{C}))$. Prove that V is always the norm closed convex hull of its extreme points (hint: use the corollary to theorem 7 in section 8).

11. Let T be a compact Hausdorff space and σ be an involuntary homeomorphism on T. Let $C_\sigma(T,\mathbb{C})=\{f\in C(T,C): f(t)=-f(\sigma(t))$ for all $t\in T\}$. Show that $C_\sigma(T,\mathbb{C})$ is the range of a contractive projection on $C(T,\mathbb{C})$ and that every $C(T,\mathbb{C})$ is some $C_\sigma(S,\mathbb{C})$.

12. Show that if $\sigma(t)\neq t$, then the functional to defined by $t(f)=f(t)$ for all $f\in C_\sigma(T,\mathbb{C})$ is an extreme point of the closed unit ball of $C_\sigma(T,\mathbb{C})$. Prove that $C_\sigma(T,\mathbb{C})$ is linearly isometric to $C_{\sigma*}(T^*,\mathbb{C})$ where $T^*=\{t_\sigma : t\in T\}$ and $\sigma^*(t_\sigma)=(\sigma(t))_\sigma$.

13. Prove that if $Y\subset C(T,\mathbb{C})$ is the range of a contractive projection, then $Y=C_\sigma(S,\mathbb{C})$ for some S,σ.

14. Prove that if $C_\sigma(T,\mathbb{R})$ has an extreme point in its closed unit ball then $C_\sigma(T,\mathbb{C})$ is linearly isometric to $C(S,\mathbb{C})$ for some compact Hausdorff space S. What if $C_\sigma(T,\mathbb{C})$ has an extreme point in its closed unit ball?

§ 11. Banach Spaces with the Hahn-Banach Extension Property

Let X be a Banach space (real or complex) and let $\lambda\geqslant 1$. We say that X has the λ *extension property* if whenever Y and Z are Banach spaces, $\psi: Y\to Z$ is a linear isometry into and $L: Y\to X$ is a bounded linear operator, there is a bounded linear operator $\hat{L}: Z\to X$ such that L is an extension of L (i.e. $\hat{L}\circ\psi=L$) and $\|\hat{L}\|\leqslant\lambda\|L\|$. Clearly $\mathbb{R}$ and $\mathbb{C}$ have the 1 extension property by the Hahn-Banach theorem. We shall show that X has the 1 extension property if and only if X is linearly isometric to $C(T,\mathbb{R})$ ($C(T,\mathbb{C})$ in the complex case) where T is an extremally disconnected compact Hausdorff space.

In a similiar fashion, X is said to have the λ *projection property* ($\lambda\geqslant 1$) if whenever Y is a Banach space and $\psi: X\to Y$ is a linear isometry there is a projection P of Y onto $\psi(X)$ such that $\|P\|\leqslant\lambda$. We now observe that these two notions are equivalent.

Theorem 1. *Let X be a Banach space. Then X has the λ extension property if and only if it has the λ projection property.*

Proof. Suppose X has the λ extension property and let $\psi: X\to Y$ be a linear isometry into. Let $i: X\to X$ be the identity operator. Then there is a linear operator $\hat{i}: Y\to X$ such that $\hat{i}\circ\psi=i$ and $\|i\|\leqslant\lambda$. Clearly $\psi\circ\hat{i}=P$ is a projection and $\|P\|\leqslant\lambda$.

The converse follows from theorem 3 below as follows. Let $\phi: X\to C(\beta T,\mathbb{R})$ be an into linear isometry for some discrete space T. Suppose $\psi: Y\to Z$ is a linear isometry into and $L: Y\to X$ is a bounded linear operator. There is a projection P of $C(\beta T,\mathbb{R})$ onto $\phi(X)$ with $\|P\|\leqslant\lambda$ and a norm preserving extension $\hat{L}$ of $\phi\circ L$ to Z. Thus $\phi^{-1}\circ P\circ\hat{L}$ is the required extension of L. ∎

The above mentioned characterization of spaces with the 1 extension property was the result of several papers. The reader is especially referred to [114], [128], [152], and [214]. The development here follows [151] and [164]. We note that the characterization of those Banach spaces with the λ extension property for $\lambda > 1$ is still an unsolved problem. The reader is referred to [190] and [245] for some partial results concerning these spaces.

We shall denote the class of all real (resp. complex) Banach spaces with the λ extension property by $\mathscr{P}_\lambda(\mathbb{R})$ (resp. $\mathscr{P}_\lambda(\mathbb{C})$). By $\mathscr{P}(\mathbb{R})$ we shall mean the union over all $\lambda \geqslant 1$ of the classes $\mathscr{P}_\lambda(\mathbb{R})$ (similiarly for $\mathscr{P}(\mathbb{C})$). The Banach spaces in $\mathscr{P}(\mathbb{R})$ or $\mathscr{P}(\mathbb{C})$ are called *injective* Banach spaces.

The following theorem has elementary proof.

Theorem 2. (a) $\mathbb{R}$ *is in* $\mathscr{P}_1(\mathbb{R})$ *and* $\mathbb{C}$ *is in* $\mathscr{P}_1(\mathbb{C})$.

(b) *If* X *is in* $\mathscr{P}_\lambda(R)$ *(resp.* $\mathscr{P}_\lambda(C)$*) and* $\psi: X \to Y$ *is an onto linear isomorphism with* $\lambda' = \|\psi\| \|\psi^{-1}\|$, *then* Y *is in* $\mathscr{P}_{\lambda\lambda'}(\mathbb{R})$ *(resp.* $\mathscr{P}_{\lambda\lambda'}(\mathbb{C})$*). In particular, if* $\lambda' = 1$, *then* Y *is in* $\mathscr{P}_\lambda(\mathbb{R})$ *(resp.* $\mathscr{P}_\lambda(\mathbb{R})$ *(resp.* $\mathscr{P}_\lambda(\mathbb{C})$*).*

(c) *If* X *is in* $\mathscr{P}_\lambda(\mathbb{R})$ *(resp.* $\mathscr{P}_\lambda(\mathbb{C})$*) and* P *is a projection of norm* λ' *on* X, *then* $Y = P(X)$ *is in* $\mathscr{P}_{\lambda\lambda'}(\mathbb{R})$ *(resp.* $\mathscr{P}_{\lambda\lambda'}(\mathbb{C})$*). In particular, if* $\lambda' = 1$, *then* Y *is in* $\mathscr{P}_\lambda(\mathbb{R})$ *(resp.* $\mathscr{P}_\lambda(\mathbb{C})$*).* ∎

The following theorem is due to Phillips [300].

Theorem 3. *Let* T *be a discrete topological space (i.e. each point of* T *is open). Then* $C(\beta T, \mathbb{R})$ *and* $C(\beta T, \mathbb{C})$ *have the* 1 *extension property.*

Proof. The same proof works in both cases so we give it for $C(\beta T, \mathbb{R})$. Let X and Y be Banach spaces and suppose $\psi: X \to Y$ is a linear isometry into and $L: X \to C(\beta T, \mathbb{R})$ is a bounded linear operator. By the Hahn-Banach theorem, for each $t \in T$ there is a $y_t^* \in Y^*$ such that $y_t^* | X = L^*(\varepsilon_t) = \varepsilon_t \circ L$ and $\|y_t^*\| = \|L^*(\varepsilon_t)\|$. We let $\hat{L}: Y \to C(\beta T, \mathbb{R})$ be defined by $(\hat{L} y)(t) = y_t^*(y)$ for $t \in T$ (recall that by definition of βT and the fact that T is discrete, the elements of $C(\beta T, \mathbb{R})$ are precisely the bounded functions on T). Now $\|\hat{L}(y)\| = \sup\{|y_t^*(y)| : t \in T\} \leqslant \sup\{\|y_t^*\| \|y\| : t \in T\} = \sup\{\|L^*(\varepsilon_t)\| \|y\| : t \in T\} = \|L^*\| \|y\| = \|L\| \|y\|$ for all $y \in Y$. Thus $C(\beta T, \mathbb{R})$ has the 1 extension property. ∎

Corollary 1. *Suppose* X *is a real (resp. complex) Banach space with the property that whenever* X *is linearly isomorphic to a subspace* Y *of a Banach space* Z, *then* Y *is the range of a bounded projection in* Z *(i.e.* Y *is complemented in* Z*). Then* X *is an injective space.*

Proof. Recall that there is a discrete space T and a linear isometric embedding of X into $C(\beta T, \mathbb{R})$. For example, let T be the set of extreme points of the closed unit ball of X^* and let $\psi: X \to C(\beta T, \mathbb{R})$ be defined by $\psi(x)(x^*) = x^*(x)$ for all $x \in X$ and $x^* \in T$. Thus by theorems 1 and 2, X is injective. ∎

Corollary 2. *Let T be an extremally disconnected compact Hausdorff space. Then $C(T,\mathbb{R})$ (resp. $C(T,\mathbb{C})$) has the 1 extension property.*

Proof. By lemma 3 of section 7 there is a discrete topological space S and continuous mappings $r:\beta S\to T$ and $s:T\to\beta S$ such that $r\circ s$ is the identity on T. Let $L_1:C(T,\mathbb{R})\to C(\beta S,\mathbb{R})$ and $L_2:C(\beta S,\mathbb{R})\to C(T,\mathbb{R})$ be defined by $L_1(f)=f\circ r$ and $L_2(g)=g\circ s$ for $f\in C(T,\mathbb{R})$ and $g\in C(\beta S,\mathbb{R})$. Then L_1 is an isometry and L_2 is onto. Moreover, $L_2\circ L_1$ is the identity on $C(T,\mathbb{R})$. Thus $P=L_1\circ L_2$ is a contractive projection of $C(\beta S,\mathbb{R})$ onto the range of L_1. Thus by theorems 1 and 2, $C(T,\mathbb{R})$ has the 1 extension property. The same proof clearly works for $C(T,\mathbb{C})$. ∎

We now wish to establish the converse of corollary 2 and thus obtain the stated characterization of Banach spaces with the 1 extension property. For convenience we shall give the proof and statements in the complex case only. The real case follows with obvious modifications.

Definition 1. Let X and Y be Banach spaces and suppose $\psi:X\to Y$ is a linear isometry into. The pair (Y,ψ) is said to be an *essential extension* of X if for each Banach space Z and each contractive linear operator $L:Y\to Z$ (i.e. $\|L\|\leqslant 1$), we have that L is an isometry if $L\circ\psi$ is. The pair (Y,ψ) is said to be *rigid* if for each contractive linear operator $L:Y\to Y$ which is not the identity, $L\circ\psi\neq\psi$.

We shall now construct an essential extension (Y,ψ) of X such that Y has the 1 extension property.

Lemma 1. *Let X and Y be Banach spaces and L a linear isometry of X into Y. If (Y,L) is an essential extension of X, then it is also rigid.*

Proof. Suppose (Y,L) is an extension of X which is not rigid. Then there is a contractive linear operator $A:Y\to Y$ such that $A\circ L=L$ and A is not the identity. Hence there is a $y\in Y$ such that $Ay\neq y$. Let M denote the span of $y-Ay$ and $Z=Y/M$. Then the natural map $\pi:Y\to Z$ is not an isometry but $\pi\circ L$ is. For, suppose $x\in X$ and $\|x\|=1$. If there is a $\delta\in\mathbb{R}$ such that $\|Lx+\delta(y-Ay)\|<1$, whenever z is close enough to $L(x)$ we have that $\|z+\delta(y-Ay)\|\leqslant\|z\|$. Then for such a z, $\|z+\varepsilon(y-Ay)\|<\|z\|$ and $\|z-\varepsilon(y-Ay)\|>\|z\|$ for all ε between 0 and δ. Now $A(Lx+\varepsilon y)=Lx+\varepsilon Ay=(Lx+\varepsilon y)-\varepsilon(y-Ay)$. Thus for $|\varepsilon|$ small enough, $\|A(Lx+\varepsilon y)\|>\|Lx+\varepsilon y\|$ which is a contradiction to A being contractive. ∎

Let X and Y be Banach spaces with $X\subset Y$, a subset F of the closed unit ball of Y^* is said to be an X *boundary* if $\|x\|=\sup\{|y^*(x)|:y^*\in F\}$ for all $x\in X$. Moreover, Y is said to be a *bounded extension* of X if for each X boundary F, $\|y\|=\sup\{|y^*(y)|:y^*\in F\}$ for all $y\in Y$.

By a standard application of Zorn's lemma it is easy to see that minimal weak* closed X boundaries exist.

Lemma 2. *Let X and Y be Banach spaces with $X \subset Y$. Then Y is a bound extension of X if and only if the only seminorm on Y which is dominated by the norm on Y and equal to the norm on X is the norm on Y itself.*

Proof. Let F be an X boundary and let p be defined by $p(y) = \sup\{|y^*(y)| : y^* \in F\}$ for all $y \in Y$. Then p is a seminorm on $Y, p(y) \leqslant \|y\|$ for all $y \in Y$ and $p(x) = \|x\|$ for all $x \in X$. Thus if the only such seminorm is the norm on Y, it follows that Y is a bound extension of X.

Conversely, suppose Y is a bound extension of X and p is a seminorm on Y such that $p(y) \leqslant \|y\|$ for all $y \in Y$ and $p(x) = \|x\|$ for all $x \in X$. Let $F = \{y^* \in Y^* : |y^*(y)| \leqslant p(y)$ for all $y \in Y\}$. Then F is an X boundary and since Y is a bound extension of X, $p = \| \ \|$. ∎

Corollary. *Let X and Y be Banach spaces with $X \subset Y$. Then Y is a bound extension of X if and only if (Y, i) is an essential extension of X, where i is the natural embedding of X into Y.*

Proof. Suppose Y is a bound extension of X and $L: Y \to Z$ is a contractive linear operator such that $L|X$ is an isometry. Let p be defined by $p(y) = \|Ly\|$ for all $y \in Y$. Then since Y is a bound extension of X, $p = \| \ \|$ and L is an isometry. That is, (Y, i) is an essential extension of X.

Conversely suppose (Y, i) is an essential extension of X. Let p be a seminorm on Y with $p(y) \leqslant \|y\|$ for all $y \in Y$ and $p(x) = \|x\|$ for all $x \in X$. Let M be the null manifold of p and π the natural map of Y onto the completion of Y/M in the norm induced by p. Then $\pi \circ i$ is an isometry and, thus, π is an isometry. That is, $p = \| \ \|$. ∎

In corollary 1 to theorem 3 we observed that any (complex) Banach space X is linearly isometric to a subspace of some $\mathscr{P}_1(\mathbb{C})$ space.

Theorem 4. *Let X and Y be Banach spaces with $X \subset Y$ and suppose Y is a $\mathscr{P}_1(\mathbb{C})$ space. Then there is a $\mathscr{P}_1(\mathbb{C})$ space Z with $X \subset Z \subset Y$ and Z is a bound extension of X.*

Proof. Let $\mathscr{F}$ denote the family of all seminorms p on Y such that $p(y) \leqslant \|y\|$ for all $y \in Y$ and $p(x) = \|x\|$ for all $x \in X$. Clearly $\mathscr{F}$ is a partially ordered set under the relation $p \leqslant q$ if and only if $p(y) \leqslant q(y)$ for all $y \in Y$. If $\{p_d\}$ is a downwards directed chain in $\mathscr{F}$, then a lower bound in $\mathscr{F}$ for this chain is defined by $p(y) = \lim p_d(y)$ for all $y \in Y$. Thus by Zorn's lemma $\mathscr{F}$ has a minimal element p_0.

By passing to the completion of Y/M in the norm induced by p_0 where M is the null manifold of p_0 and using the fact that Y is a $\mathscr{P}_1(\mathbb{C})$ space, there is a linear operator $L:Y\to Y$ such that $Lx=x$ for all $x\in X$ and $\|Ly\|\leqslant p_0(y)$ for all $y\in Y$. Let p_1 be defined by $p_1(y)=\|Ly\|$ for all $y\in Y$. Then $p_1\leqslant p_0$ and $p_1\in\mathscr{F}$ so by the minimality of $p_0, p_1=p_0$. Let p_2 be defined by $p_2(y)=\limsup\left\|\frac{1}{n}\sum_{i=1}^{n}L^i(y)\right\|$ for all $y\in Y$. Then $p_2\leqslant p_0$ and $p_2\in\mathscr{F}$ so again $p_2=p_0$. Moreover, $p_2(y-Ly)=\limsup\left\|\frac{1}{n}(Ly-L^{n+1}y)\right\|=0$ for all $y\in Y$. That is, $p_2(y-Ly)$ $=p_0(y-Ly)=p_1(y-Ly)=\|Ly-L^2y\|=0$ for all $g\in Y$ and it follows that $L=L^2$, that is, L is a contractive projection and $X\subset L(Y)$. Let $Z=L(Y)$. Then clearly Z is a $\mathscr{P}_1(\mathbb{C})$ space by theorem 2(c) and it only remains to show that Z is a bound extension of X. Let p be a semi-norm on Z with $p(z)\leqslant\|z\|$ for all $z\in Z$ and $p(x)=\|x\|$ for all $x\in X$. Then $p\circ L$ is in $\mathscr{F}$ and $p\circ L\leqslant p_0$. Thus $p\circ L=p_0$ and it follows that Z is a bound extension for X. ∎

Corollary. *Let X and Y be Banach spaces with $X\subset Y$ and suppose Y is a $\mathscr{P}_1(\mathbb{C})$ space and (Y,i) an essential extension of X. If Z is a Banach space and L is a linear operator from X into Z so that (Z,L) is an essential extension of X, then there is a unique linear contractive operator $A:Z\to Y$ with $A\circ L$ equal to the identity on X. Moreover A is an isometry.*

Proof. Since Y is a $\mathscr{P}_1(\mathbb{C})$ space there is a contractive linear operator $A:Z\to Y$ such that $A\circ L$ is the identity on X. Since (Z,L) is an essential extension of X, A is an isometry.

Now suppose there is a contractive linear operator $A_1:Z\to Y$ such that $A_1\circ L$ is the identity on X. Then A_1 is also an isometry. Let $L_1:A(Z)\to A_1(Z)$ be defined by $L_1(Az)=A_1(z)$ for all $z\in Z$. Then L_1 is contractive and since Y is a $\mathscr{P}_1(\mathbb{C})$ space, there is a contractive $\hat{L}_1$ on Y such that $\hat{L}_1=L_1$ on $A(Z)$. But, $\hat{L}_1x=x$ for all $x\in X$ implies $\hat{L}_1$ is the identity on Y by lemma 1. Thus $A=A_1$. ∎

For the next two lemmas and theorem we suppose T is a compact Hausdorff space and X is a closed linear subspace of $C(T,\mathbb{C})$ such that for each closed proper subset F of T, F is not an X boundary.

Lemma 3. *Let V be a nonempty open set in T. Then for each $\varepsilon>0$ $(\pi/4>\varepsilon>0)$ there is an $f\in X$ such that for all $t\in T\setminus V\cap\{s:\operatorname{Re}f(s)>\|f\|\cos\varepsilon\}$ we have that $|f(t)|<\|f\|$.*

Proof. We clearly may assume that $V\neq T$. By the hypothesis on X there is an $f_1\in X$ which attains its maximum modulus only in V. Moreover, we may take f_1 so that $\|f_1\|=\max\{\operatorname{Re}f_1(t):t\in T\}$. Let $0<\delta\leqslant\varepsilon$ be such that $|f_1(t)|<\|f_1\|\cos\delta$ if $t\notin V$. Similiarly we may choose an $f_2\in X$ which attains its maximum modulus only in $V\cap\{t:\operatorname{Re}f_1(t)>\|f_1\|\cos\delta/3\}$

and such that $\|f_2\| = \max\{\operatorname{Re} f_2(t): t \in T\}$. Let $g_n = nf_2 + f_1$. We shall show that for sufficiently large n, g_n satisfies the lemma.

Suppose $t_n \in T$ is such that $|g_n(t_n)| = \|g_n\|$ for $n = 1, 2, \ldots$. Note that $\|g_n\| \geq n\|f_2\| + \|f_1\| \cos\delta/3$. Thus $n^2\|f_2\|^2 + 2n\|f_2\|\|f_1\|\cos\delta/3 + \|f_1\|^2\cos^2\delta/3 \leq n^2|f_2(t_n)|^2 + 2n\operatorname{Re}[f_2(t_n)\overline{f_1(t_n)}] + |f_1(t_n)|^2 \leq n^2|f_2(t_n)|^2 + 4n\|f_2\|\|f_1\| + \|f_1\|^2$. In particular,

$$|f_2(t_n)|^2 \geq \|f_2\|^2 + (2/n)[\|f_2\|\|f_1\|\cos\delta/3 - 2\|f_2\|\|f_1\|] + (1/n^2)\|f_1\|^2(\cos 2\delta/3 - 1).$$

Thus $\lim|f_2(t_n)| = \|f_2\|$ and note that the right hand side of the above inequality does not depend on t_n so that $|f_2(t)|$ can be made uniformly close to $\|f_2\|$ for any t such that $|g_n(t)| = \|g_n\|$. Now, since $|f_2(t_n)|^2 \leq \|f_2\|^2$, we also obtain that $2n\|f_2\|\|f_1\|\cos\delta/3 + \|f_1\|^2\cos^2\delta/3 \leq 2n\operatorname{Re}[f_2(t_n)\overline{f_1(t_n)}] + |f_1(t_n)|^2 \leq 2n\operatorname{Re}[f_2(t_n)\overline{f_1(t_n)}] + \|f_1\|^2$. By a similar argument we see that $\liminf\operatorname{Re}[f_2(t_n)\overline{f_1(t_n)}] \geq \|f_1\|\|f_2\|\cos\delta/3$. Now if $t \notin V$, then for any n, $|g_n(t)| \leq n|f_2(t)| + |f_1(t)| < n\|f_2\| + \|f_1\|\cos\delta \leq \|g_n\|$.

By the above there is an n_0 such that for $n \geq n_0$, $\operatorname{Re} f_1(t) > \|f_1\|\cos\delta/3$ for all t such that $|g_n(t)| = \|g_n\|$. That is, the argument of $f_1(t)$ is in $[-\delta/3, \delta/3]$. Since $\liminf\operatorname{Re}[f_2(t_n)\overline{f_1(t_n)}] \geq \|f_1\|\|f_2\|\cos\delta/3$ (and this does not depend on the way the t_n's are chosen), we can find an $n \geq n_0$ such that for each $t \in T$ with $|g_n(t)| = \|g_n\|$. We have that $\operatorname{Re} f_1(t) > \|f_1\|\cos\delta/3$ and $\operatorname{Re} f_2(t) > \|f_2\|\cos 2\delta/3$. Thus $|g_n(t)| \geq \operatorname{Re} g_n(t) = n\operatorname{Re} f_2(t) + \operatorname{Re} f_1(t) > n\|f_2\|\cos 2\delta/3 + \|f_1\|\cos\delta/3 \geq (n\|f_2\| + \|f_1\|)\cos\varepsilon$. Thus g_n satisfies the lemma. ∎

Lemma 4. *Let $f \in C(T, \mathbb{C})$ and suppose $0 \leq d < \|f\|$. Then there is a $g \in X$ such that $\|g + f\| < \|g\| - d$.*

Proof. Clearly we may assume that there is a $t \in T$ such that $f(t) = \|f\|$. Let $d < a < \|f\|$ and suppose $S(\delta) = \{a \in \mathbb{C}: |a + 1| < \delta\}$. By lemma 3 we can choose a $g \in X$ such that $\|g\| = 1$, $g(t) = -1$ for some $t \in T$, and g attains its maximum modulus only in $\{t: \operatorname{Re} f(t) > a\} \cap \{t: g(t) \in S(\delta)\}$. Now $|f(t) + ng(t)|^2 = |f(t)|^2 + 2n\operatorname{Re}[f(t)\overline{g(t)}] + n^2|g(t)|^2 \leq n^2 + 2nM(\delta) + \|f\|^2$, where $M(\delta) = \sup\{\operatorname{Re}\bar{b} f(t): \operatorname{Re} f(t) > a \text{ and } b \in S(\delta)\}$. Suppose the lemma is not valid for ng. Then $\|f + ng\| \geq n - d$ so that $(n-d)^2 \leq \|f + ng\|^2 \leq n^2 + 2nM(\delta) + \|f\|^2$. Hence $M(\delta) \geq -d$. But, taking the limit as $\delta \to 0$, we obtain a $t \in T$ such that $-\operatorname{Re} f(t) \geq -d$ and $\operatorname{Re} f(t) \geq a$ which is impossible. Thus, the lemma is true for some $ng \in X$. ∎

Theorem 5 (Kaufman). *$(C(T, \mathbb{C}), i)$ is an essential extension of X, where i is the natural embedding of X into $C(T, \mathbb{C})$.*

Proof. Let p be a seminorm on $C(T,\mathbb{C})$ with $p\leqslant\|\ \|$ and $p=\|\ \|$ on X. Thus for $f\in C(T,\mathbb{C})$ and $g\in X$, $\|f+g\|\geqslant p(f+g)\geqslant p(g)-p(f)=\|g\|-p(f)$. Thus by lemma 4 we obtain that $p=\|\ \|$. ∎

We can now prove the characterization of $\mathscr{P}_1(\mathbb{C})$ and $\mathscr{P}_1(\mathbb{R})$ spaces.

Theorem 6 (Nachbin, Kelley, Goodner, Hasumi). *Let X be a $\mathscr{P}_1(\mathbb{C})$ (resp. $\mathscr{P}_1(\mathbb{R})$) space. Then X is linearly isometric to $C(T,\mathbb{C})$ (resp. $C(T,\mathbb{R})$) for some extremally disconnected compact Hausdorff space T.*

Proof. Let T be a minimal weak* closed subset of the closed unit ball of X^* such that $\|x\|=\sup\{|x^*(x)|:x^*\in T\}$ for all $x\in X$. By theorem 5, $(C(T,\mathbb{C}),L)$ is an essential extension of X where $L(x)(x^*)=x^*(x)$ for all $x^*\in T$ and $x\in X$.

Since X is a $\mathscr{P}_1(\mathbb{C})$ space there is a contractive projection P of $C(T,\mathbb{C})$ onto the range of L. Since the extension is essential and $P\circ L=L$, it follows that P is the identity and L is onto.

Let Ω be an extremally disconnected compact Hausdorff space and let $h:\Omega\to T$ be a minimal continuous mapping of Ω onto T (see theorem 2 of section 7). Then the operator $A:C(T,\mathbb{C})\to C(\Omega,\mathbb{C})$ defined by $A(f)=f\circ h$ for $f\in C(T,\mathbb{C})$ is a linear isometry. By theorem 4, $(C(\Omega,\mathbb{C}),A)$ is an essential extension of $C(T,\mathbb{C})$. Since $C(T,\mathbb{C})$ is a $\mathscr{P}_1(\mathbb{C})$ space, the above argument shows that A is onto and, thus, that h is a homeomorphism. ∎

We now present two special characterizations of $\mathscr{P}_1(\mathbb{R})$ spaces.

Theorem 7. *Let T be a compact Hausdorff space. Then T is extremally disconnected if and only if $C(T,\mathbb{R})$ is order complete.*

Proof. Suppose $C(T,\mathbb{R})$ is order complete and let V be a nonempty open set in T. For each nonempty closed set $F\subset V$ choose $f_F\in C(T,\mathbb{R})$ such that $0\leqslant f_F\leqslant 1$, $f_F=1$ on F, and $f_F=0$ on $T\setminus V$. Then the family $\{f_F\}$ is bounded from above by the constant function 1 and, hence, $\sup f_F=f$ exists in $C(T,\mathbb{R})$. Clearly $f=1$ on $\bar{V}$. Moreover, for any $t\in T\setminus\bar{V}$ there is a $g\in C(T,\mathbb{R})$ with $0\leqslant g\leqslant 1$, $g(t)=0$ and $g=1$ on $\bar{V}$. Thus $f\leqslant g$ and f is necessarily the characteristic function of $\bar{V}$. Therefore $\bar{V}$ is open and T is extremally disconnected.

Now suppose T is extremally disconnected and $\{f_d\}$ is a norm bounded set in $C(T,\mathbb{R})$. Let g be the pointwise supremum of the f_d's. Then g is lower semicontinuous and bounded. Let $M>0$ be such that $g(T)\subset(-M,M)$ and for any partition $\Pi=\{-M=x_0<x_1<\cdots<x_n=M\}$ of $[-M,M]$ let $U_i=g^{-1}(x,\infty)$ for $i=0,\dots,n$. Then each U_i is open, $T=U_0\supset\cdots\supset U_n=\emptyset$ and $T=\bigcup_{i=0}^{n-1}[\bar{U}_i\cap(T\setminus\bar{U}_{i+1})]$ determines a parti-

tion of T into pairwise disjoint open and closed sets. Let $f_\Pi = \sum_{i=1}^{n} x_i f_i$ where f_i is the characteristic function of $\bar{U}_{i-1} \setminus \bar{U}_i$ for $i=1,\ldots,n$. Then for $f_\Pi \in C(T,\mathbb{R})$ and for $t \notin \bigcup_{i=0}^{n-1} \bar{U}_i \setminus U_i$, $0 \leqslant g(t) - f_\Pi(t) \leqslant \operatorname{mesh}(\Pi)$. Let $\{\Pi_n\}$ be a sequence of partitions with Π_{n+1} a refinement of Π_n and $\operatorname{mesh}\Pi_n \to 0$. Then $0 \leqslant f_{\Pi_{n+p}}(t) - f_{\Pi_n}(t) \leqslant \operatorname{mesh}\Pi_n$ for all $t \in T$ and $\{f_{\Pi_n}\}$ is a Cauchy sequence in $C(T,\mathbb{R})$. It follows that $\sup f_a = \lim f_{\Pi_n}$. ∎

A collection of sets is said to have the *binary intersection property* if every pair of sets in the collection has nonempty intersection.

Theorem 8. *Let X be a real Banach space. Then X is a $\mathscr{P}_1(\mathbb{R})$ space if and only if for every collection of closed balls in X with the binary intersection property, the collection has nonempty intersection.*

Proof. Suppose B is a $\mathscr{P}_1(\mathbb{R})$ space. Then B is linearly isometric to $C(T,\mathbb{R})$ for some extremally disconnected compact Hausdorff space T and by theorem 7 $C(T,\mathbb{R})$ is order complete. Let $S_a = S_a(f_a, r_a)$ be the closed ball with radius r_a and center f_a. If a family $\{S_a\}$ has the binary intersection property, then for any a_1 and a_2 there is an f with $f_{a_i} - r_{a_i} \leqslant f \leqslant f_{a_i} + r_{a_i}$ for $i=1,2$. Hence $\sup(f_a - r_a) \leqslant \inf(f_a + r_a)$ and since $C(T,\mathbb{R})$ is order complete there is an $f \in \bigcap S_a$.

Suppose X has the binary intersection property. Let Y, Z be Banach spaces suppose $\psi: Y \to Z$ is a linear isometry into and $L: Y \to X$ a bounded linear operator. For $z \in Z \setminus \psi(Y)$, $L \circ \psi^{-1}$ is a bounded linear extension to $\psi(Y) \oplus \operatorname{span}(z)$ of norm $\|L\|$ if and only if there is an $x \in X$ such that for all $y \in Y$ and all $a \neq 0$, $\|L(y/a) + x\| \leqslant \|L\| \, \|\psi(y/a) + z\|$. That is, if and only if for each $y \in Y$, $\|Ly - x\| \leqslant \|L\| \, \|\psi(y) - z\|$. This is clearly equivalent to $\bigcap_{y \in Y} S(Ly, \|L\| \, \|\psi(y) - z\|) \neq \emptyset$. From the binary intersection property this is true. Zorn's lemma finishes the proof. Thus X is a $\mathscr{P}_1(\mathbb{R})$ space. ∎

The study turns to $C(T,\mathbb{R})$ for particular extremally disconnected compact Hausdorff spaces T. The main goal is to completely characterize those compact Hausdorff spaces T such that $C(T,\mathbb{R})$ is linearly isometric to X^* for some Banach space X. This will require the study of certain types of regular Borel measures on T.

A positive regular Borel measure μ on a compact Hausdorff space T is said to be *normal* if $\mu(B) = 0$ for each Borel set B of first category in T. Let $N(T,\mathbb{R})^+$ denote the set of all positive normal regular Borel measures on T. Clearly $N(T,\mathbb{R})^+$ is a norm closed proper cone in $M(T,\mathbb{R})$ and moreover, $N(T,\mathbb{R}) = N(T,\mathbb{R})^+ - N(T,\mathbb{R})^+$ is a closed ideal of $M(T,\mathbb{R})$. Note that if t is an atom of $\mu\in N(T,\mathbb{R})$, then $\{t\}$ is an open set in T, i.e., t is an isolated point in T.

The symbol f_B denotes the *characteristic function* of a set B.

Lemma 5. *Let T be an extremally disconnected compact Hausdorff space. A positive regular Borel measure μ on T is normal if and only if for every bounded upwards directed set $\{f_d\}$ in $C(T,\mathbb{R})$, $\lim\int f_d\,d\mu=\int f\,d\mu$ where $f=\sup f_d$. Furthermore, if σ is a regular Borel measure on T and $\lim\int f_d\,d\mu=\int f\,d\sigma$ for f_d, f as above, then σ^+ and σ^- are normal.*

Proof. Let B be a closed nowhere dense Borel set in T and $D=\{f\in C(T,\mathbb{R}): f\geqslant f_B\}$. Then D is downwards directed and by the regularity of μ and the hypothesis, $\mu(B)=\lim_{f\in D}\int f\,d\mu=\int f_0\,d\mu$ where $f_0=\inf\{f: f\in D\}$. But f_0 is 0 on $T\setminus B$ which is a dense open set in T. Thus $f_0=0$ and it follows that $\mu(B)=0$. Hence μ is normal.

Suppose μ is normal and let $\{f_d\}_{d\in D}$ be a bounded upwards directed set in $C(T,\mathbb{R})$. Put $f_0=\sup f_d$ in $C(T,\mathbb{R})$ and let f_1 be the pointwise supremum of the f_d's. Then f_1 is lower semicontinuous and if $\varepsilon>0$ there is a closed set $F\subset T$ such that $f_1|F$ is continuous and $\mu(T\setminus F)<\varepsilon$. Let $f_2=f_1 f_A+f_0 f_{T\setminus A}$ where A is the closure of the interior of F. Then f_2 is continuous and is an upper bound for the f_d's. Thus $f_2\geqslant f$ but clearly $f_2\leqslant f_0$. Hence $f_0=f_1$ on A and since $F\setminus A$ is nowhere dense, $\mu(F\setminus A)=0$. By compactness we have that $\{f_d\}$ converges uniformly to f_1 on A. Thus there is a $d_0\in D$ such that for $d\geqslant d_0$, $\sup_{t\in A}|(f_1-f_d)(t)|<\varepsilon$. Hence $\int(f_1-f_d)\,d\mu=\int_A(f_1-f_d)\,d\mu+\int_{T\setminus A}(f_1-f_d)\,d\mu \leqslant \varepsilon\mu(A)+2\|f_0\|\varepsilon$. Therefore $\lim\int f_d\,d\mu=\int f_1\,d\mu=\int f_0\,d\mu$.

Suppose σ has the stated property and $\{f_d\}$ is a bounded upwards directed set in $C(T,\mathbb{R})$ with $f=\sup f_d$. Let (T^+,T^-) be the Hahn-decomposition of σ. For $\varepsilon>0$ let $F\subset T^+\subset U$ with F closed and U open and $|\sigma|(U\setminus F)<\varepsilon/2$. Let V be open and closed with $F\subset V\subset U$. Then $\int(f-f_d)\,d\sigma^+=\int_{T^+}(f-f_d)\,d\sigma=\int_{T^+\setminus V}(f-f_d)\,d\sigma+\int_V(f-f_d)\,d\sigma -\int_{V\setminus T^+}(f-f_d)\,d\sigma\leqslant|\int_V(f-f_d)\,d\sigma|+2\varepsilon\|f\|$. Similarly for σ^-. ∎

Theorem 9. *Let T be an extremally disconnected compact Hausdorff space. Suppose μ is a positive normal regular Borel measure on T and F is a bounded Borel function on T which is 0 on the complement of the support $S(\mu)$ of μ. Then there is a $g\in C(T,\mathbb{R})$ such that $g=f$ on $S(\mu)$, except perhaps for a nowhere dense set.*

Proof. Suppose $f\geqslant 0$. For each n let F_n be a closed subset of T with $f|F_n$ continuous and $\mu(T\setminus F_n)<1/2^n$. Since T is extremally disconnected F_n can be taken to be open and closed. Hence f is continuous on $\bigcup_{n=1}^{\infty}F_n$.

Let f_n be defined by $f_n = f \cdot f_{A_n}$ where $A_n = \bigcup_{i=1}^{n} F_i$. Then $\{f_n\}$ increases and is bounded from above. Let $g = \sup f_n$ in $C(T, \mathbb{R})$. Then $\int (f - f_n) d\mu = \int_{T \setminus F_n} |f - f_n| d\mu \leqslant (1/2^{n-1}) \|f\|$. Thus $\int f d\mu = \int g d\mu$ and $\{t \in S(\mu) : f(t) \neq g(t)\}$ is nowhere dense. The general case follows readily. ∎

The right condition in order to insure that $C(T, \mathbb{R})$ is a conjugate space must be imposed. The condition is not intuitive and, in fact, there is no known topological characterization of it. The condition is the following: *the union of the supports of the positive normal measures is dense in T.* Following Dixmier [89] we call extremally disconnected compact Hausdorff spaces satisfying this condition *hyperstonian* spaces.

Note that if T is the Stone-Čech compactification of its isolated points, then it is clearly hyperstonian. Moreover, if T is hyperstonian, then every Borel set of first category in T (i.e. a union of nowhere dense sets) is, in fact, nowhere dense.

If μ is a measure on T and $g \in C(T, \mathbb{R})$, we denote by $g \cdot \mu$ the measure defined by $(g \cdot \mu)(B) = \int_B g \, d\mu$ for all Borel sets B.

Theorem 10 (Dixmier). *If T is a hyperstonian space, then $C(T, \mathbb{R})$ is linearly isometric to $N(T, \mathbb{R})^*$ under the operator defined by $(Lf)(\mu) = \int f d\mu$ for all $f \in C(T, \mathbb{R})$ and all $\mu \in N(T, \mathbb{R})$.*

Proof. Clearly L is linear and $\|L\| \leqslant 1$. We first show that L is an isometry. Let $0 \neq f \in C(T, \mathbb{R})$ and suppose $0 < \varepsilon < \|f\|$. For $V = \{t : \|f\| - f(t) < \varepsilon\}$ there is a $\mu \in N(T, \mathbb{R})^+$ such that $\|\mu\| = 1$ and $S(\mu) \subset V$. Thus $|\int f d\mu| = \int f d\mu = \|f\| - \int [\|f\| - f] d\mu \geqslant \|f\| - \varepsilon$. That is, $\|Lf\| \geqslant \|f\| - \varepsilon$. Hence L is an isometric embedding of $C(T, \mathbb{R})$ into $N(T, \mathbb{R})^*$ and we need only show that it is onto.

Let $\mu \in N(T, \mathbb{R})^+$ and suppose that x^* is a positive linear functional on $N(T, \mathbb{R})$. Since $N(T, \mathbb{R})$ is a closed ideal in $M(T, \mathbb{R})$, $L_1(\mu, \mathbb{R}) \subset N(T, \mathbb{R})$ and, thus, $x^* | L_1(\mu, \mathbb{R})$ is a positive linear functional on $L_1(\mu, \mathbb{R})$. In particular, there is a positive bounded Borel function g on T such that $x^*(v) = \int g \, dv$ for all $v \in L_1(\mu, \mathbb{R})$. By theorem 9 there is a (necessarily unique) positive $f_\mu \in C(T, \mathbb{R})$ such that $f_\mu = 0$ on $T \setminus S(\mu)$ and $x^*(v) = \int f_\mu \, dv$ for $v \in L_1(\mu, \mathbb{R})$. For $\sigma \in L_1(\mu, \mathbb{R})$ we define v_σ by $v_\sigma(B) = \int_B f_\mu \, d\mu$ for all Borel sets B. Clearly $v_\sigma \in L_1(\mu, \mathbb{R})$ and $\int f \, dv_\sigma = \int f f_\mu \, d\mu$ for all $f \in C(T, \mathbb{R})$.

For each $h \in C(T, \mathbb{R})^+$, $h \cdot v_\mu = v_{h \cdot \mu}$. For, let V be open and closed in T and let $h = f_V$. Then $h \cdot v_\mu(B) = v_\mu(B \cap V) = v_{h \cdot \mu}(B)$ for all Borel sets B. By linearity the above is true for all positive simple functions in $C(T, \mathbb{R})$. Now for any positive g in $C(T, \mathbb{R})$, $H = \{h \in C(T, \mathbb{R}) : 0 \leqslant h \leqslant g$, h is a simple function$\}$ is upwards directed and $g = \sup H$. Thus $\lim_H h_\mu = g$ and $v_{g_\mu} = \lim_H v_{h_\mu} = \lim_H h \, v_\mu = g \cdot v_\mu$.

For any μ_1, μ_2 in $N(T,\mathbb{R})^+$, $f_{\mu_1} = f_{\mu_2}$ on $S(\mu_1) \cap S(\mu_2)$. Without loss of generality it can be assumed that $S(\mu_1) = S(\mu_2)$. Thus μ_1, μ_2 are mutually absolutely continuous and $(f_{\mu_2})\mu_2 = \nu_{\mu_2}$, $\nu_{h\cdot\mu_1} = h\cdot\nu_{\mu_1} = (h\cdot f_{\mu_1})\mu_1 = (f_{\mu_1})\mu_2$, where h is the Radon-Nikodym derivative of μ_2 with respect to μ_1. Hence $f_{\mu_1} = f_{\mu_2}$ by the uniqueness of f_μ.

Thus if $g\colon \bigcup_{\mu\in N(T,\mathbb{R})^+} S(\mu) \to \mathbb{R}$ is defined by $g|S(\mu) = f_\mu$, then g is well defined and continuous. Moreover, g is bounded. For, suppose g is not bounded from above. Then for each n there is a $\mu_n \in N(T,\mathbb{R})^+$ such that $\|\mu_n\| = 1$ and $S(\mu_n) \subset \{t : g(t) > n\}$. Now $\|x^*\| \geqslant \int g\, d\mu_n \geqslant n$ which is clearly impossible. A similiar argument shows that g is bounded from below. Hence $\{f_\mu : \mu \in N(T,\mathbb{R})^+\}$ is bounded and upwards directed and $f = \sup\{f_\mu : \mu\in N(T,\mathbb{R})^+\}$ exists in $C(T,\mathbb{R})$. For $\mu_0 \in N(T,\mathbb{R})^+$, $\int f\, d\mu_0 = \lim \int f_\mu\, d\mu_0 = \int f_{\mu_0} = \int f_{\mu_0}\, d\mu_0 = x^*(\mu_0)$.

Therefore $L(f) = x^*$ and it follows that L is onto. ∎

We now wish to prove a strong converse to theorem 10 due to Grothendieck [120]. In the following theorem T is a compact Hausdorff space, X is a Banach space, L is a linear isometry of $C(T,\mathbb{R})$ onto X^*, and J is the natural map of X into X^{**}.

Theorem 11 (Grothendieck). *T is hyperstonian and $A = L^* \circ J$ is a linear isometry of X onto $N(T,\mathbb{R})$.*

Proof. For each $x\in X$ let $\mu_x = L^*(Jx)$. Let U be a nonempty open set in T and $\varepsilon > 0$ be given. There is an $x \in X$ such that $\|x\| = 1$, $\mu_x^-(U) < \varepsilon$ and $\mu_x^+(U) \geqslant 1-\varepsilon$. For, let $t\in U$ and $f\in C(T,\mathbb{R})$ be such that $f(t) = 1$, $f = 0$ on $T\setminus U$, and $0 \leqslant f \leqslant 1$. Now $\|f\| = \|Lf\| = \sup\{\int f\, d\mu_x : \|x\| = 1\}$. Let $\|x\| = 1$ be such that $1 \geqslant \int f\, d\mu_x \geqslant 1-\varepsilon$. Then $1-\varepsilon \leqslant \int_U f\, d\mu_x \leqslant \mu_x^+(U)$ and hence $\mu_x^-(U) = |\mu_x|(U) - \mu_x^+(U) \leqslant 1-(1-\varepsilon) = \varepsilon$.

Let $\{f_d\}$ be an upwards directed bounded set in $C(T,\mathbb{R})$ and let x^* be defined by $x^*(x) = \lim \int f_d\, d\mu_x$ for all $x\in X$. Then $x^* \in X^*$ and there is an $f\in C(T,\mathbb{R})$ such that $(Lf)(x) = \int f\, d\mu_x = \lim \int f_d\, d\mu_x$ for all $x\in X$. Suppose there is a d_0 and a $t_0 \in T$ such that $f(t_0) < f_{d_0}(t_0)$. Then there is a $\delta > 0$ and an open set U such that $f_{d_0}(t) - f(t) > \delta$ for all $t\in U$. Let $\varepsilon > 0$ be given and put $M = \max(\|f\|, \sup\|f_d\|)$. Choose $x\in X$ with $\|x\| = 1$, $\mu_x^-(U) < \varepsilon$ and $\mu_x^+(U) \geqslant 1-\varepsilon$. Then for $d \geqslant d_0$, $\int (f_d - f)\, d\mu_x = \int_U (f_d - f)\, d\mu_x^+ + \int_{T\setminus U} (f_d - f)\, d\mu_x - \int_U (f_d - f)\, d\mu_x^- \geqslant \delta\mu_x^+(U) - M\varepsilon + \int_{T\setminus U}(f_d - f)\, d\mu_x \geqslant \delta\mu_x^+(U) - M\varepsilon - 2M|\mu_x|(T\setminus U) \geqslant \delta(1-\varepsilon) - 2M\varepsilon = \delta - \varepsilon(\delta + 2M)$. Thus for $\varepsilon < \delta/2(\delta+2M)$, $\int (f_d - f)\, d\mu_x \geqslant \delta/2$ which is a contradiction. Hence $f \geqslant f_d$ for all d.

Suppose there is a $g\in C(T,\mathbb{R})$ such that $f_d \leqslant g < f$ for all d. Let $t\in T$ be such that $g(t_0) < f(t_0)$. Then there is a $\delta > 0$ and an open set U such that $f(t) - g(t) > \delta$ for all $t\in U$. Let $\varepsilon > 0$ be given and choose $x\in X$ such that $\|x\| = 1$, $\mu_x^-(U) < \varepsilon$ and $\mu_x^+(U) \geqslant 1-\varepsilon$.

Then $\int f\,d\mu_x - \int g\,d\mu_x = \int_U (f-g)\,d\mu_x^+ + \int_{T\setminus U}(f-g)\,d\mu_x^+ - \int_U (f-g)\,d\mu_x^- - \int_{T\setminus U}(f-g)\,d\mu_x^- \geqslant \delta\mu_x^+(U) - M\mu_x^-(U) - 2M|\mu_x|(T\setminus U) \geqslant \delta(1-\varepsilon) - M\varepsilon - 2M\varepsilon = \delta - \varepsilon(\delta+3M)$. Hence for $\varepsilon < \delta/2(\delta+3M)$, $\int f\,d\mu_x - \int g\,d\mu_x > \delta/2$. But, $f_d \leqslant g < f$ and $\int f\,d\mu_x = \lim \int f_d\,d\mu_x$ implies $\int f\,d\mu_x = \int g\,d\mu_x$. Thus $g < f$ is impossible and $f = \sup f_d$. Hence $C(T,\mathbb{R})$ is order complete and it follows that T is extremally disconnected.

Clearly the above shows that μ_x^+ and μ_x^- are normal so $\mu_x \in N(T,\mathbb{R})$. The first part of the proof shows that T must be hyperstonian.

Now by theorem 10, the operator A defined by $(Bf)(\mu) = \int f\,d\mu$ is a linear isometry of $C(T,\mathbb{R})$ onto $N(T,\mathbb{R})^*$. By the above $L^* \circ J$ is a linear isometry of X into $N(T,\mathbb{R})$. Moreover, $(Bf)(L^*Jx) = \mu_x(f) = (Lf)(x)$ for all $f \in C(T,\mathbb{R})$ and $x \in X$. Hence $L^* \circ J$ is onto $N(T,\mathbb{R})$. ∎

If X is an abstract L_1 space, then X^* is an abstract M space with strong order unit and so $X^* = C(T,\mathbb{R})$ for some compact Hausdorff space T. By the above T is hyperstonian and the natural map J carries X onto $N(T,\mathbb{R})$. Since J is also a lattice isomorphism, X is completely identified with $N(T,\mathbb{R})$. That is, any abstract L_1 space can be realized as the space of all normal measures on some hyperstonian space T.

Theorems 10 and 11 are also valid in the complex cases. We leave the proofs to the reader as exercises. The two steps which are necessary are outlined below. As before, suppose T is a compact Hausdorff space and L is a linear isometry of $C(T,\mathbb{C})$ onto a (complex) Banach space X^* so that we may speak of the weak* topology on $C(T,\mathbb{C})$. The two steps are:

(a) *$C(T,\mathbb{R})$ and $C(T,\mathbb{R})^+$ are weak* closed in $C(T,\mathbb{C})$.*

(b) *For each $f \in C(T,\mathbb{R}) \setminus C(T,\mathbb{R})^+$ there is a positive weak* continuous linear functional ψ on $C(T,\mathbb{C})$ such that $\psi(f) < 0$.*

We close this section with some remarks on the separable λ extension property. A separable Banach space X is said to have the *separable λ extension property* if it satisfies the condition of λ extension whenever all Banach spaces under consideration are separable. Clearly X satisfies the separable λ extension property if and only if it satisfies the separable λ projection property (which is defined in the obvious manner). The only known real spaces which satisfy the separable λ extension property are finite dimensional spaces and spaces isomorphic to $c_0(\mathbb{N},\mathbb{R})$. Sobczyk proved in [263] that $c_0(\mathbb{N},\mathbb{R})$ has the separable 2 projection property. We give a proof of this due to Veech [274].

Theorem 12 (Sobczyk). *$c_0(\mathbb{N},\mathbb{R})$ has the separable 2 projection property.*

Proof. Let X and Y be separable Banach spaces and suppose $\psi: X \to Y$ is a linear isometry into and $L: X \to c_0(\mathbb{N},\mathbb{R})$ is a bounded linear operator. Let V^* be the closed unit ball of Y^* and suppose d is a metric on V^* which is compatible with the weak* topology. Choose $x_n^* \in V^*$ such that for each $x \in X$, $x_n^*(\psi(x))$ is the nth coordinate of $L(x)$ and choose $y_n^* \in V^* \cap \psi(x)^\perp$ such that $d(x_n^*, y_n^*)$ minimizes the distance from x_n^* to $\psi(x)^\perp$ (recall that if $S \subset Y$, $S^\perp = \{y^* \in Y^* : y^*|S = 0\}$). We define $\hat{L}: Y \to c_0(\mathbb{N}, \mathbb{R})$ by $\hat{L}(y) = \{x_n^*(y) - y_n^*(y)\}$ for all $y \in Y$. Then it is easy to see that indeed $\hat{L}(y) \in c_0(\mathbb{N}, \mathbb{R})$, $\|L\| \leqslant 2\|L\|$, and $\hat{L} \circ \psi = L$. ∎

In [5] Amir proved that if T is a compact metric space, then $C(T, \mathbb{R})$ has the separable λ extension property (for some $\lambda \geqslant 1$) if and only if $C(T, \mathbb{R})$ is isomorphic to $c_0(\mathbb{N}, \mathbb{R})$. In particular, by the result of Bessaga and Pełczyński [39] on the isomorphic classification of $C(T, \mathbb{R})$ when T is countable we obtain that $C(T, \mathbb{R})$ has the separable λ extension property if and only if T is homeomorphic to $\Gamma(\alpha)$ for some ordinal α with $\alpha < \omega^\omega$. The general question, however, is still open.

Question. *If X has the separable λ extension property (for some $\lambda \geqslant 1$), is it true that X is either finite dimensional or linearly isomorphic to $c_0(\mathbb{N}, \mathbb{R})$.*

Finally we note that Baker [26] has recently shown that for each n, the infimum of the λ's such that $C(\Gamma(\omega^n k), \mathbb{R})$ has the λ separable extension property is $2n+1$.

In [300] Phillips showed that $c_0(\mathbb{N}, \mathbb{R})$ is not complemented in $l_\infty(\mathbb{N}, \mathbb{R})$. Recently Lindenstrauss [186] settled completely the problem of which subspaces of $l_\infty(\mathbb{N}, \mathbb{R})$ are complemented by showing that an infinite dimensional subspace of $l_\infty(\mathbb{N}, \mathbb{R})$ is complemented if and only if it is linearly isomorphic to $l_\infty(\mathbb{N}, \mathbb{R})$.

Exercises. 1. Suppose X and Y are Banach spaces with Y a $\mathscr{P}_1(\mathbb{C})$ space and $X \subset Y$. Show that (Y, i) is an essential extension of X if and only if for any $\mathscr{P}_1(\mathbb{C})$ space Z with $X \subset Z \subset Y$, $Z = Y$.

2. (Cohen). Let X be a Banach space and let S^* be the closed unit sphere of X^*. Show that there is a set T of extreme points of S^* such that the weak* closure of $\{a x^* : |a| = 1, x^* \in T\}$ is equal to the weak* closure of the set of extreme points of S^* and such that $a x^* \notin T^*$ for all $x^* \in T$ if $|a| = 1$ and $a \neq 1$, where T^* is the closure of T with respect to the weak* topology. (The real and complex cases should be considered separately.)

Let P be an extremally disconnected compact Hausdorff space and $f: P \to T^*$ a minimal onto continuous map. Show that $(C(P, \mathbb{C}), L)$ is an essential extension of X, where $L(x)(p) = f(p)(x)$ for all $x \in X$, $p \in P$.

3. Prove theorem 7 for the complex case and show that theorem 8 is not valid in the complex case.

4. Let X be a real order complete vector lattice and let $Y = \{L: X \to X : L$ is a linear map and there is an $M > 0$ such that

$-Mx \leqslant Lx \leqslant Mx$ for all $x \geqslant 0\}$. Show that Y is a Banach space under $\|L\| = \inf\{M > 0 : -Mx \leqslant Lx \leqslant Mx$ for all $x \geqslant 0\}$. Show further that Y can be ordered in such a way to form a partially ordered Banach space with a strong order unit and order unit norm under $L \geqslant 0$ if and only if $Lx \geqslant 0$ for all $x \geqslant 0$. Prove that Y is a commutative Banach algebra under composition and that, in fact, Y is linearly isometric order isomorphic and algebraically isomorphic to $C(T, \mathbb{R})$ for some extremally disconnected compact Hausdorff space T.

5. Prove theorem 10 for the complex case.

6. Prove theorem 11 for the complex case.

In the next few problems we shall be concerned with the space Y. Y is a $\mathscr{P}_1(\mathbb{C})$ space (resp. $\mathscr{P}(\mathbb{R})$) and (Y, ψ) is an essential extension of X. We shall refer to Y as the Hahn-Banach space of X and suppress the embedding ψ. These problems are taken from Lacey-Cohen [164].

7. Let T be a compact Hausdorff space. Show that the Hahn-Banach space of $C(T, \mathbb{C})$ is $l_\infty(\Gamma, \mathbb{C})$ if and only if T has a dense set of isolated points of cardinal number the cardinal of Γ.

8. Let Γ be an index set. Show that the Hahn-Banach space of $c_0(\Gamma, \mathbb{C})$ is $l_\infty(\Gamma, \mathbb{C})$.

9. Let T be a compact Hausdorff space. Show that the following are equivalent.

(1) T is dispersed.

(2) For every closed subspace X of $C(T, \mathbb{C})$ the Hahn-Banach space of X is of the form $l_\infty(\Gamma, \mathbb{C})$.

(3) For every separable closed subspace X of $C(T, \mathbb{C})$ the Hahn-Banach space of X is of the form $l_\infty(\Gamma, \mathbb{C})$ for some countable index set Γ.

(4) For every two dimensional subspace X of $C(T, \mathbb{C})$, the Hahn-Banach space of X is of the form $l_\infty(\Gamma, \mathbb{C})$ for some countable set Γ.

10. In the real case show that the Hahn-Banach space of $l_1(k, \mathbb{R})$ is $l_\infty(2^{k-1}, \mathbb{R})$ for each positive integer k. In the complex case show that the Hahn-Banach space of $l_1(k, \mathbb{C})$ is infinite dimensional for $k > 1$.

11. Show that there are exactly five possible Hahn-Banach spaces of separable Banach spaces.

12. (Dixmier). Let T be an extremally disconnected compact Hausdorff space. Show that there is one and only one partition T_1, T_2, T_3 of T into pairwise disjoint closed and open sets such that

(a) T_1 contains a Borel set of the first category which is dense in T_1 (and every normal measure is 0 on T_1),

(b) T_2 is hyperstonian,

(c) Every Borel set of the first category in T_3 is nowhere dense and the support of each regular Borel measure on T_3 is nowhere dense (so each normal measure is 0 on T_3).

13. Let X be an abstract L_1 space such that there is an $x \in X$ with $|x| \wedge |y| = 0 \Rightarrow y = 0$. Show that there is a compact Hausdorff space T and a positive normalized regular Borel measure μ on T such that X is linearly isometric and lattice isomorphic to $L_1(\mu, \mathbb{R})$.

14. Give an example of a rigid extension which is not essential.

Chapter 4

Classical Sequence Spaces

By the term classical sequence space we shall mean the spaces $l_p(\mathbb{N}, \mathbb{R})$, $c(\mathbb{N}, \mathbb{R})$, and $c_0(\mathbb{N}, \mathbb{R})$ and their complex analogues. In section 12 we briefly develop the notion of a Schauder basis and study these bases in classical sequence spaces. In particular, we use basis theory to show that each infinite dimensional complemented subspace of a classical sequence space X is linearly isomorphic to X and that each infinite dimensional closed subspace of X contains an infinite dimensional complemented subspace.

Section 13 is devoted to the study of linear isomorphic embeddings of classical Banach spaces into $C(T, \mathbb{C})$ (and $C(T, \mathbb{R})$). We show that $c_0(\mathbb{N}, \mathbb{C})$ can always be embedded into $C(T, \mathbb{C})$ when T is an infinite compact Hausdorff space and that when T is not dispersed, all separable Banach spaces can be embedded into $C(T, \mathbb{C})$.

§ 12. Schauder Bases in Classical Sequence Spaces

The results in this section are true for both the real and complex case and the proofs given will work in either case. Consequently we only state them for the complex case.

Definition 1. Let X be a (complex) Banach space and $\{x_n\}$ be a sequence in X. We say that $\{x_n\}$ is a *Schauder basis* for X if for each $x \in X$ there is a unique sequence $\{a_n\}$ of scalars such that $x = \sum_{n=1}^{\infty} a_n x_n$ (where the convergence is taken in norm).

Clearly if X has a Schauder basis, then X is separable. The converse has been a celebrated question for many years. Recently Per Enflo [290] has shown that there is indeed a separable Banach space without a Schauder basis (in fact, his example shows there is a Banach space which fails to possess a weaker property the *Grothendieck approximation property*).

We shall only be concerned with Schauder basis as they apply to classical sequence spaces. For a complete reference on the subject the reader may check [304].

Theorem 1. *Let X be a Banach space and suppose that $\{x_n\}$ is a Schauder basis for X. Then for each n, the functional x_n^* defined by $x_n^*\left(\sum_{k=1}^{\infty} a_k x_k\right) = a_n$ is continuous.*

Proof. Let $P_n: X \to X$ be defined by $P_n\left(\sum_{k=1}^{\infty} a_k x_k\right) = \sum_{k=1}^{n} a_k x_k$. Clearly P_n is a linear projection on X and for each $x \in X$, $\rho(x) = \sup_n \|P_n(x)\| < \infty$. Moreover, ρ is a norm on X and clearly $\rho(x) \geqslant \|x\|$ for all $x \in X$. We shall first show that ρ is an equivalent norm on X. To do this it obviously suffices to show that ρ is a complete norm. Let $\{y_m\}$ be a Cauchy sequence with respect to ρ. Then for $j > k$, $\|(P_j - P_k)(y_m - y_n)\| \leqslant \|P_j(y_m - y_n)\| + \|P_k(y_m - y_n)\| \leqslant 2\rho(y_m - y_n)$. In particular, for $j = k+1$ we get that $\|x_j^*(y_m - y_n)x_j\| = |x_j^*(y_m - y_n)|\,\|x_j\| \leqslant 2\rho(y_m - y_n)$. Thus $\lim_{m\to\infty} x_j^*(y_m)x_j = a_j x_j$ for some a_j.

Now, let $\varepsilon > 0$ be given and choose m_0 such that if $n > m \geqslant m_0$, then $\rho(y_m - y_n) < \varepsilon$. Then by taking the limit on n we see that $\left\|(P_j - P_k)y_m - \sum_{i=k+1}^{j} a_i x_i\right\| \leqslant 2\varepsilon$. Moreover, $\lim_{k\to\infty} \|P_k(y_m) - y_m\| = 0$ implies that there is a k_0 such that if $j > k \geqslant k_0$, then $\|(P_j - P_k)y_m\| < \varepsilon (m \geqslant m_0)$. Thus we get that $\left\|\sum_{i=k+1}^{j} a_i x_i\right\| < 3\varepsilon$ when $j > k \geqslant k_0$. By the completeness of the original norm on X, $\sum_{i=1}^{\infty} a_i x_i$ converges to some element $y \in X$. By the uniqueness of the expansion of y, $a_i = x_i(y)$ and $(P_j - P_k)y = \sum_{i=k+1}^{j} a_i x_i$. Thus $\rho(y_m - y) \leqslant \varepsilon$ when $m \geqslant m_0$ and it follows that ρ is complete.

To see that each x_n^* is continuous note that $|x_n^*(x)| = \|P_n(x) - P_{n-1}(x)\| \leqslant 2\rho(x)$ when $\|x\| = 1$. ∎

We say that the sequence $\{x_n^*\}$ is *biorthogonal* to $\{x_n\}$.

Definition 2. Let X be a Banach space. A sequence $\{x_n\}$ in X is said to be a *basic sequence* if $\{x_n\}$ is a Schauder basis for the closed linear span $[x_n]$ of $\{x_n\}$.

The following theorem is very useful for determining when sequences are basic.

Theorem 2 (Nikolskii). *Let $\{x_n\}$ be a sequence of nonzero vectors in a Banach space X. Then $\{x_n\}$ is a basic sequence if and only if there is a $K \geqslant 1$ such that for each finite sequence $a_1, \ldots, a_n$ of scalars, $\left\| \sum_{i=1}^{m} a_i x_i \right\| \leqslant K \left\| \sum_{i=1}^{n} a_i x_i \right\|$ for all $m \leqslant n$.*

Proof. Suppose $\{x_n\}$ is a basic sequence. Then the norm defined in theorem 1 is an equivalent norm to the original norm on $[x_n]$. In particular, there is a $K \geqslant 1$ such that for each sequence $\{a_n\}$ such that $\sum_{n=1}^{\infty} a_n x_n$ exists in X, $\sup_n \left\| \sum_{k=1}^{n} a_k x_k \right\| \leqslant K \left\| \sum_{k=1}^{\infty} a_k x_k \right\|$.

Conversely, if the condition holds, then by induction it is easy to see that expansions in terms of the x_n's are unique. Suppose $x \in [x_n]$. That is, $x = \lim_n \sum_{i=1}^{m_n} a_{i,n} x_i$ for some $a_{i,n}$'s. Then for $n > m$, $\left\| \sum_{i=1}^{m} (a_{i,n} - a_{i,m}) x_i \right\| \leqslant K \left\| \sum_{i=1}^{n} a_{i,n} x_i - \sum_{i=1}^{m} a_{i,m} x_i \right\|$. In particular, for each i the sequence $\{a_{i,n}\}$ is Cauchy. An argument similiar to the one in theorem 1 shows that $x = \sum_{n=1}^{\infty} a_n x_n$, where $a_n = \lim_i a_{i,n}$ for all n. ∎

Let X and Y be Banach spaces and suppose that $\{x_n\}$ and $\{y_n\}$ are basic sequences in X and Y respectively. We say that $\{x_n\}$ and $\{y_n\}$ are *equivalent* if $\sum_{n=1}^{\infty} a_n x_n$ converges in X if and only if $\sum_{n=1}^{\infty} a_n y_n$ converges in Y. That is, $\{x_n\}$ and $\{y_n\}$ are equivalent if and only if there is a linear isomorphism A of $[x_n]$ onto $[y_n]$ such that $A(x_n) = y_n$ for all n.

Theorem 3. *Let X be a Banach space and $\{x_n\}$ a basic sequence in X with $\{x_n^*\}$ its corresponding biorthogonal sequence. If $\{y_n\}$ is a sequence of nonzero elements in X, and $\sum_{n=1}^{\infty} \|x_n - y_n\| \, \|x_n^*\| = a < 1$, then $\{y_n\}$ is a basic sequence equivalent to $\{x_n\}$.*

Proof. Let $a_1, \ldots, a_{n+m}$ be a finite sequence of scalars. Then

$$|a_i| = \left| x_i^* \left(\sum_{j=1}^{n} a_j x_j \right) \right| \leqslant \left\| \sum_{j=1}^{n} a_j x_j \right\| \|x_i^*\| \quad \text{for } i = 1, \ldots, n$$

and

$$\left\| \sum_{j=1}^{n} a_j y_j \right\| = \left\| \sum_{j=1}^{n} a_j x_j + \sum_{j=1}^{n} a_j (y_j - x_j) \right\| \leqslant \left\| \sum_{j=1}^{n} a_j x_j \right\| + \sum_{j=1}^{n} |a_j| \, \|x_j - y_j\|$$

$$\leqslant \left\| \sum_{j=1}^{n} a_j x_j \right\| + \sum_{j=1}^{n} \left\| \sum_{j=1}^{n} a_j x_j \right\| \cdot \|x_j - y_j\| \cdot \|x_j^*\| \leqslant (1+a) \left\| \sum_{j=1}^{n} a_j x_j \right\|.$$

Moreover,

$$\left\|\sum_{i=1}^{n+m} a_i y_i\right\| = \left\|\sum_{i=1}^{n+m} a_i(x_i+y_i-x_i)\right\| \geqslant \left\|\sum_{i=1}^{n+m} a_i x_i\right\| - \left\|\sum_{i=1}^{n+m} a_i(y_i-x_i)\right\|$$

$$\geqslant \left\|\sum_{i=1}^{n+m} a_i x_i\right\| - \sum_{i=1}^{n+m} |a_i|\,\|y_i-x_i\|$$

$$\geqslant \left\|\sum_{i=1}^{n+m} a_i x_i\right\| - \sum_{i=1}^{n+m}\left(\left\|\sum_{i=1}^{n+m} a_j x_j\right\|\|x_i^*\|\,\|y_i-x_i\|\right) \geqslant (1-a)\left\|\sum_{i=1}^{n+m} a_i x_i\right\|.$$

Hence $\left\|\sum_{j=1}^{n} a_i y_i\right\| \leqslant K\left(\frac{1+a}{1-a}\right)\left\|\sum_{j=1}^{n+m} a_j y_j\right\|$ where K is the constant in theorem 2 for $\{x_n\}$. Thus $\{y_n\}$ is a basic sequence and it is clearly equivalent to $\{x_n\}$. ∎

The following theorem is of the same type as theorem 3.

Theorem 4 (Bessaga-Pełczyński). *Let X be a Banach space and suppose $\{x_n\}$ and $\{y_n\}$ are two basic sequences in X. If there is a bounded linear projection P of X onto $[x_n]$ such that $\|P\| \sum_{n=1}^{\infty} \|x_n^*\|\,\|x_n-y_n\|<1$, then $[y_n]$ is complemented in X.*

Proof. Let $L: X \to X$ be defined by $L(x)=x-P(x)+\sum_{n=1}^{\infty} x_n^*(Px)y_n$. Clearly L is a bounded linear operator. Moreover $\|I-L\| = \sup\{\|x-Lx\| : \|x\| \leqslant 1\}$ $= \sup\left\{\left\|\sum_{n=1}^{\infty} x_n^*(Px)(x_n-y_n)\right\| : \|x\| \leqslant 1\right\} \leqslant \|P\| \sum_{n=1}^{\infty} \|x_n^*\|\,\|x_n-y_n\| < 1$. Thus L is an isomorphism of X onto X and clearly $L(x_n)=y_n$ so L maps $[x_n]$ onto $[y_n]$. The mapping $Q=LPL^{-1}$ is a projection of X onto $[y_n]$. ∎

Let $\{x_n\}$ be a basic sequence in X, $\{p_n\}$ a strictly increasing sequence of positive integers, and $\{a_n\}$ a sequence of scalars. The sequence $\{y_n\}$ defined by $y_n = \sum_{i=p_n+1}^{p_{n+1}} a_i x_i$ is called a *block basis* with respect to $\{x_n\}$. The following theorem is also due to Bessaga and Pełczyński [36].

Theorem 5. *Let X be a Banach space, $\{x_n\}$ a Schauder basis for X, and $\{x_n^*\}$ the biorthogonal sequence to $\{x_n\}$. If $\{y_n\}$ is a sequence satisfying $\inf\|y_n\| = \varepsilon > 0$ and $\lim_n x_i^*(y_n)=0$ for all i, then there is a subsequence of $\{y_n\}$, which is equivalent to a block basis with respect to $\{x_n\}$.*

Proof. Let K be the constant of theorem 2 for $\{x_n\}$. Choose strictly increasing sequences $\{p_n\}$, $\{q_n\}$ of positive integers such that

$$\frac{4K}{\varepsilon}\left\|\sum_{i=q_n+1}^{\infty} x_i^*(y_{p_n})x_i\right\| \leqslant \frac{1}{2^{n+2}} \quad\text{and}\quad \frac{4K}{\varepsilon}\left\|\sum_{i=1}^{q_n} x_i^*(y_{p_{n+1}})x_i\right\| \leqslant \frac{1}{2^{n+2}}$$

which is possible since $\lim_n x_i^*(y_n)=0$. Let $z_n = \sum_{i=q_n+1}^{q_n+1} x_i^*(y_{p_{n+1}})x_i$. Then $\|z_n\| \geqslant \varepsilon/2$ and $\sum_{n=1}^{\infty} \frac{4K}{\varepsilon}\|z_n - y_{p_{n+1}}\| < \frac{1}{2}$. Let $\{z_n^*\}$ in $[z_n]^*$ be the biorthogonal sequence to $\{z_n\}$. Then $\|z_n^*\| \leqslant 4K/\varepsilon$ and it follows that $\{y_{p_{n+1}}\}$ is a basic sequence equivalent to $\{z_n\}$. ∎

We shall need the following notation here and at various other places throughout the text. Let Γ be a nonempty index set and suppose $\{X_\gamma\}_{\gamma\in\Gamma}$ is a collection of Banach spaces. By $\left(\oplus \sum_{\gamma\in\Gamma} X_\gamma\right)_\lambda$ (where $\lambda=0$ or ∞ or $1\leqslant\lambda<\infty$) we mean the Banach space of all $\{x_\gamma\}$ such that $\{\|x_\gamma\|\}$ is in $c_0(\Gamma,\mathbb{R})$, $l_\infty(\Gamma,\mathbb{R})$, or $l_p(\Gamma,\mathbb{R})$ for the case $\lambda=0$ or $\lambda=\infty$ or $1\leqslant p<\infty$ respectively and $x_\gamma\in X_\gamma$ for each $\gamma\in\Gamma$. The norm on X is defined to be the norm of $\{\|x_\gamma\|\}$. It is easily checked that X is indeed a Banach space.

The following theorem is left as an exercise.

Theorem 6. *Let $X_n = l_p(\mathbb{N},\mathbb{C})$ (resp. $c_0(\mathbb{N},\mathbb{C})$) for each n. Then $\left(\oplus \sum_{n=1}^{\infty} X_n\right)_p$ $\left(\text{resp. } \left(\oplus \sum_{n=1}^{\infty} X_n\right)_0\right)$ is linearly isometric to $l_p(\mathbb{N},\mathbb{C})$ (resp. $c_0(\mathbb{N},\mathbb{C})$).* ∎

Another useful lemma due to Pełczyński is now given.

Lemma 1. *Let X be Banach spaces and suppose that X is isomorphic to $(X\oplus X\oplus\cdots)_p$ for $p=0$ or $1\leqslant p<\infty$. If X is isomorphic to a complemented subspace of Y and Y is isomorphic to a complemented subspace of X, then X is isomorphic to Y.*

Proof. We shall use the symbol $\sim$ to mean "is isomorphic to". By hypothesis there are Banach spaces Z and W such that $X\sim Z\times Y$ and $Y\sim X\times W$. Hence $X\sim(X\oplus X\oplus\cdots)_p\sim((X\oplus Z\oplus W)\oplus(X\oplus Z\oplus W)\oplus\cdots)_p$ $\sim(X\oplus X\oplus\cdots)_p\oplus(Z\oplus Z\oplus\cdots)_p\oplus(W\oplus W\oplus\cdots)_p\oplus W\sim X\oplus W\sim Y$. ∎

The next three theorems are valid for both $l_p(\mathbb{N},\mathbb{C})$ and $c_0(\mathbb{N},\mathbb{C})$, but we shall only give proofs based on $l_p(\mathbb{N},\mathbb{C})$. Each of these spaces have natural Schauder bases called the *natural unit vector basis*. It is the sequence $\{e_n\}$ where $e_n=(0,\ldots,1,0,\ldots)$ where the 1 occurs in the nth place.

We shall use the symbol X to stand for $l_p(\mathbb{N},\mathbb{C})$ or $c_0(\mathbb{N},\mathbb{C})$ in the statements of the theorems due to Pełczyński [233].

Theorem 7. *Let $\{x_n\}$ be a sequence of nonzero terms in X such that there is a strictly increasing sequence $\{p_n\}$ of positive integers with $x_m = \sum_{i=p_m+1}^{p_{m+1}} a_i^m e_i$. Then*

(1) $[x_n]$ *is isometrically isomorphic to* X *and*
(2) $[x_n]$ *is complemented in* X.

Proof. Let $\{b_n\}$ be a sequence of scalars. Then $\left\|\sum_{m=1}^{k} b_m x_m\right\|$ $=\left\|\sum_{m=1}^{k}\sum_{i=p_m+1}^{p_{m+1}} b_m a_i^m e_i\right\| = \left(\sum_{m=1}^{k}\sum_{i=p_m+1}^{p_{m+1}} |b_m a_i^m|^p\right)^{1/p} = \left(\sum_{m=1}^{k} |b_m|^p \|x_m\|^p\right)^{1/p}$. Hence $\{x_m\}$ is a basic sequence and $\sum_{m=1}^{\infty} b_m \frac{x_m}{\|x_m\|}$ converges if and only if $\sum_{m=1}^{\infty} |b_m|^p < \infty$. Let L be defined by $L(\{b_m\}) = \sum_{m=1}^{\infty} b_m \frac{x_m}{\|x_m\|}$ for $\{b_m\}$ in $l_p(\mathbb{N},\mathbb{C})$. Then L is a linear isometry of $l_p(\mathbb{N},\mathbb{C})$ onto $[x_m]$.

Let $X_m = \operatorname{span}(e_{p_m+1},\ldots,e_{p_{m+1}})$. Then $x_m \in X_m$ and there is an $x_m^* \in X_m^*$ such that $x_m^*(x_m)=1$ and $\|x_m^*\| = \frac{1}{\|x_m\|}$. If $x = \sum_{n=1}^{\infty} b_n e_n$ is in $l_p(\mathbb{N},\mathbb{C})$, then P defined by $Px = \sum_{m=1}^{\infty} x_m^*\left(\sum_{i=p_m+1}^{p_{m+1}} b_i e_i\right) x_m$ maps $l_p(\mathbb{N},\mathbb{C})$ into $[x_m]$ since $\left|x_m^*\left(\sum_{i=p_m+1}^{p_{m+1}} b_i e_i\right)\right| \leqslant \frac{1}{\|x_m\|}\left(\sum_{i=p_m+1}^{p_{m+1}} |b_i|^p\right)^{1/p}$. Since $\{x_m\}$ is a basis for $[x_m]$ and $Px_m = x_m$ for all m, P is a projection of $l_p(\mathbb{N},\mathbb{C})$ onto $[x_m]$. Moreover, $\|Px\| \leqslant \|x\|$ for all x in $l_p(\mathbb{N},\mathbb{C})$. ∎

Theorem 8. *Let* Y *be an infinite dimensional closed subspace of* X. *Then* Y *has a subspace complemented in* X *and isomorphic to* X.

Proof. Since Y is infinite dimensional it is possible to choose $\{y_n\}$ in Y and a strictly increasing sequence $\{p_n\}$ of positive integers such that $y_m = \sum_{i=p_m+1}^{\infty} a_i^m e_i$, $\|y_m\|=1$, and $\left\|\sum_{i=p_{m+1}+1}^{\infty} a_i^m e_i\right\| \leqslant 1/2^{m+1}$ for $m=1,2,\ldots$. If $x_m = \sum_{i=p_m+1}^{p_{m+1}} a_i^m e_i$, then $\|x_m - y_m\| \leqslant 1/2^{m+1}$ and $x_m \neq 0$. Let P be a projection of $l_p(\mathbb{N},\mathbb{C})$ onto $[x_m]$ and $\{x_m^*\}$ be the biorthogonal sequence to $\{x_m\}$ in $[x_m]^*$. Then $\|x_m^*\| = \frac{1}{\|x_m\|} \leqslant \frac{1}{\|y_m\| - \|y_m - x_m\|} \leqslant \frac{1}{1 - 1/2^{m+1}}$.

Since $\|P\| \sum_{n=1}^{\infty} \|x_m^*\|\,\|y_m - x_m\| \leqslant \sum_{m=1}^{\infty} \frac{1}{2^{m+1}-1} < 1$, by theorem 5 there is a subsequence of $\{y_m\}$, say $\{y_{m_k}\}$, equivalent to a basis with respect to $\{e_i\}$ and $[y_{m_k}]$ is complemented in X. By theorem 7, $[y_{m_k}]$ is isometrically isomorphic to X. ∎

Theorem 9. *Each infinite dimensional closed subspace of* X *complemented in* X *is isomorphic to* X.

Proof. Let Y be an infinite dimensional complemented subspace of X. Then Y contains a subspace Z complemented in X and isomorphic to X. Thus by lemma 1, $X \sim Y$. ∎

We shall need below the fact that a separable Banach space can be embedded into a space with a Schauder basis. To see this let X be a separable Banach space and let T be the closed unit ball of X^* in the weak* topology. Then T is metrizable and $C(T, \mathbb{C})$ is, in fact, embeddable in $C(\Delta, \mathbb{C})$ where Δ is the Cantor set. For, there is a continuous map h of Δ onto T (see theorem 3 of section 6) and $L: C(T, \mathbb{C}) \to C(\Delta, \mathbb{C})$ defined by $L(f) = f \circ h$ is a linear isometric embedding (see exercise 1 of section 9). Clearly X can be embedded into $C(T, \mathbb{C})$ by $(Ax)(x^*) = x^*(x)$ for $x \in X$ and $x^* \in T$. By exercise 2 at the end of this section, $C(\Delta, \mathbb{C})$ has a Schauder basis.

We now discuss when a Banach space X contains an isomorphic copy of $c_0(\mathbb{N}, \mathbb{C})$.

A series $\sum_{n=1} x_n$ of elements in a Banach space X is said to be *weakly unconditionally convergent* if $\sum_{n=1}^{\infty} |x^*(x_n)| < \infty$ for all $x^* \in X^*$.

The following lemma gives the basic characterizations of weakly unconditionally convergent series.

Lemma 2. *Let X be a Banach space and $\{x_n\}$ a sequence in X. Then the following are equivalent.*

(1) $\sum_{n=1}^{\infty} x_n$ *is weakly unconditionally convergent,*

(2) *there is a constant $K > 0$ such that for each bounded sequence $\{a_n\}$ of scalars,* $\sup_n \left\| \sum_{i=1}^{n} a_i x_i \right\| \leqslant K \sup_n |a_n|$.

Proof. Let $Z_n = \left\{ x^* \in X^* : \sum_{i=1}^{\infty} |x^*(x_i)| \leqslant n \right\}$. Then each Z_n is norm closed and $X^* = \bigcup_{n=1}^{\infty} Z_n$. Thus by the Baire category theorem there is a $K > 0$ such that $\sum_{n=1}^{\infty} |x^*(x_n)| \leqslant K$ for all x^* with $\|x^*\| \leqslant 1$. Hence if $\{a_n\}$ is a bounded sequence of scalars, $\left\| \sum_{i=1}^{m} a_i x_i \right\| = \sup_{\|x^*\| \leqslant 1} \left| x^* \left(\sum_{i=1}^{n} a_i x_i \right) \right|$ $\leqslant \sup_{\|x^*\| \leqslant 1} \left| \sum_{i=1}^{n} x^*(x_i) \right| \sup_n |a_n| \leqslant K \sup_n |a_n|$ and (1) implies (2).

Let $x^*\in X^*$ and a_n be the scalar of modulus 1 such that $a_n x^*(x_n) = |x^*(x_n)|$ for all n. Then $\left|\sum_{i=1}^{n} x^*(a_n x_n)\right| \leqslant \|x^*\| \left\|\sum_{i=1}^{n} a_i x_i\right\| \leqslant \|x^*\| K$ and it follows that $\sum_{n=1}^{\infty} |x^*(x_n)| < \infty$ and (2) implies (1). ∎

Corollary. *Let T be a compact Hausdorff space. Then the series $\sum_{n=1}^{\infty} f_n$ of elements $f_n\in C(T,\mathbb{C})$ is weakly unconditionally convergent if and only if there is an $M>0$ such that $\sum_{n=1}^{\infty} |f_n(t)| \leqslant M$ for all $t\in T$.* ∎

Theorem 9. *Let X be a Banach space. Then the following conditions are equivalent.*

(1) *There is a weakly unconditionally convergent series $\sum_{n=1}^{\infty} x_n$ which is not unconditionally convergent (in norm).*

(2) *There is a weakly unconditionally convergent series $\sum_{n=1}^{\infty} x_n$ in X such that $\inf_n \|x_n\| > 0$.*

(3) *X contains a subspace isomorphic to $c_0(\mathbb{N},\mathbb{C})$.*

Lemma 3. *Let X be a Banach space and suppose that $\{x_n\}$ is a basic sequence in X such that $\inf_n \|x_n\| > 0$ and $\sum_{n=1}^{\infty} x_n$ is weakly unconditionally convergent. Then $\{x_n\}$ is equivalent to the natural basis $\{e_n\}$ in $c_0(\mathbb{N},\mathbb{C})$ and thus $[x_n]$ is isomorphic to $c_0(\mathbb{N},\mathbb{C})$.*

Proof. Clearly $\sum_{n=1}^{\infty} a_n x_n$ converges if and only if $\lim a_n = 0$. ∎

Proof of theorem 9. (1) implies (2). If $\sum_{n=1}^{\infty} x_n$ is not unconditionally convergent, then for some permutation $\{k_n\}$ of the indices, $\sum_{n=1}^{\infty} x_{k_n}$ does not converge. Thus there is an increasing sequence $\{q_n\}$ of indices such $\inf_n \left\|\sum_{i=q_n+1}^{q_{n+1}} x_{k_i}\right\| > 0$. Thus $y_n = \sum_{i=q_n+1}^{q_{n+1}} x_{k_i}$ defines a weakly unconditionally convergent series which satisfies (2).

(2) implies (3). Let $\sum_{n=1}^{\infty} x_n$ be a weakly unconditionally series such that $\inf_n \|x_n\| = \delta > 0$. Thus $x_n \to 0$ weakly and by theorem 5 there is a subsequence $\{x_{n_k}\}$ which is a basic sequence. Hence $[x_{n_k}]$ is isomorphic to $c_0(\mathbb{N},\mathbb{C})$ by lemma 3. (It is here that we need the fact that X is contained in a space with a Schauder basis in order to apply theorem 5.)

The series $\sum_{n=1}^{\infty} e_n$ in $c_0(\mathbb{N}, \mathbb{C})$ is weakly unconditionally convergent but not convergent. Hence (3) implies (1) is clear. ∎

Exercises. 1. (Schauder basis for $C([0,1], \mathbb{C})$). Let $f_0 = 1$, $f_1(t) = t$, $f^{(t)}_{2^n+k+1} = 2^{n+1}(t - k2^{-n})$ if $k2^{-n} \leqslant t \leqslant (2k+1)2^{-n-1}$, $f^{(t)}_{2^n+k+1} = 2^{n+1}((k+1)2^{-n} - t)$ if $(2k+1)2^{-n-1} \leqslant t \leqslant (k+1)2^{-n}$ and $f^{(t)}_{2^n+k+1} = 0$ elsewhere for $n = 0, \ldots$ and $1 \leqslant k \leqslant 2^n$. Then this sequence is a Schauder basis for $C([0,1], \mathbb{C})$.

2. (Schauder basis for $C(\Delta, \mathbb{C})$). We consider that $\Delta \subset [0,1]$ in the classical fashion. Let $h_0 = 1$, $h_{2^{n-1}+k} = f_{G(n,2k)} - f_{G(n,2k+1)}$ for $n = 1, 2, \ldots$ and $0 \leqslant k < 2^{n-1}$, where $G(n,k) = \Delta \cap \left[\frac{k}{2}, \frac{k+1}{2}\right]$. Then this sequence is a Schauder basis for $C(\Delta, \mathbb{C})$.

3. Find a Schauder basis for $L_p([0,1], \mathbb{C})$ for $1 \leqslant p < \infty$.

4. Let X be a complex Banach space. Show that there is an index set Γ and a bounded linear map of $l_1(\Gamma, \mathbb{C})$ onto X. In particular, if X is separable, Γ can be taken to be $\mathbb{N}$.

5. Let T be a compact Hausdorff space which has a sequence $\{t_n\}$ of distinct terms which converges. Show that there is a bounded linear map from $C(T, \mathbb{C})$ onto $c_0(\mathbb{N}, \mathbb{C})$. Find an example of a compact Hausdorff space which does not have a sequence of distinct terms which converges. Prove that every infinite dispersed compact Hausdorff space has a sequence of distinct terms which converges. Can you find an infinite compact Hausdorff space T such that $C(T, \mathbb{C})$ cannot be mapped in a linear continuous fashion onto $c_0(\mathbb{N}, \mathbb{C})$?

6. Let X be a Banach space and suppose that $\{x_n\}$ is a bounded sequence in X. Let $A: l_1(\mathbb{N}, \mathbb{C}) \to X$ be the unique linear operator such that $A(e_n) = x_n$ for all n.

a) When is A weakly compact?

b) When is A compact?

c) Show that X is finite dimensional if and only if every bounded linear operator from $l_1(\mathbb{N}, \mathbb{C})$ to X is compact.

d) Show that X is reflexive if and only if every bounded linear operator from $l_1(\mathbb{N}, \mathbb{C})$ to X is weakly compact.

7. Let X be a complex Banach space and suppose $A: X \to c_0(\mathbb{N}, \mathbb{R})$ is a bounded linear operator and $x_n^* = A^*(e_n)$ for each n.

a) Show that A is compact if and only if $\lim x_n^* = 0$ (in norm).

b) Show that A is weakly compact if and only if $\lim x_n^* = 0$ (in the weak topology on X^* from X^{**}).

c) Prove that weak* and norm sequential convergence coincide in X^* if and only if every bounded linear operator from X to $c_0(\mathbb{N}, \mathbb{R})$ is compact.

d) Prove that weak* and weak sequential convergence coincide in X^* if and only if every bounded linear operator from X to $c_0(\mathbb{N},\mathbb{R})$ is weakly compact.

8. Prove that a block basis with respect to a basic sequence is indeed a basic sequence.

9. Prove that if $X = l_p(\mathbb{N},\mathbb{C})$ or $c_0(\mathbb{N},\mathbb{C})$ and Y is a closed linear subspace of X linearly isometric to X, then Y is complemented in X with a projection of norm one. What can you say if Y is only linearly isomorphic to X?

10. Prove that $l_p(\mathbb{N},\mathbb{C})$ can be linearly isometrically embedded into $L_p([0,1],\mathbb{C})$ for $1 \leqslant p < \infty$.

§ 13. Embedding of Classical Sequence Spaces into Continuous Function Spaces

We investigate the possibility of embedding the classical sequence spaces into spaces of the type $C(T,\mathbb{R})$ (and $C(T,\mathbb{C})$). It was noted in section 12 that any separable Banach space can be embedded in $C(\Delta,\mathbb{R})$ (or $C(\Delta,\mathbb{C})$ in the complex case). Since one can embedd $C(\Delta,\mathbb{R})$ (resp. $C(\Delta,\mathbb{C})$) into $C([0,1],\mathbb{R})$ (resp. $C([0,1],\mathbb{C})$), we get that these spaces are universal for all separable Banach spaces (see exercise 11). We shall see that $C(T,\mathbb{R})$ is such a universal space only when T is not dispersed and that, in fact, if T is dispersed, then every closed infinite dimensional subspace of $C(T,\mathbb{R})$ contains a subspace isomorphic to $c_0(\mathbb{N},\mathbb{R})$.

By the *linear dimension* of a Banach space X we mean the smallest cardinal m for which there is a set of cardinal m whose linear span is dense in X. It is denoted by $\dim X$ since the linear space dimension *(Hamel dimension)* of X is not used in this book except to distinguish between finite dimensional and infinite dimensional spaces. In particular, if X is infinite dimensional, then $\aleph_0 \leqslant \dim X \leqslant$ the Hamel dimension of X.

The first theorem concerns the relationship between $\dim C(T,\mathbb{C})$ and the *weight* $\omega(T)$ of T for compact Hausdorff spaces. It is easily seen that $\dim C(T,\mathbb{R}) = \dim C(T,\mathbb{C})$ for each compact Hausdorff space T.

Theorem 1. *Let T be an infinite compact Hausdorff space. Then $\omega(T) = \dim C(T,\mathbb{C})$.*

Proof. If $\{f_i\}_{i\in I}$ is a set in $C(T,\mathbb{C})$ whose linear span is dense in $C(T,\mathbb{C})$, then it separates the points of T and thus the weakest topology on T under which all the f_i are continuous is the original topology of T. Therefore $\dim C(T,\mathbb{C}) \geqslant \omega(T)$.

On the other hand, let $\{U_i\}_{i\in I}$ be a base for the topology of T. Then card $I \geqslant \aleph_0$. Let $t_i \in U_i$ and $f_i \in C(T,\mathbb{C})$ with $f_i(t_i)=1$ and $f_i=0$ on $T\setminus U_i$. Now card I is the same as the cardinal of the set S of all finite products of f_i's together with the constant function 1. Now, the linear span of S is an algebra containing 1 and separating the points of T and by the Stone-Weierstrass theorem it is dense in $C(T,\mathbb{C})$. Therefore $\omega(T) \geqslant \dim C(T,\mathbb{C})$. ∎

Corollary. *Let T be a compact Hausdorff space. Then T is metrizable if and only if $C(T,\mathbb{C})$ is separable, that is, $\dim C(T,\mathbf{C})=\aleph_0$.*

Proof. This is clear if $C(T,\mathbb{C})$ is finite dimensional. Suppose T is infinite. Then $\omega(T)\geqslant\aleph_0$ and $\omega(T)=\dim C(T,\mathbb{C})$. Since $\omega(T)=\aleph_0$ if and only if T is metrizable, the proof is complete. ∎

Note that if T is an infinite compact Hausdorff space and $X\subset C(T,\mathbb{C})$ is a linear subspace with $\dim X=m$, then there is a closed linear subspace $Y\supset X$ with $\dim Y=m$ and $Y=\{f\circ\alpha : f\in C(S,\mathbb{C})\}$ where S is some compact Hausdorff space and α is a continuous map of T onto S. For, take Y to be the closed subalgebra generated by X and 1.

Theorem 2. *Let T be an infinite compact Hausdorff space and let Γ be an infinite index set. Then $l_1(\Gamma,\mathbb{R})$ can be embedded into $C(T,\mathbb{R})$ in a linear isometric fashion if and only if T admits an interlocking family $\{A_i,B_i\}_{i\in\Gamma}$ of closed sets.*

Proof. Suppose T admits an interlocking family $\{A_i,B_i\}_{i\in\Gamma}$ of closed sets. Let $f_i\in C(T,\mathbb{R})$ with $-1\leqslant f_i(t)\leqslant 1$ for all $t\in T$, $i\in\Gamma$, and $f_i=-1$ on A_i, $f_i=1$ on B_i. For any finite nonempty set $F\subset I$ and any set of scalars a_i for $i\in F$ let s_i be defined by $s_i=1$ if $a_i\leqslant 0$ and $s_i=-1$ if $a_i>0$. Then by assumption there is a $t\in\bigcap_{i\in F} s_i A_i$ (recall that $1\cdot A_i=A_i$ and $-1\cdot A_i=B_i$) and thus $\sum_{i\in F} a_i f_i(t)=\sum_{i\in F}|a_i|$. Hence the operator L defined on the linear span M of the f_i's by $L\left(\sum_{i\in F} a_i f_i\right)=\sum_{i\in F} a_i\varepsilon_i$ is a linear isometry of M onto a dense set in $l_1(\Gamma,\mathbb{R})$ and thus has a unique linear isometric extension from $\bar{M}$ onto $l_1(\Gamma,\mathbb{R})$.

Suppose $L: l_1(\Gamma,\mathbb{R})\to C(T,\mathbb{R})$ is a linear isometric embedding and $f_i=L(\varepsilon_i)$ for all $i\in\Gamma$. Then $\|f_i\|=1$ and for any nonempty finite set $F\subset I$ and any $a_i\in\mathbb{R}$, $\left\|\sum_{i\in F} a_i f_i\right\|=\sum_{i\in F}|a_i|$. For each $i\in\Gamma$, let $A_i=f_i^{-1}(-1)$ and $B_i=f_i^{-1}(1)$. Then each A_i, B_i are closed and $A_i\cap B_i=\emptyset$ for all $i\in\Gamma$. If $\{s_i : i\in F\}$ be a sequence of 1's and -1's. Then for $s_i'=-s_i$ for $i\in F$, either $\bigcap_{i\in F} s_i A_i\neq\emptyset$ or $\bigcap_{i\in F} s_i' A_i\neq\emptyset$. For, if $a_i=s_i$, then $\left\|\sum_{i\in F} a_i f_i\right\|=\left\|\sum_{i\in F} a_i\varepsilon_i\right\|=\sum_{i\in F}|a_i|=\operatorname{card} F$. Therefore for some $t\in T$,

$\sum_{i\in F} a_i f_i(t) = \operatorname{card} F$. But $|a_i f_i(t)| \leqslant 1$ for all $i \in F$ and either $\bigcap_{i\in F} s_i A_i$ or $\bigcap_{i\in F} s_i' A_i$ contains t.

For convenience we now assume that $\Gamma = \{\alpha : \alpha < \alpha_0\}$ for some initial ordinal α_0. Now suppose that for some sequence $\alpha_1 < \cdots < \alpha_n$ in Γ and $s_i \in \{-1, 1\}$, we have that $\bigcap_{i=1}^{n} s_i A_{\alpha_i} = \emptyset$. Let $\tilde{A}_\alpha = A_{\alpha+\alpha_n}$ and $\tilde{B}_\alpha = B_{\alpha+\alpha_n}$ for all $a \in \Gamma$. Then $\tilde{A}_\alpha, \tilde{B}_\alpha$ are closed and disjoint for all $\alpha \in \Gamma$. Suppose there is a sequence $\beta_1 < \cdots < \beta_k$ in Γ and $t_j \in \{-1, 1\}$ such that $\bigcap_{j=1}^{k} t_j \tilde{A}_{\beta_j} = \emptyset$. Then $\bigcap_{j=1}^{k} t_j' \tilde{A}_{\beta_j} \neq \emptyset$ where $t_j' = -t_j$, for $j = 1, \ldots, k$. Now, either $\left(\bigcap_{i=1}^{n} s_i A_{\alpha_i}\right) \cap \left(\bigcap_{j=1}^{k} t_j' \tilde{A}_{\beta_j}\right) \neq \emptyset$ or $\left(\bigcap_{i=1}^{n} s_i' A_{\alpha_i}\right) \cap \left(\bigcap_{j=1}^{k} t_j \tilde{A}_{\beta_j}\right) \neq \emptyset$ but this is impossible and it follows that $(\tilde{A}_\alpha, \tilde{B}_\alpha\}_{\alpha\in\Gamma}$ is interlocking. ∎

Corollary. *A compact Hausdorff space T is dispersed if and only if $l_1(\mathbb{N}, \mathbb{R})$ cannot be linearly and isometrically embedded in $C(T, \mathbb{R})$.* ∎

We shall see later that isometry can be replaced with isomorphism in the above corollary.

On the other hand, the space $c(\mathbb{N}, \mathbb{C})$ of all convergent sequences can be embedded into $C(T, \mathbb{C})$ for each infinite compact Hausdorff space T. For, let $\{U_n\}$ be a sequence of nonempty pairwise disjoint open sets in T and for each n let $t_n \in U_n$. Then there is an $f_n \in C(T, \mathbb{C})$ with $0 \leqslant f_n \leqslant 1$, $f_n(t_n) = 1$ and $f_n = 0$ on $T \setminus U_n$. Let $L: c(\mathbb{N}, \mathbb{C}) \to C(T, \mathbb{C})$ be defined by $L(\{a_n\}) = \sum_{n=1}^{\infty} (a_n - \lim a_n)\, \mathrm{f}_n + (\lim a_n) \cdot 1 = f$. Then L is a linear map and $\|f\| = \sup |a_n|$.

However, only spaces of a special type can be linearly isometrically embedded into $c(N, \mathbb{C})$ (the proof also works for $c(N, \mathbb{R})$).

Theorem 3 (Lacey-Morris). *Let X be a Banach space, U^* the closed unit ball of X^*, and F the set of extreme points of U^*. Then X can be linearly isometrically embedded into $c(N, \mathbb{C})$ if and only if $F = \bigcup_{t\in T} F_t$ where T is the set of scalars of modulus 1, $F_t = t F_1$ for all $t \in T$ and $F_t \cap F_1 = \emptyset$ for $t \neq 1$, and F_1 is countable and has at most one weak* limit point.*

Proof. We shall use the fact that $c(\mathbb{N}, \mathbb{C}) = C(\Gamma(\omega), \mathbb{C})$ where $S = \{\alpha : 1 \leqslant \alpha \leqslant \omega\}$. Suppose $L: X \to c(\mathbb{N}, \mathbb{C})$ is a linear isometric embedding. Then L^* maps the closed unit ball V^* of $c(\mathbb{N}, \mathbb{C})^*$ onto U^* and $L^*(E) \supset F$ where E is the set of extreme points of V^*. Let k_1 be the smallest integer such that $L^*(\varepsilon_{k_1}) \in F$. Suppose $1 \leqslant k_1 < \cdots < k_n$ have been chosen so that $L^*(\varepsilon_{k_j}) \in F$ and $L^*(\varepsilon_j) \notin F$ for j in $\{1, \ldots, k_n\} \setminus \{k_1, \ldots, k_n\}$

and that $L^*(\varepsilon_{k_i}) \neq a L^*(\varepsilon_{k_j})$ for all $|a|=1$ and $i \neq j$. If $F = \bigcup_{t \in T} t\{\varepsilon_{k_1}, \ldots, \varepsilon_{k_n}\}$, then $F_1 = \{\varepsilon_{k_1}, \ldots, \varepsilon_{k_n}\}$. Otherwise let k_{n+1} be the smallest integer $k > k_n$ such that $L^*(\varepsilon_{k_{n+1}}) \in F$ and $L^*(\varepsilon_{k_{n+1}}) \neq a L^*(\varepsilon_{k_i})$ for all $|a|=1$ and $i=1, \ldots, n$. By induction $F_1 = \{L^*(\varepsilon_{k_n})\}$. Moreover F_1 has at most one weak* limit point and exactly one when it is infinite. In such a case it is $L^*(\varepsilon_\omega)$.

Suppose such an F_1 exists. If $F_1 = \{x_1^*, \ldots, x_n^*\}$, then let L be defined by $L(x) = \{(x_1^*(x)\}, \ldots, x_n^*(x), 0, \ldots)$. If $F_1 = \{x_1^*, \ldots, x_n^*, \ldots\}$ with weak* limit x^*, then let L be defined by $L(x) = \{x_i^*(x)\}$. ∎

Corollary 1. *Let X be a Banach space. Then X can be linearly isometrically embedded into $c_0(\mathbb{N}, \mathbb{C})$ if and only if there is an F_1 (as in the theorem) such that either F_1 is finite or F_1 forms a sequence converging to 0 in the weak* topology.* ∎

Corollary 2 (Klee). *Let X be a finite dimensional real Banach space. Then X can be linearly isometrically embedded into $c_0(\mathbb{N}, \mathbb{R})$ if and only if the closed unit ball of X has only finitely many extreme points (i.e. is polyhedral).* ∎

The next theorem is due to Pełczyński and Semadeni [235]. It is valid for both the real and complex case.

Theorem 4. *Let T be a compact Hausdorff space. Then T is dispersed if and only if every closed infinite dimensional subspace of $C(T, \mathbb{C})$ contains an isomorphic copy of $c_0(\mathbb{N}, \mathbb{C})$.*

Proof. Clearly the theorem is true if T is finite. Suppose T is not dispersed. Then $C(T, \mathbb{C})$ contains a subspace isometrically isomorphic to $C([0,1], \mathbb{C})$ since T can be mapped continuously onto $[0,1]$. Since $C([0,1], \mathbb{C})$ contains isometric copies of all separable Banach spaces and, for example, $l_1(\mathbb{N}, \mathbb{C})$ does not contain a subspace isomorphic to $c_0(\mathbb{N}, \mathbb{C})$, the sufficiency is established.

Suppose T is dispersed and X is an infinite dimensional closed subspace of $C(T, \mathbb{C})$. Without loss of generality X can be taken to be separable. Thus there is a subspace Y of $C(T, \mathbb{C})$ containing X such that $Y = \{f \circ \alpha : f \in C(S, \mathbb{C})\}$ where S is a compact metrizable space and α is a map of T onto S. Since T is dispersed, so is S and thus by theorem 3 of section 5, $S = \{\beta : \beta \leqslant \beta_0\} = \Gamma(\beta_0)$ for some countable ordinal β_0. Hence it suffices to prove the theorem for spaces of the type $C(\Gamma(\beta_0), \mathbb{C})$ for a countable ordinal β_0. When $\beta_0 = \omega$, then $C(\Gamma(\beta_0), \mathbb{C}) = c(\mathbb{N}, \mathbb{C})$ and the theorem follows from theorem 8 of section 12.

Suppose for some countable ordinal $\beta_0 > \omega$ there is an infinite dimensional closed subspace X of $C(\Gamma(\beta_0), \mathbb{C})$ which does not contain a subspace isomorphic to $c(\mathbb{N},\mathbb{C})$. Then β_0 can be taken to be the smallest such ordinal and must then be a limit ordinal. Let $X_0=\{f\in X: f(\beta_0)=0\}$. Then X_0 is infinite dimensional. For $\beta<\beta_0$ let $Y_\beta=\{f\in C(\Gamma(\beta),\mathbb{C}): f = g|\Gamma(\beta)$ for some $g\in X_0\}$. For each $\varepsilon>0$ there is an $f\in X_0$ such that $\|f\|=1$ and $|f(\gamma)|<\varepsilon$ for all $\gamma\in\Gamma(\beta)$. For, we have the following two cases.

Case 1. The space Y_β is finite dimensional. Since X_0 is infinite dimensional there are f_1, f_2 in X_0 such that $f_1\neq f_2$ and $f_1|\Gamma(\beta)=f_2|\Gamma(\beta)$. Then for $f=(f_1-f_2)/\|f_1-f_2\|$, $\|f\|=1$ and $|f(\gamma)|<\varepsilon$ for all $\gamma\in\Gamma(\beta)$.

Case 2. The space Y_β is infinite dimensional. Then since $Y_\beta\subset C(\Gamma(\beta),\mathbb{C})$ there is an isomorphism of $c(\mathbb{N},\mathbb{C})$ into Y. Thus by the corollary to lemma 2 of section 12 and theorem 9 of section 12 there is a sequence $f_n\in Y_\beta$ such that $\|f_n\|=1$ and $1\leqslant\sup_{\gamma\leqslant\beta}\sum_{n=1}^{\infty}|f_n(\gamma)|=K<\infty$.

Let $g_n\in X_0$ such that $g_n|\Gamma(\beta)=f_n$. Then $\|g_n\|\geqslant\|f_n\|=1$ and $\sum_{n=1}^{\infty} g_n$ is not weakly unconditionally convergent (since otherwise X_0 would contain a copy of $c_0(\mathbb{N},\mathbb{C})$). Thus there is a γ_0 with $\beta<\gamma_0\leqslant\beta_0$ and integers $n_1<\cdots<n_k$ such that $|g_{n_1}(\gamma_0)+\cdots+g_{n_k}(\gamma_0)|>K/\varepsilon$ and for $f=(g_{n_1}+\cdots+g_{n_k})/\|g_{n_1}+\cdots+g_{n_k}\|$, $\|f\|=1$ and $|f(\gamma)|<\varepsilon$ for all $\gamma\leqslant\beta$.

Now let $h_1\in X_0$ with $\|h_1\|=1$. Since $\lim_{\gamma\to\beta_0} h_1(\gamma)=h_1(\beta_0)=0$, there is a $\gamma_1<\beta_0$ such that $|h_1(\gamma)|<1/2$ for all $\gamma_1<\gamma\leqslant\beta_0$. By the above there is an $h_2\in X_0$ such that $\|h_2\|=1$ and $|h_2(\gamma)|<1/2$ for all $\gamma\leqslant\gamma_1$. Assume $h_1,\ldots,h_k$ have been chosen so that $\|h_i\|=1$ and $\gamma_1<\cdots<\gamma_{k-1}<\beta_0$ are such that $|h_i(\gamma)|<1/2^{k-1}$ for $\gamma>\gamma_{i-1}$, $i=1,\ldots,k$. Let $\gamma_k>\gamma_{k-1}$ be such that $|h_i(\gamma)|<1/2^k$ for $i=1,\ldots,k$ and $\gamma>\gamma_k$. Choose $h_{k+1}\in X_0$ such that $\|h_{k+1}\|=1$ and $|h_{k+1}(\gamma)|<1/2^k$ for $\gamma\leqslant\gamma_k$. Then $\sum_{n=1}^{\infty}|h_n(\gamma)|\leqslant 2$ for all $\gamma\in\Gamma(\beta_0)$ and $\sum_{n=1}^{\infty} h_n$ is weakly unconditionally convergent. Thus X_0 contains a copy of $c_0(\mathbb{N},\mathbb{C})$ which is a contradiction. ∎

Corollary. *Let T be a compact Hausdorff space. Then T is dispersed if and only if $C(T,\mathbb{C})$ does not contain a subspace isomorphic to $l_1(\mathbb{N},\mathbb{C})$.* ∎

The final theorem is due to Lindenstrauss and Pełczyński [187].

Theorem 5. *Let X be a closed subspace of $c_0(\mathbb{N},\mathbb{R})$ and T be a compact Hausdorff space. If $L: X\to C(T,\mathbb{R})$ is a bounded linear transformation, then for each $\varepsilon>0$ there is a bounded linear extension L_ε of L from $c_0(\mathbb{N},\mathbb{R})$ to $C(T,\mathbb{R})$ such that $\|L_\varepsilon\|\leqslant(1+\varepsilon)\|L\|$.*

Proof. Clearly it may be assumed that $\|L\|=1$. Since $L(X)$ is separable it is contained in a subspace of $C(T,\mathbb{R})$ linearly isometric to $C(S,\mathbb{R})$ for some compact metrizable space S. Thus without loss of generality, T is taken to be metrizable. Finally, by Zorn's lemma it is enough to show that for each y in $c_0(\mathbb{N},\mathbb{R})\setminus X$, L can be extended to an operator of norm $\leqslant 1+\varepsilon$ on $X\oplus\mathbb{R}\,y$. That is, it must be shown that there is an $f\in C(T,\mathbb{R})$ with $\|f-Lx\|\leqslant(1+\varepsilon)\|y-x\|$ for all $x\in X$. Thus for every $t\in T$, f must satisfy $\sup_{x\in X}(Lx(t)-(1-\varepsilon)\|y-x\|)=G(t)\leqslant f(t)\leqslant F(t)$ $=\sup_{x\in X}(Lx(t)+(1+\varepsilon)\|y-x\|)$. By theorem 1 of section 4 it suffices to show that $\bar{G}\leqslant\underline{F}$ where $\bar{G}$ is the smallest upper semicontinuous function $\geqslant G$ and $\underline{F}$ is the largest lower semicontinuous function $\leqslant F$. Suppose to the contrary that $\underline{F}(t)<\bar{G}(t)$ for some $t\in T$. Then there are sequences $\{s_n\}$ and $\{t_n\}$ in T with $t=\lim s=\lim t_n$ and $\lim F(s_n)<\lim G(t_n)$. From the definition of F and G it follows that there are sequences $\{x_n\}$ and $\{y_n\}$ in X such that $\lim(Lx_n(s_n)+(1+\varepsilon)\|y-x_n\|)<\lim(Ly_n(t_n)-(1+\varepsilon)\|y-y_n\|)$.

Let $x_n^*=L^*(\varepsilon_{s_n})$ and $y_n^*=L^*(\varepsilon_{t_n})$ and μ_n,ν_n be norm preserving extensions of x_n^*,y_n^* respectively to all of $c_0(\mathbb{N},\mathbb{R})$ for $n=1,2,\ldots$. By passing to subsequences it can be assumed that $\mu_n\to\mu$ and $\nu_n\to\nu$ in the weak* topology. Moreover $\mu|X=\nu|X=L^*(\varepsilon_t)$. Now since the norm is additive in $c_0(\mathbb{N},\mathbb{R})^*$, it follows that $\lim(\|\mu_n\|-\|\mu_n-\mu\|-\|\mu\|)$ $=\lim(\|\nu_n\|-\|\nu_n-\nu\|-\|\nu\|)=0$.

By restricting to X and the fact that $\|\mu_n\|=\|x_n^*\|$, $\|\nu_n\|=\|y_n^*\|$, it follows that $\|\mu\|=\|\nu\|=\|L^*\varepsilon_t\|=r$ and $\limsup\|\mu_n-\mu\|\leqslant 1-r$, $\limsup\|\nu_n-\nu\|\leqslant 1-r$. Since $\lim(\mu_n-\mu)(y)=\lim(\nu_n-\nu)(y)=0$, from the definitions of x_n^*,y_n^*,μ_n,ν_n and $\mu|X=\nu|X=L^*\varepsilon_t$, it follows that $\lim(Lx_n(s_n)+(1+\varepsilon)\|y-x_n\|-Ly_n(t_n)+(1+\varepsilon)\|y-y_n\|)$

$$\begin{aligned}&=\lim(\mu_n(x_n)-\nu_n(y_n)+(1+\varepsilon)\|y-x_n\|+(1+\varepsilon)\|y-y_n\|)\\&=\lim((\mu_n-\mu)(x_n)-(\nu_n-\nu)(y_n)+L^*\varepsilon_t(x_n-y_n)+(1+\varepsilon)(\|y-x_n\|+\|y-y_n\|))\\&=\lim((\mu_n-\mu)(x_n-y)-(\nu_n-\nu)(y_n-y)+L^*\varepsilon_t(x_n-y_n)\\&\quad+(1+\varepsilon)(\|y-x_n\|+\|y-y_n\|))=\lim(S_1(n)+S_2(n)+S_3(n))\end{aligned}$$

where

$$\begin{aligned}S_1(n)&=(\mu_n-\mu)(x_n-y)+(1-r+\varepsilon)\|y-x_n\|\\S_2(n)&=-(\nu_n-\nu)(y_n-y)+(1-r+\varepsilon)\|y-y_n\|\\S_3(n)&=L^*\varepsilon_t(x_n-y_n)+r(\|x_n-y\|+\|y_n-y\|).\end{aligned}$$

But, $S_1(n)$, $S_2(n)\geqslant 0$ for n large enough and $S_3(n)\geqslant 0$ for all n which is a contradiction. ∎

Corollary. Let X be an infinite dimensional complemented subspace of $C(T,\mathbb{R})$. If X is isomorphic to a subspace of $c_0(\mathbb{N},\mathbb{R})$, then it is isomorphic to $c_0(\mathbb{N},\mathbb{R})$. ∎

Exercises. 1. Let X be a real Banach space and V^* the closed unit ball of X^*. Show that X can be linearly isometrically embedded in $C(T,\mathbb{R})$ for some compact Hausdorff dispersed space T if and only if the weak* closure of the set of extreme points of V^* is dispersed.

2. Give an example of an abstract M space X such that $X^* = l_1(\mathbb{N},\mathbb{R})$ and X is not linearly isometrically embeddable in $C(T,\mathbb{R})$ for any compact Hausdorff dispersed space T.

3. Let X be a separable real Banach space and Y a closed subspace of X which is also linearly isometric to a subspace of $c_0(\mathbb{N},\mathbb{R})$. Show that for every compact Hausdorff space T, every bounded linear operator $L: Y \to C(T,\mathbb{R})$ and every $\varepsilon > 0$, there is an extension L of L to X with $\|L\| \leqslant (2+\varepsilon)\|L\|$.

4. Let X be a subspace of $c_0(\mathbb{N},\mathbb{R})$ and V^* the closed unit ball of X^*. Then for every separable Banach space $Y \supset X$ there is a weak* continuous mapping $\psi: V^* \to Y^*$ such that $\psi(x^*)(x) = x^*(x)$ for all $x^* \in X^*$ and $x \in X$.

5. Prove that $c_0(\mathbb{N},\mathbb{R})$ cannot be embedded into $C([0,1],\mathbb{R})$ as a sublattice.

6. Prove that $l_1(\mathbb{N},\mathbb{C})$ does not contain a subspace isomorphic to $c_0(\mathbb{N},\mathbb{C})$.

7. Prove that $c(\mathbb{N},\mathbb{R})$ contains a two dimensional subspace X which is not polyhedral (i.e. the closed unit ball of X has an infinite number of extreme points).

8. Let T be a dispersed compact Hausdorff space. Prove that every reflexive quotient space of $C(T,\mathbb{R})$ is finite dimensional.

9. Let T be an infinite compact Hausdorff space. Prove that there is a closed subspace X of $C(T,\mathbb{R})$ such that the quotient space of $C(T,\mathbb{R})$ by X is infinite dimensional and separable (Note: *it is an unsolved problem as to whether or not every infinite dimensional Banach space has an infinite dimensional separable quotient*).

10. Let X be a finite dimensional real Banach space with polyhedral closed unit ball. Show that the closed unit ball in X^* is polyhedral.

Chapter 5

Representation Theorems for Spaces of the Type $L_p(T, \Sigma, \mu, \mathbb{C})$

In section 14 we shall show that when μ is a finite measure $L_p(T, \Sigma, \mu, \mathbb{C})$ can be faithfully represented as a Banach lattice in terms of countable direct sums of spaces $L_p([0,1]^m, \mathbb{C})$ where m is an infinite cardinal number and $[0,1]^m$ is m products of $[0,1]$ with product Lebesgue measure being considered. We also discuss the isomorphic classification of these spaces.

Section 15 is devoted to a proof that a Banach lattice X with the property that $\|x+y\|^p = \|x\|^p + \|y\|^p$ whenever $x \wedge y = 0$ is an $L_p(\mu, \mathbb{C})$ space for some measure μ. We also prove that these spaces together with abstract M spaces are exactly the Banach lattices which have the property that $\|x+y\|$ is a function of $\|x\|$ and $\|y\|$ whenever $x \wedge y = 0$.

§ 14. Measure Algebras and the Representation of $L_p(T, \Sigma, \mu, \mathbb{C})$ when μ is a Finite Measure

We will require the notion of a Boolean algebra and the important Maharam classification of measure algebras. The proof of this classification occupies a major portion of this section. The details are given in a series of steps below.

Definition 1. A nonempty set $\mathscr{A}$ with three operations $\vee$ $\wedge$, $-$ (join, meet, and complement) satisfying the following relations is called a *Boolean Algebra.*

For each A, B, C in $\mathscr{A}$.

(1) $A \vee B = B \vee A$ and $A \wedge B = B \wedge A$,

(2) $A \vee (B \vee C) = (A \vee B) \vee C$ and $A \wedge (B \wedge C) = (A \wedge B) \wedge C$,

(3) $(A \wedge B) \vee B = B$ and $(A \vee B) \wedge B = B$,

(4) $A \wedge (B \vee C) = (A \wedge B) \vee (A \wedge C)$ and $A \vee (B \wedge C) = (A \vee B) \wedge (A \vee C)$,

(5) $(-A \vee A) \wedge B = B$ and $(-A \wedge A) \vee B = B$.

The following results are elementary consequences of these axioms. We use the notation $-A \vee B$ to mean $(-A) \vee B$ and similiarly for $-A \wedge B$.

(a) There is a unique element $0\in\mathscr{A}$ such that $-A\wedge A=0$ for all $A\in\mathscr{A}$.

(b) There is a unique element $1\in\mathscr{A}$ such that $-A\vee A=1$ for all $A\in\mathscr{A}$.

(c) For all $A\in\mathscr{A}$, $0\vee A=A$, $0\wedge A=0$, $1\vee A=1$, and $A\wedge A=A$.

(d) For all A,B in $\mathscr{A}$, $A=-(-A)$, $-(A\vee B)=(-A)\wedge(-B)$, and $-(A\wedge B)=(-A)\vee(-B)$.

(e) $-0=1$.

A Boolean algebra $\mathscr{A}$ has a natural partial order defined by $A\leqslant B$ if and only if $A\wedge B=A$. Moreover, $\mathscr{A}$ is said to be σ *complete* if every sequence of elements in $\mathscr{A}$ has a least upper bound in this ordering and $\mathscr{A}$ is said to be *complete* if every nonempty set of elements of $\mathscr{A}$ has a least upper bound in this ordering.

The least upper bound (if it exists) of a set S of elements of $\mathscr{A}$ is denoted by $\bigvee S$.

A function $\Phi:\mathscr{A}_1\to\mathscr{A}_2$, where $\mathscr{A}_1$ and $\mathscr{A}_2$ are Boolean algebras, is said to be a *homomorphism* if for each A,B in $\mathscr{A}_1$, $\Phi(A\vee B)=\Phi(A)\vee\Phi(B)$, $\Phi(A\wedge B)=\Phi(A)\wedge\Phi(B)$, and $\Phi(-A)=-\Phi(A)$. It is said to be an *isomorphism* if, in addition, it is one-to-one and onto. If Φ is an isomorphism, then for any set S of elements of $\mathscr{A}_1$ such that $\bigvee S$ exists, $\bigvee\Phi(S)$ exists and $\Phi(\bigvee S)=\bigvee\Phi(S)$.

A model for Boolean algebras which, in fact, characterizes them is the Boolean algebra of open and closed sets in a topological space T where the operations are set-theoretic union, intersection, and complementation. Indeed, it will be seen that T can be taken to be compact, Hausdorff and totally disconnected. The following elementary lemma is easily established and its proof is left to the reader.

Lemma 1. *Let T_1,T_2 be two totally disconnected compact Hausdorff spaces and $\mathscr{A}_1,\mathscr{A}_2$ the Boolean algebras of open and closed sets in T_1 and T_2 respectively. Then $\mathscr{A}_1$ and $\mathscr{A}_2$ are isomorphic if and only if T_1 and T_2 are homeomorphic.* ∎

Let us note that any Boolean algebra $\mathscr{A}$ is a linear algebra over the integers mod(2), Z_2, under the natural operations given by $A+B=(-A\wedge B)\vee(-B\wedge A)$, $A\cdot B=A\wedge B$, and $0\cdot A=0$, $1\cdot A=A$ for all A and B in $\mathscr{A}$.

Theorem 1 (Stone). *Let $\mathscr{A}$ be a Boolean algebra. Then $\mathscr{A}$ is isomorphic to the Boolean algebra of open and closed sets of a unique totally disconnected compact Hausdorff space.*

Proof. The uniqueness is an immediate consequence of lemma 1.

For motivation note that if T is a totally disconnected compact Hausdorff space and $\mathscr{B}$ is the Boolean algebra of all open and closed sets in T, then $\mathscr{B}$ is isomorphic to the Boolean algebra of all continuous characteristic functions on T, where $\wedge$ is min, $\vee$ is max, and the negation of f is $1-f$. Moreover, for each $t\in T$, the point evaluation functional ε_t is a ring homomorphism on $\mathscr{B}$ and since there are enough elements of $\mathscr{B}$ to separate the points of T and T is compact, the topology on T is the weakest topology which makes each element of $\mathscr{B}$ continuous.

Now, let T denote the set of ring homomorphisms from $\mathscr{A}$ into Z and for each $A\in\mathscr{A}$, let $\hat{A}:T\to Z$ be defined by $\hat{A}(t)=t(A)$ for all $t\in T$. If H is a Hamel basis for $\mathscr{A}$ over Z, then the map $f:T\to\prod_H Z$ (Z_2 producted with itself H times) by $f(t)(A)=t(A)$ for all $t\in T$ and $A\in H$ is clearly a one-to-one embedding of T onto a closed subset of $\prod_H Z_2$ where Z_2 has the discrete topology and $\prod_H Z_2$ has the product topology. If T has the topology induced from this embedding, then T is a totally disconnected compact Hausdorff space and, in fact, T has the weakest topology such that each $\hat{A}$ is continuous for $A\in\mathscr{A}$. Moreover, for each $A\in\mathscr{A}$, $\{t\in T:t(A)=0\}$ is both open and closed in T and since $\{t\in T:t(A_1)=t(A_2)=0\}=\{t\in T:t(A_1A_2+A_1+A_2)=0\}$ for A_1,A_2 in $\mathscr{A}$, it follows that the sets of the form $\{t\in T:t(A)=0\}$ $(A\in\mathscr{A})$ form a base for the topology of T.

Let $U\subset T$ be an open and closed set. Then there are $A_1,\ldots,A_n$ in such that $U=\bigcup_{i=1}^{n}\{t\in T:t(A_i)=0\}=\{t\in T:t(A_1\cdots A_n)=0\}$. Thus $A\to\hat{A}$ is an algebra homomorphism of $\mathscr{A}$ onto the continuous functions from T to Z_2. Let $A\in\mathscr{A}$ with $A\neq 0$ and M be a maximal ideal in $\mathscr{A}$ which contains $\{AB-B:B\in\mathscr{A}\}$. Then $\mathscr{A}/M$ is a field and for each $B\in\mathscr{A}$, $B^2+M=B+M$. Thus $\mathscr{A}/M=Z_2$ and $A\notin M$. If t is defined by $t(B)=0$ for $B\in M$ and $t(B)=1$ for $B\notin M$, then $t\in T$ and $t(A)=1$. Therefore $\mathscr{A}$ is isomorphic to the Boolean algebra of continuous functions from T to Z_2. ∎

We can now give a characterization of complete Boolean algebras also due to Stone.

Theorem 2 (Stone). *Let $\mathscr{A}$ be a Boolean algebra. Then $\mathscr{A}$ is complete if and only if it is isomorphic to the Boolean algebra of open and closed sets of a (unique) extremally disconnected compact Hausdorff space.*

Proof. Let $\mathscr{A}$ be isomorphic to the open and closed sets of a compact Hausdorff totally disconnected space T as in theorem 1. If T is extremally disconnected and S is a class of subsets of $\mathscr{A}$, then it is easy to see that $\bigvee S$ is the closure of the union of the elements of S (the elements of S are open and closed sets in T) so that $\mathscr{A}$ is complete.

Now suppose $\mathscr{A}$ is complete. Let $U\subset T$ be open and $S=\{B\in\mathscr{A}:B\subset U\}$. Then $\bigvee S$ exists and is equal to $\bar{U}$, that is, $\bar{U}$ is open in T and T is extremally disconnected. ∎

The Boolean algebras of interest in the study of classical Banach spaces arise mainly as measure algebras. If $\mathscr{A}$ is a σ complete Boolean algebra a *measure* μ on $\mathscr{A}$ is a real valued function on $\mathscr{A}$ such that for all pairwise disjoint sequences $\{A_n\}$ in $\mathscr{A}$, $\mu(\bigvee A_n)=\sum_{n=1}^{\infty}\mu(A_n)$. The measure μ is said to be *strictly positive* if $\mu(A)>0$ for all $A\neq 0$ and it is said to be *normalized* if $\mu(1)=1$. A *measure algebra* is a pair $(\mathscr{A},\mu)$ where $\mathscr{A}$ is a σ complete Boolean algebra and μ is a strictly positive normalized measure on $\mathscr{A}$. In such a case it can easily be seen that $\mathscr{A}$ is a complete Boolean algebra.

Theorem 3. *Let T be a totally disconnected compact Hausdorff space, $\mathscr{A}$ the Boolean algebra of open and closed sets in T, and Φ a positive additive function on $\mathscr{A}$. Then there is a unique regular Borel measure μ on T such that $\mu(A)=\Phi(A)$ for all $A\in\mathscr{A}$ and $\|\mu\|=\Phi(1)$.*

Proof. Since T is totally disconnected, the closed linear span of the characteristic functions f_A for $A\in\mathscr{A}$ is all of $C(T,\mathbb{R})$. Thus there is a unique linear functional x^* on $C(T,\mathbb{R})$ such that $x^*(f_A)=\Phi(A)$ for all $A\in\mathscr{A}$. Hence μ is given by the Riesz representation theorem (theorem 4 of section 8). ∎

We shall also need the notion of an infinite product measure. The main theorem we need is stated below. For a proof of it the reader is referred to [130] or [131].

Theorem 4. *Let $\{T_i\}_{i\in I}$ be a family of compact Hausdorff spaces and suppose that μ_i is a positive normalized regular Borel measure on T_i for each $i\in I$. Then there is a unique positive normalized regular Borel measure on $T=\prod_{i\in I}T_i$ such that*

(1) *if $A_i\subset T_i$ is μ_i measurable and only countably many A_i's are not equal to T_i, then $\prod_{i\in I}A_i$ is μ measurable and $\mu\left(\prod_{i\in I}A_i\right)=\prod_{i\in I}\mu(A_i)$ $\equiv\inf\left\{\prod_{i\in F}\mu(A_i): F\subset I \text{ is a finite set}\right\}$,*

(2) *if $\mu_i(A_i)<1$ for uncountably many $A_i\subset T_i$, then $\mu\left(\prod_{i\in I}A_i\right)=0$.* ∎

We now briefly give some terminology we shall need for the classification of measure algebras.

An *atom* in a Boolean algebra is an element $A \neq 0$ such that if $B \leqslant A$, then $B=0$ or $B=A$. A measure μ on a Boolean algebra $\mathscr{A}$ is said to be *purely nonatomic* if $\mathscr{A}$ has no atoms.

Note that if (T,Σ,μ) is a measure space where μ is a positive normalized measure on Σ, then there is a natural corresponding measure algebra $(\mathscr{A},\hat{\mu})$ where $\mathscr{A}$ is the set of equivalence classes of elements of Σ under the equivalence relation equal μ-almost everywhere and $\hat{\mu}(\hat{A})=\mu(A)$ for $A \in \Sigma\,(\hat{A}=\{B \in \Sigma : \mu[(A \setminus B) \cup (B \setminus A)]\}=0)$.

Two measure algebras $(\mathscr{A},\mu)$ and $(\mathscr{B},\nu)$ are said to be *isomorphic* if there is a Boolean algebra isomorphism h of $\mathscr{A}$ onto $\mathscr{B}$ such that $\nu(h(A))=\mu(A)$ for all $A \in \mathscr{A}$.

Any measure algebra has a natural metric as follows. If $(\mathscr{A},\mu)$ is a measure algebra, then by theorem 3 there is a totally disconnected compact Hausdorff space T and a regular Borel measure ν on T such that $\mathscr{A}$ is isomorphic to the closed and open sets in T under an isomorphism Φ and $\nu(\Phi(A))=\mu(A)$ for all $A \in \mathscr{A}$. Thus $A \to f_{(A)}$ = the characteristic function of $\Phi(A)$ is also an isomorphism. Now $f_{(A)} \in L_1(\nu,R)$ and if the metric on $\mathscr{A}$ is defined by $d(A,B)=\|f_{(A)}-f_{(B)}\|$, then it is easy to see that $\mathscr{A}$ is a complete space under this metric.

The measure algebra $(\mathscr{A},\mu)$ is called *separable* if $\mathscr{A}$ is separable in the above metric, that is, if $L_1(\nu,R)$ is a separable Banach space.

The classification of separable purely nonatomic measure algebras was accomplished by Caratheodory.

A nice exposition of the proof of the following theorem can be found in [253].

Theorem 5 (Caratheodory). *Let $(\mathscr{A},\mu)$ be a separable measure algebra. Then there is an isomorphism of $(\mathscr{A},\mu)$ into the measure algebra of the measure space $([0,1],\Lambda,\lambda)$ where Λ is the Lebesgue measurable sets and λ is Lebesgue measure. Moreover, the isomorphism can be taken to be onto if and only if μ is purely nonatomic.* ∎

Basically the classification of general purely nonatomic measure algebras is a generalization of theorem 5 where product Lebesgue measure is used instead of Lebesgue measure.

Let α be an infinite ordinal and let P_α denote the measure algebra generated by product Lebesgue measure on $I^\alpha=\prod_{\beta<\alpha} I_\beta$ where $I_\beta=[0,1]$ for all $\beta<\alpha$. Similarly, let Q_α denote the measure algebra on $D^\alpha=\prod_{\beta<\alpha} D_\beta$ where $D_\beta=\{0,1\}$ and the measure is given by $\mu_\beta\{0\}=\mu_\beta\{1\}=1/2$. Clearly by theorem 5, P_α and Q_α are isomorphic when α is a countable ordinal since $\prod_{\beta<\alpha} I_\beta$ and $\prod_{\beta<\alpha} D_\beta$ are metrizable compact Hausdorff spaces.

It is easy to see that they are always isomorphic. For, let J be an index set with card $J = \operatorname{card}\alpha$ and let $J = \bigcup_{\beta<\alpha} J_\beta$ with $J_\beta \cap J_\gamma = \emptyset$ for $\beta \neq \gamma$ and card $J_\beta = \aleph_0$. Then there is an isomorphism of the measure algebra on $[0,1]$ onto the measure algebra on $\prod_{\gamma\in J_\beta} D_\gamma$. This can be used to construct an isomorphism between P_α and Q_α.

We shall use Q_α as our model in the classification theorem.

Let $\mathscr{A}$ be a Boolean algebra and $A \in \mathscr{A}$. The *principle ideal* generated by A is given by $\mathscr{A}_A = \{A \wedge B : B \in \mathscr{A}\}$. A subset of $\mathscr{A}$ is called a *generating set* if the smallest σ subalgebra of $\mathscr{A}$ containing it is $\mathscr{A}$ itself. Let $\bar{\mathscr{A}}$ denote the smallest cardinal possible for a generating set for $\mathscr{A}$. The algebra $\mathscr{A}$ is said to be *homogeneous* if $\bar{\mathscr{A}}_A = \bar{\mathscr{A}}$ for each $A \in \mathscr{A}$.

The proof of the following theorem is left as an exercise.

Theorem 6. Let α be an infinite ordinal. Then Q_α is homogeneous and $\bar{Q}_\alpha = \operatorname{card}\alpha$. ∎

We first prove the following decomposition theorem, also due to Maharam, which shows that it is only necessary to consider homogeneous measure algebras.

Theorem 7. Let $(\mathscr{A}, \mu)$ be a purely nonatomic measure algebra. Then there is a unique decomposition $1 = \bigvee_{i\in I} 1_i$ where I is finite or countably infinite, $1_i \neq 0$, $1_i \wedge 1_j = 0$ for $i \neq j$, and $\mathscr{A}_{1_i}$ is homogeneous for all $i \in I$.

Proof. Let $I = \{\bar{\mathscr{A}}_A : A \in \mathscr{A}, A \neq 0\}$. For each $A \in \mathscr{A}$, $A \neq 0$, there is a largest element B of $\mathscr{A}$ such that $\bar{\mathscr{A}}_B = \bar{\mathscr{A}}_A$, namely $\bigvee\{C : \bar{\mathscr{A}}_C = \bar{\mathscr{A}}_A\}$ is such an element. Let $\{m_\beta : \beta < \alpha\}$ be a well ordering of I. The 1_i's are obtained as follows. Let 1_0 be the largest element of $\mathscr{A}$ such that $\bar{\mathscr{A}}_{1_0} = m_0$ and if 1_β has been defined for $\beta < \gamma < \alpha$ let 1_γ be the largest element of $\mathscr{A}$ less than or equal to $-\left(\bigvee_{\beta<\gamma} 1_\beta\right)$ and $\bar{\mathscr{A}}_{1_\gamma} = m_\gamma$. Then $1 = \bigvee_{\beta<\alpha} 1_\beta$ and since μ is finite, α is necessarily a countable ordinal. Clearly $\mu(1_\beta) > 0$ and $1 = \sum_{\beta<\alpha} \mu(1_\beta)$. ∎

The main classification theorem is now stated and discussed briefly. Its proof is accomplished in a series of numbered steps and lemmas given below.

Theorem 8 (Maharam). *Let $(\mathscr{A}, \mu)$ be a homogeneous purely nonatomic measure algebra and suppose $m = \bar{\mathscr{A}}$. Then $(\mathscr{A}, \mu)$ is isomorphic to Q_α where α is the initial ordinal of cardinality m.*

If α and β are ordinals with $\beta<\alpha$, then Q_β is naturally isomorphic to a σ subalgebra of Q_α, namely, if B is a measurable set in D^β, we correspond B to $B \prod_{\beta\leqslant\gamma<\alpha} D_r$, where $D_\gamma=\{0,1\}$ for all γ. The main step in the proof of theorem 8 is to establish the following: if $\mathscr{B}$ is a σ subalgebra of $\mathscr{A}$ and $\mathscr{B}$ is isomorphic to Q_β where $\operatorname{card}\beta<m$, then for any $A\in\mathscr{A}\setminus\mathscr{B}$, there is a σ subalgebra $\mathscr{B}_1$ of $\mathscr{A}$ containing A and $\mathscr{B}$ such that $\mathscr{B}_1$ is isomorphic to $Q_\beta\times Q_\omega$ in such a way that when the isomorphism is restricted to $\mathscr{B}$ it is the original one onto Q_β. Once this is accomplished, the result follows readily by transfinite induction.

Proof of theorem 8. Let $\mathscr{B}$ be a σ subalgebra of $\mathscr{A}$ containing 1 and suppose $\mathscr{B}$ is isomorphic to Q_β where $\operatorname{card}\beta<m$. In particular, for any nonzero A in $\mathscr{A}$, $\mathscr{A}_A=\{A\wedge C: C\in\mathscr{A}\}\neq\{A\wedge B: B\in\mathscr{B}\}=\mathscr{B}_A$ since $(\mathscr{A},\mu)$ is homogeneous of order m and $\mathscr{B}_A$ is not. Let $T=\prod_{\gamma<\beta} D_\gamma$ where $D_\gamma=\{0,1\}$ for all γ and h be the isomorphism of $\mathscr{B}$ onto Q_β. We will think of $h(B)$ as an element of the collection Σ of measurable subsets of T. Also, all statements which follow should be interpreted as holding ν almost everywhere if necessary, where ν is the product measure on T.

For each $A\in\mathscr{A}$ let ν_A be defined on Σ by $\nu_A(C)=\mu[A\wedge h^{-1}(C)]$ for all $C\in\Sigma$. Then ν_A is a measure on Σ and clearly ν_A is absolutely continuous with respect to ν. Thus by the Radon-Nikodym theorem there is a measurable function g_A on T with $0\leqslant g_A\leqslant 1$ and $\nu_A(C)=\int_C g_A\,d\nu$ for all $C\in\Sigma$. The following properties of g_A are easily verified.

(1) *If A,B are in $\mathscr{A}$ with $A\leqslant B$, then $g_A\leqslant g_B$.*

(2) *For each pair A,B in $\mathscr{A}$, $g_{A\wedge B}\leqslant\min(g_A,g_B)\leqslant\max(g_A,g_B)\leqslant g_{A\vee B}$.*

(3) *If $A\wedge B=0$, then $g_A+g_B=g_{A\vee B}$.*

(4) *If $A\in\mathscr{B}$, then g_A is the characteristic function of $h(A)$.*

(5) *For each $A\in\mathscr{A}$, g_A is strictly positive on T if and only if $A\wedge B\neq 0$ for all $0\neq B\in\mathscr{B}$.*

Some further properties of g_A are established below

(6) *If $A\in\mathscr{A}$ and $B\in\mathscr{B}$, then $g_{A\wedge B}=g_A g_B$.*

Proof. For $C\in\Sigma$, $\nu_{A\wedge B}(C)=\mu[A\wedge B\wedge h^{-1}(C)]=\mu[A\wedge h^{-1}(h(B)\wedge C)]$ $=\nu_A((h(B)\wedge C))=\int_{h(B)\cap C} g_A\,d\nu=\int_C g_A\cdot f_{h(B)}\,d\nu=\int_C g_A\cdot g_B\,d\nu$. Thus $g_{A\wedge B}$ $=g_A g_B$. ∎

(7) *If A and B are in $\mathscr{A}$ and $A\leqslant B$, then $g_A=g_B$ implies $A=B$.*

Proof. Clearly $g_B=g_A+g_{-A\wedge B}$ and thus $g_{-A\wedge B}=0$. Hence $-A\wedge B=0$ and $A=B$. ∎

(8) *If A_1,A_2 are in $\mathscr{A}$ with $A_1\leqslant A_2$ and $B\in\mathscr{B}$ with $g_{A_1}=g_{A_2}\cdot g_B$, then $A_1=A_2\wedge B$.*

Proof. For $C\in\mathscr{B}$, $\mu(A_1\wedge C)=\int_{h(C)} g_{A_1}dv=\int_{h(C)} g_{A_2}\cdot g_B dv=\int_{h(C\wedge B)} g_{A_2}dv$ $=\mu(A_2\wedge B\wedge C)$. Thus for $C=B$, $\mu(A_1\wedge B)=\mu(A_2\wedge B)$ and since $A_1\wedge B\leqslant A_2\wedge B$, it follows that $A_1\wedge B=A_2\wedge B$. For $C=1$, $\mu(A_1)=\mu(A_2\wedge B)$ and since $A_1\geqslant A_1\wedge B=A_2\wedge B$, it follows that $A_1=A_2\wedge B$. ∎

Lemma 2. *Let $A\in\mathscr{A}$ be such that g_A is strictly positive on $U\in\Sigma$. Then there is a $B\in\mathscr{A}$ and $V\in\Sigma$ such that*

(a) *$B\leqslant A$, $V\subset U$, and $v(V)>0$,*

(b) *$0<g_B<g_A$ on V,*

(c) *$g_B=0$ on $T\setminus V$.*

Proof. Let $C=A\wedge h^{-1}(U)$. Then $g_C=g_A\cdot f_U$ is strictly positive on U. From the observation at the beginning of the proof of theorem 8 there is a $0<D\leqslant C$ such that $D\neq C\wedge B$ for all $B\in\mathscr{B}$. Let $V=\{t\in T:0<g_D(t)<g_C(t)\}$. If $v(V)=0$, then $g_D=0$ or $g_D=g_C$. Clearly $g_D=0$ is impossible. If $g_D=g_C$, then $g_D=g_C\cdot f_W$ for $W=T\setminus V$ and by (8) $D=C\wedge h^{-1}(W)$ which is again impossible. Thus $v(V)>0$ and for $B=D\wedge h^{-1}(V)$, $g_B=g_D\cdot f_V$ (see (6)). ∎

Lemma 3. *Let $A\neq 0$ be in $\mathscr{A}$ and $U=\{t\in T:g_A(t)>0\}$. Then there is a $B\in\mathscr{A}$ such that*

(i) *$B\leqslant A$ and $g_B=0$ on $T\setminus U$,*

(ii) *$0<g_B<g_A$ on U.*

Proof. By lemma 2 there are $B_1\in\mathscr{A}$ and $U_1\in\Sigma$ such that $v(U_1)>0$ and $g_{B_1}=0$ on $T\setminus U_1$, $U_1\subset U$, $B_1\leqslant A$ and $0<g_{B_1}<g_A$ on U_1. Let α be a countable ordinal and suppose $B_\beta\in\mathscr{A}$, $U_\beta\in\Sigma$ have been chosen for $\beta<\alpha$ so that

(1) *$\{B_\beta\}_{\beta<\alpha}$ is a pairwise disjoint family of elements $B_\beta\leqslant A$, and $\{U_\beta\}_{\beta<\alpha}$, is a pairwise disjoint family of subsets of U,*

(2) *$v(U_\beta)>0$ and $g_{B_\beta}=0$ on $T\setminus U_\beta$ for all $\beta<\alpha$,*

(3) *$0<g_{B_\beta}<g_A$ on U_β for all $\beta<\alpha$.*

Let $C=\bigvee_{\beta<\alpha} B_\beta$, $D=-C\wedge A$ and $V=U\setminus\bigcup_{\beta<\alpha} U_\beta$. Then $g_B=\sum_{\beta<\alpha} g_{B_\beta}$. If $v(V)=0$ let $B=C$ and if $v(V)>0$, $g_D=g_A-g_C>0$ on V. By lemma 2 there are $B_\alpha\in\mathscr{A}$ and $U_\alpha\in\Sigma$ such that

(4) *$B_\alpha\leqslant D\leqslant A$, $U_\alpha\subset V\subset U$, and $v(U_\alpha)>0$,*

(5) *$0\leqslant g_{A_\alpha}<g_D<g_A$ on U_α,*

(6) *$g_{B_\alpha}=0$ on $T\setminus U_\alpha$.*

Since $v(U)$ is finite, this process terminates at some countable ordinal α_0. Thus $B=\bigvee_{\beta<\alpha_0} B_\beta$ will do. ∎

Lemma 4. *Let A,B be in $\mathscr{A}$ and $U\in\Sigma$ such that $\nu(U)>0$ and $g_A=0$ on $T\setminus U$, $A\leqslant B$, and $0<g_A<g_B$ on U. Then there is a $C\in\mathscr{A}$ such that*

(i) *$C\leqslant B$ and $g_C=0$ on $T\setminus U$,*

and

(ii) *$0<g_C<\frac{1}{2}g_B$ on U.*

Proof. Let $V=\{t:0<g_A(t)<\frac{1}{2}g_B(t)\}$ and $W=\{t\in U:g_A(t)\geqslant\frac{1}{2}g_B(t)\}=U\setminus V$. If $C_1=[h^{-1}(V)\wedge A]\vee[h^{-1}(W)\wedge(-A\wedge B)]$, then $g_{C_1}=g_{h^{-1}(V)\wedge A}+g_{h^{-1}(W)\wedge(-A\wedge B)}\leqslant\min(f_V,g_A)+\min(f_W,g_{-A\wedge B})\leqslant\frac{1}{2}g_B$. Also $g_{C_1}\geqslant g_{h^{-1}(V)\wedge A}=f_V\cdot g_A>0$ on V and $g_{C_1}\geqslant g_{h^{-1}(W)\wedge(-A\wedge B)}=f_W g_{-A\wedge B}>0$ on W. Hence $g_{C_1}>0$ on U. Let $D=C_1\wedge h^{-1}(U)$. Then $U=\{t:g_D(t)>0\}$ and by lemma 3 there is a $C\leqslant D$ such that $g_C=0$ on $T\setminus U$ and $0<g_C<g_D$ on U. ∎

Lemma 5. *Let $U\in\Sigma$ and $\varepsilon>0$. If $A\in\mathscr{A}$ is such that $0<g_A$ on U, then there is a $B\leqslant A\wedge h^{-1}(U)$ such that*

(i) *$g_B=0$ on $T\setminus U$,*

and

(ii) *$0<g_B<\varepsilon$ on U.*

Proof. Let $C=h^{-1}(U)\wedge A$. Then $g_C=g_A\cdot f_U>0$ and U and $g_C=0$ on $T\setminus U$. By lemma 3 there is an $A_1\in\mathscr{A}$ such that $A_1\leqslant C$, $g_{A_1}=0$ on $T\setminus U$, and $0<g_{A_1}<g_C$ on U. By lemma 4 there is a $B_1\in\mathscr{A}$ such that $B_1\leqslant C$, $g_{B_1}=0$ on $T\setminus U$ and $0<g_{B_1}<\frac{1}{2}g_C$ on U. By applying lemmas 3 and 4 repeatedly we obtain A_n,B_n such that $A_{n+1}\leqslant B_n$, $g_{A_{n+1}}=0$ on $T\setminus U$, $0<g_{A_{n+1}}<g_{B_n}$ on U, $B_{n+1}\leqslant B_n$, $g_{A_{n+1}}=0$ on $T\setminus U$, and $0<g_{B_{n+1}}\cdot\frac{1}{2}g_{B_n}<(1/2)^{n+1}g_C$ on U. Thus let $B=B_n$ for n so that $(\frac{1}{2})^{n+1}<\varepsilon$. ∎

Lemma 6. *Let f be a real valued measurable function on T with $0<f<1$. Let $U=\{t:0<f(t)\}$ and $A\in\mathscr{A}$ be such that $0<g_A$ on U. Then there is a $B\in\mathscr{A}$ such that*

(i) *$B\leqslant A$ and $g_B=0$ on $T\setminus U$,*

and

(ii) *$0<g_B\leqslant f$ on U.*

Proof. Let $U_n=\{t:1/2^n<f(t)<1/2^{n-1}\}$ and $\varepsilon=1/2^n$ in the preceeding lemma. Then there is an $A_n\in\mathscr{A}$ such that $A_n\leqslant A\wedge h^{-1}(U_n)$, $g_{A_n}=0$ on $T\setminus U_n$, and $0<g_{A_n}<1/2^n$ on U_n. Let $B=\bigvee_n A_n$. ∎

The following lemma is lemma 2 of [199] and it provides the key to the proof of theorem 8.

Lemma 7. *Let f be a real valued measurable function on T and $A\in\mathscr{A}$ such that $0\leqslant f\leqslant g_A$. Then there is a $B\in\mathscr{A}$ such that $B\leqslant A$ and $g_B=f$.*

Proof. Let $U_1=\{t:f(t)>0\}$. By lemma 6 there is a $B_1\in\mathscr{A}$ with $B_1\leqslant A$, $g_{B_1}=0$ on $T\setminus U_1$, and $0<g_{B_1}\leqslant f$ on U_1. Suppose α is a countable ordinal and $\{B_\beta\}_{\beta<\alpha}$ have been defined so that

(i) $B_\beta\neq 0$ for $\beta<\alpha$,

and

(ii) the B_β's are pairwise disjoint, $B_\beta\leqslant A$, and $g_{B_\beta}\leqslant f$ for all $\beta<\alpha$.

Let $D=\bigvee_{\beta<\alpha} B$. Then $0\leqslant g_D\leqslant f$ and let $U_\alpha=\{t:g_D(t)<f(t)\}$. If $v(U_\alpha)=0$, then we are done. If $v(U_\alpha)>0$, let $C=-D\wedge A$. Then $g_C=g_A-g_D\geqslant f-g_D>0$ on U_α. By lemma 6 there is a $B_\alpha\in\mathscr{A}$ such that $C_\alpha\leqslant -D\wedge A$ with $0<g_B\leqslant f-g_D$ on U_α. Then

(iii) $B_\alpha\neq 0$ and $B_\alpha\wedge B_\beta=0$ for $\beta<\alpha$.

(iv) $B_\alpha\leqslant A$,

(v) $g_{B_\alpha\wedge D}=g_D+g_{B_\alpha}\leqslant g_D+f-g_D=f$. Since $v(T)<\infty$, this process terminates with some countable ordinal α_0 and $B=B_{\alpha_0}$ will do. ∎

Recall that the problem is to construct a σ subalgebra $\mathscr{B}_1$ containing $\mathscr{B}$ in such a fashion that $\mathscr{B}_1$ is isomorphic to $Q_\beta\times Q_\omega$ in such a way that it extends the isomorphism h of $\mathscr{B}$ onto Q_β. The construction of $\mathscr{B}_1$ is accomplished below.

Let Δ denote all finite sequences of 0's and 1's. If $A\in\mathscr{A}$ and $B=-A$, then for each $\delta\in\Delta(\delta=(\varepsilon_1,\ldots,\varepsilon_n))$, there are A_δ, B_δ in $\mathscr{A}$ such that

(1) $A=A_0\vee A_1$ and $A_0\wedge A_1=0$,

(2) $B=B_0\vee B_1$ and $B_0\wedge B_1=0$,

(3) $A_\delta=A_{\delta,0}\vee A_{\delta,1}$ and $A_{\delta,0}\wedge A_{\delta,1}=0$,

(4) $B_\delta=B_{\delta,0}\vee B_{\delta,1}$ and $B_{\delta,0}\wedge B_{\delta,1}=0$,

(5) $$g_{A_\delta}(t)=\min\left(g_A(t),\frac{\varepsilon_1}{2}+\cdots+\frac{\varepsilon_n}{2^n}+\frac{1}{2^n}\right)-\min\left(g_A(t),\frac{\varepsilon_1}{2}+\cdots+\frac{\varepsilon_n}{2^n}\right),$$

(6) $$g_{B_\delta}(t)=\max\left(g_B(t),\frac{\varepsilon_1}{2}+\cdots+\frac{\varepsilon_n}{2^n}+\frac{1}{2^n}\right)-\max\left(g_B(t),\frac{\varepsilon_1}{2}+\cdots+\frac{\varepsilon_n}{2^n}\right),$$

In the above $A_0=A_{\{0\}}$, similarly for A_1,B_0,B_1 and $\delta,0=\{\varepsilon_1,\ldots,\varepsilon_n,0\}$.

Now $g_A\wedge\frac{1}{2}\leqslant g_A$ and by lemma 7 there is an $A_0\in\mathscr{A}$ such that $A_0\leqslant A$ and $g_{A_0}=g_A\wedge\frac{1}{2}$. Let $A_1=-A_0\wedge A$. Then $g_{A_1}=g_A-g_{A_0}=(g_A\wedge 1)-(g_A\wedge\frac{1}{2})$. Clearly B_0,B_1 can be obtained similarly and (1) and (2) are satisfied. The construction then proceeds as above by induction since $g_A\wedge\frac{1}{2}\leqslant g_A\wedge\frac{1}{2}=g_{A_0}$ and lemma 7 can be applied again.

Thus if $A\in\mathscr{A}$ and $A\notin\mathscr{B}$ and $\mathscr{B}_1$ is the σ algebra generated by $\{C_\delta\}_{\delta\in\Delta}$ and $\mathscr{B}$ where $C_\delta=A_\delta\vee B_\delta$ for all $\delta\in\Delta$, then we show that $A\in\mathscr{B}_1$ and $\mathscr{B}_1$ is isomorphic to $Q_\beta\times Q_\omega$. Let $\Delta_n=\{\delta\in\Delta:\delta$ has n coordinates$\}$ and for $\delta=(\varepsilon_1,\ldots,\varepsilon_n)$ let $a_\delta=\frac{\varepsilon_1}{2}+\cdots+\frac{\varepsilon_n}{2^n}$. For each $\delta\in\Delta_n$

let $U_\delta=\{t\colon g_A(t)\geqslant\alpha_\delta+1/2^n\}$ and $D_\delta=h^{-1}(U_\delta)$. Then each D_δ is in $\mathscr{B}$ and for $E_n=\bigvee_{\delta\in\Delta_n}(C_\delta\wedge D_\delta)$ we have

(a) $E_N\leqslant E_{n+1}$,

(b) $E_n\leqslant A$,

(c) $\mu(A\wedge(-E_n))\leqslant 1/2^n$.

From this $A=\bigvee_n E_n$ and $A\in\mathscr{B}_1$. The proofs of (a), (b) and (c) are as follows.

(a) For each $\delta\in\Delta$, $U_\delta=U_{\delta,1}$ so $U_\delta=U_{\delta,1}\subset U_{\delta,0}$. Now $C_\delta\wedge D_\delta=[C_{\delta,0}\vee C_{\delta,1}]\wedge D_{\delta,1}=(C_{\delta,0}\wedge D_{\delta,0})\vee(C_{\delta,1}\wedge D_{\delta,1})\leqslant(C_{\delta,0}\wedge D_{\delta,0})\vee(C_{\delta,1}\wedge D_{\delta,1})$. Thus $\bigvee_{\delta\in\Delta_n}(C_\delta\wedge D_\delta)\leqslant\bigvee_{\delta\in\Delta_n}[(C_{\delta,0}\wedge D_{\delta,0})\vee(C_{\delta,1}\wedge D_{\delta,1})]=E_{n+1}$.

(b) For $\delta\in\Delta$, $g_B=0$ on U_δ and thus $g_{B_\delta\wedge D_\delta}=g_{B_\delta}\cdot f_{D_\delta}=0$. Hence $B_\delta\wedge D_\delta=0$ and $C_\delta\wedge D_\delta=(A_\delta\vee B_\delta)\wedge D_\delta=A_\delta\wedge D_\delta\leqslant A_\delta$ and $E_n\leqslant A$.

Hence $B_\delta\wedge D_\delta=0$ and $C_\delta\wedge D_\delta=(A_\delta\vee B_\delta)\wedge D_\delta=A_\delta\wedge D_\delta\leqslant A_\delta$ and $E_n\leqslant A$.

(c) Now $-E_n\wedge A=\bigvee_{\delta\in\Delta_n}A_\delta\wedge\Big[-\bigvee_{\delta\in\Delta_n}(A_\delta\wedge D_\delta)\Big]=\bigvee_{\delta\in\Delta_n}(-D_\delta\wedge A_\delta)$.

Therefore $\mu(-E_n\wedge A)=\sum_{\delta\in\Delta_n}\mu(-D_\delta\wedge A_\delta)$.

For each $\delta\in\Delta_n$ let $T_\delta=\{t\colon a_\delta<g_A(t)<a_\delta+1/2^n\}$. Then $\mu(-D_\delta\wedge A_\delta)=\int_{T\setminus U_\delta}g_{A_\delta}\,dv=\int_{T\setminus U_\delta}(g_A\wedge a_\delta+1/2^n)-(g_A\wedge a_\delta)\;dv=\int_{T_\delta}[(g_A\wedge a_\delta+1/2^n)-(g_A\wedge a_\delta)]\,dv\leqslant v(T_\delta)1/2^n$.

Since $\delta\neq\delta'$ implies $T_\delta\cap T_\delta=\emptyset$, $\mu(-E_n\wedge A)=\sum_{\delta\in\Delta_n}\mu(-D_\delta\wedge A_\delta)\leqslant 1/2^n$.

Finally $\mathscr{B}_1$ is isomorphic to $B\times Q_\omega$ in such a way that the isomorphism extends h. To see this let $I_n=\bigvee\{C_\delta\colon\delta\in\Delta_n$ and $\varepsilon_n=0$ where $\delta=(\varepsilon_1,\dots,\varepsilon_n)\}$. Then for each $B\in\mathscr{B}$ and $I_{n_1},\dots,I_{n_k}$, $\mu(B\wedge I_{n_1}\wedge\cdots\wedge I_{n_k})=\mu(B_1)\mu(I_{n_1})\cdots\mu(I_{n_k})$ and the smallest σ-algebra containing the I_n's also contains the C_δ's. For if $F_0^n=I_n$ and $F_1^n=-I_n$, then $C_\delta=F_{\varepsilon_1}^N\vee\cdots\vee F_{\varepsilon_n}^n$ where $\delta=(\varepsilon_1,\dots,\varepsilon_n)$. It follows that $\mathscr{B}_1$ is isomorphic to $B\times Q_\omega$ in such a way that the isomorphism extends the identity of $\mathscr{B}$ onto itself. ∎

One immediate consequence of this theorem is the isometric classification of $L_p(T,\Sigma,\mu,\mathbb{C})$ and $L_p(T,\Sigma,\mu,\mathbb{R})$ when μ is a finite measure. The details are left to the reader. We state the theorem for the complex case only.

Theorem 9. *Let (T,Σ,μ) be a finite measure space. The following statements are valid for $1\leqslant p<\infty$.*

(1) *If μ is purely nonatomic, then there is a countable set $\{m_\beta\colon\beta<\alpha\}$ of distinct cardinals $(m_\beta\geqslant\aleph_0)$ such that $L_p(T,\Sigma,\mu,\mathbb{C})$ is linearly isometric and order isomorphic to $\Big[\oplus\sum_{\beta<\alpha}L_p([0,1]^{m_\beta},\mathbb{C})\Big]_p$.*

(2) *If μ is purely atomic, then $L_p(T,\Sigma,\mu,\mathbb{C})$ is linearly isometric and order isomorphic to $l_p(\Gamma,\mathbb{C})$ where $\operatorname{card}\Gamma\leqslant\aleph_0$.*

(3) *If μ has atoms, but is not purely atomic, then $L_p(T,\Sigma,\mu,\mathbb{C})$ is linearly isometric and order isomorphic to $\left(l_p(\Gamma,\mathbb{C})\oplus\left[\oplus\sum_{\beta<\alpha} L_p([0,1]^{m_\beta},\mathbb{C})\right]_p\right)_p$ for some set Γ with $\operatorname{card}\Gamma\leqslant\aleph_0$ and some countable set of distinct cardinals $m_\beta\geqslant\aleph_0$.* ∎

Corollary. *If $L_p(T,\Sigma,\mu,\mathbb{C})$ is separable, then $L_p(T,\Sigma,\mu,\mathbb{C})$ is linearly isometric and order isomorphic to exactly one of the following three spaces:*

(1) $l_p(\Gamma,\mathbb{C})$ *where* $\operatorname{card}\Gamma\leqslant\aleph_0$,

(2) $(l_p(\Gamma,\mathbb{C})\oplus L_p([0,1],\mathbb{C}))_p$ *where* $\operatorname{card}\Gamma\leqslant\aleph_0$,

(3) $L_p([0,1],\mathbb{C})$. ∎

The following theorem is also easily established.

Theorem 10. *Let $m\geqslant\aleph_0$. Then $L_p([0,1]^m,\mathbb{C})$ is linearly isometric and lattice isomorphic to $\left[\oplus\sum_{k=1}^{\infty} L_p([0,1]^{m_k},\mathbb{C})\right]_p$ where $m_k=m$ for all k and $1\leqslant p<\infty$.* ∎

We shall need the following technical result in the discussion of the isomorphic classification of $L_p(T,\Sigma,\mu,\mathbb{C})$.

Theorem 11. *Let (T_i,Σ_i,μ_i) be measure spaces where μ_i is positive and normalized for $i=1,2$. Then $L_p(T_1,\Sigma_1,\mu_1,\mathbb{C})$ is linearly isometric and order isomorphic to a sublattice X of $Y=L_p(T_1\times T_2,\Sigma_1\times\Sigma_2,\mu_1\times\mu_2,\mathbb{C})$ and there is a contractive projection of Y onto X.*

Proof. For each $f\in L_p(T_1,\Sigma_1,\mu_1,\mathbb{C})$ let $\hat{f}\in Y$ be defined by $\hat{f}(t_1,t_2)=f(t_1)$ for all $(t_1,t_2)\in T_1\times T_2$. Then $\|\hat{f}\|=[\int\int|\hat{f}(t_1,t_2)|^p d\mu_2\, d\mu_1]^{1/p}=[\int|f(t_1)|^p d\mu_1]^{1/p}=\|f\|$. Moreover, the correspondence $f\to\hat{f}$ is clearly an order preserving linear isometry onto $X=\{f\in Y: f(t_1,\#)$, is constant on T_2 for all $t_1\in T_1\}$. If P is defined by $(Pf)(t_1,t_2)=\int_{T_2} f(t_1)\,d\mu_2$ for all $f\in Y$, then it is easy to see that P is a contractive projection of Y onto X. ∎

We shall also need the following theorem. The proof for the separable case can be found in [96].

Theorem 12. *Let $m\geqslant\aleph_0$. Then $L_p([0,1]^m,\mathbb{C})$ contains a subspace linearly isomorphic to $l_2(\Gamma,\mathbb{C})$ where $\operatorname{card}\Gamma=m$ and $1\leqslant p<\infty$.*

Proof. We shall assume the theorem is true for $m=\aleph_0$. Consider the product space $\{-1,1\}^m$ where, as before, we take the measure on $\{-1,1\}$ which assigns to each point the value 1/2 and on the product space we take the corresponding product measure. By theorem 8 (see theorem 9) $L_p([0,1]^m,\mathbb{C})$ is linearly isometric and lattice isomorphic to $L_p(\{-1,1\}^m,\mathbb{C})$. Let π denote the collection of coordinate projections from $\{-1,1\}^m$ to $\{-1,1\}$ and let H be the closed linear span in $L_p(\{-1,1\}^m,\mathbb{C})$ of π.

From the separable case (see [96]) there are fixed positive constants c_p and C_p such that for each countable set $\{r_n\}$ of coordinates, $c_p\|\sum a_n f_{r_n}\| \leqslant \|\{a_n\}\|_2 \leqslant C_p\|\sum a_n f_{r_n}\|$, where $\{a_n\}\in l_2(\mathbb{N},\mathbb{C})$ and f_{r_n} is the r_n^{th} coordinate projection. Thus we clearly have an isomorphism of $l_2(\Gamma,\mathbb{C})$ onto H (where $\operatorname{card}\Gamma=m$). ∎

Finally, we need the following result from operator theory due to Pitt [240]. The proof given here is due to Rosenthal [247].

Theorem 13. *Let Γ and Δ be nonempty index sets and suppose $1\leqslant p<q<\infty$. Then every bounded linear operator from a subspace of $l_q(\Gamma,\mathbb{C})$ to $l_p(\Delta,\mathbb{C})$ is compact.*

Proof. If $p=1$, the conclusion follows from the well known fact that weak and norm sequential convergence coincide in $l_1(\Delta,\mathbb{C})$ (see theorem 4 of section 18) and the fact that every bounded linear operator from $l_q(\Gamma,\mathbb{C})$ to $l_1(\Delta,\mathbb{C})$ is weakly compact since $l_q(\Gamma,\mathbb{C})$ is reflexive ($q>1$).

Suppose $1<p<q<\infty$. If Y is a subspace of $l_q(\Gamma,\mathbb{C})$ and $A: Y\to l_p(\Delta,\mathbb{C})$ is a bounded linear operator which is not compact, then there is a sequence $\{x_n\}$ in $l_q(\Gamma,\mathbb{C})$ which converges weakly to 0 and $\{A(x_n)\}$ does not converge to 0 in norm. Thus by passing to a subsequence if necessary we may assume that $\inf\|x_n\|>0$ and $\inf\|A(x_n)\|>0$.

By theorems 5 and 7 of section 12 there is a subsequence $\{y_k\}$ of $\{x_n\}$ such that $\{y_k\}$ is a basic sequence equivalent to the natural unit vector basis in $l_q(\Gamma,\mathbb{C})$ and $\{A(y_k)\}$ is a basic sequence equivalent to the natural unit vector basis in $l_p(\Delta,\mathbb{C})$ (clearly we may assume that $\Gamma=\Delta=\mathbb{N}$).

Thus there are $K,\delta>0$ such that for all scalars $a_1,\dots,a_k$,

$$\delta\left(\sum_{i=1}^{k}|a_i|^q\right)^{1/q}\leqslant\left\|\sum_{i=1}^{k}a_i y_i\right\|\leqslant K\left(\sum_{i=1}^{k}|a_i|^q\right)^{1/q}$$

and

$$\delta\left(\sum_{i=1}^{k}|a_i|^p\right)^{1/p}\leqslant\left\|\sum_{i=1}^{k}a_i A y_i\right\|\leqslant K\left(\sum_{i=1}^{k}|a_i|^p\right)^{1/p}.$$

Hence, $\left\|\sum_{i=1}^{k}y_i\right\|\leqslant K k^{1/q}$ and $\left\|\sum_{i=1}^{k}A y_i\right\|\geqslant\delta k^{1/p}$. Therefore $\delta k^{1/p}\leqslant\|A\|K k^{1/q}$ or $k^{(1/p)-(1/q)}\leqslant\dfrac{\|A\|K}{\delta}$ which is a contradiction since $k^{(1/p)-(1/q)}\to\infty$ as $k\to\infty$. ∎

A cardinal $m\geqslant\aleph_0$ is said to be of *type I* if $m=\sum_{k=1}^{\infty}m_k$ implies $m=m_k$ for some k. Otherwise it is said to be of *type II*.

The following isomorphic classification of $L_p(T,\Sigma,\mu,\mathbb{C})$ when μ is a finite measure is an unpublished result due to Lindenstrauss (presented at [183]).

Theorem 14 (Lindenstrauss). *Let (T,Σ,μ) be a finite purely nonatomic measure space and suppose $\{m_\beta:\beta<\alpha\}$ is the set of cardinal numbers of the homogeneous parts of the measure algebra with respect to (T,Σ,μ) and put $m=\sup\{m_\beta:\beta<\alpha\}$. Then*

(1) *If m is of type I, $L_p(T,\Sigma,\mu,\mathbb{C})$ is a linearly isomorphic to $L_p([0,1]^m,\mathbb{C})$,*

(2) *If m is of type II, then $L_p([0,1],\mathbb{C})$ is linearly isomorphic to $\left(\oplus\sum_{\beta<\alpha} L_p([0,1]^{m_\beta},\mathbb{C})\right)_p$.*

Moreover, the spaces in (1) *and* (2) *are distinct when α is infinite and $p\neq 2$.*

Proof. The proofs of (1) and (2) follow readily from theorems 9 and 10.

To see that (1) and (2) are distinct, let P_n be the natural projection of $X=\left(\oplus\sum_{k=1}^{\infty} L_p([0,1]^{m_k},\mathbb{C})\right)_p$ and let Y be a subspace of X with $\dim Y=m$ where $m=\sum_{k=1}^{\infty} m_k$ and each $m_k<m$.

Choose $x_1\in Y$ with $\|x_1\|=1$ and choose n_1 so that $\|x_1-P_{n_1}x_1\|<1/2$. Suppose $x_1,\dots,x_k$ and $n_1,\dots,n_k$ have been chosen to be linearly independent with $x_i\in Y$, $\|x_i\|=1$ and $\|x_i-P_{n_i}(x)\|\leqslant 1/2^i$ for $i=1,\dots,k$. Since $\dim P_{n_k}(Y)\leqslant\sum_{i=1}^{k} m_i<m$, $P_{n_k}|Y$ is not an isomorphism. Thus there is an $x_{k+1}\in Y$ linearly independent from $x_1,\dots,x_k$ such that $\|x_{k+1}\|=1$ and $\|P_{n_k}(x_{k+1})\|\leqslant 1/2^{k+2}$. Choose n_{k+1} such that $\|P_{n_{k+1}}(x_{k+1})-x_{k+1}\|\leqslant 1/2^{k+1}$. The sequence x_n is equivalent to the unit vector basis in $l_p(\mathbb{N},\mathbb{C})$. In particular, by theorem 13, Y is not isomorphic to $l_2(\Gamma,\mathbb{C})$ where $\operatorname{card}\Gamma=m$. But, by theorem 12, $L_p([0,1]^m,\mathbb{C})$ contains such a space. Thus (1) and (2) are not isomorphic. ∎

We close with some comments on embeddings into $L_p([0,1],\mathbb{R})$. In [28] Banach presented the result that $l_q(\mathbb{N},\mathbb{R})$ cannot be linearly isomorphically embedded into $L_p([0,1],\mathbb{R})$ for $2<p<q<\infty$ and $1\leqslant q<p<2$. Paley [299] proved the same for $p>2>q$, $q>2>p$, and $2<q<p$. On the other hand, Kadec [144] showed that $l_p(\mathbb{N},\mathbb{R})$ can be isomorphically embedded into $L_p([0,1],\mathbb{R})$ for $1\leqslant p\leqslant q\leqslant 2$ and in [188] Lindenstrauss and Pełczyński went onto to prove that $L_q([0,1],\mathbb{R})$ can, in fact, be isomorphically embedded into $L_p([0,1],\mathbb{R})$ for $1\leqslant p\leqslant q\leqslant 2$. For the case $p=\infty$, Rosenthal [245] has shown that $L_\infty(\mu,\mathbb{R})$ is isomorphic to $L_\infty([0,1]^m,\mathbb{R})$ where $m=\dim L_1(\mu,\mathbb{R})$ where μ is a finite measure. He also proved that $\dim L_\infty([0,1]^m,\mathbb{R})=m^{\aleph_0}$ (here m is an infinite cardinal). In particular, there is no distinction between type I and type II cardinals in the case $p=\infty$.

§ 15. Abstract L_p Spaces

We define an abstract L_p space and prove that a real or complex abstract L_p space is linearly isometric and lattice isomorphic to $L_p(\mu, \mathbb{R})$ or $L_p(\mu, \mathbb{C})$ respectively for some measure μ.

If X is a real or complex Banach lattice, we say that the norm on X is *p additive* (for $1 \leqslant p \leqslant \infty$) if whenever $x \wedge y = 0$, we have that $\|x+y\|^p = \|x\|^p + \|y\|^p$ in case $1 \leqslant p < \infty$ and $\|x+y\| = \max(\|x\|, \|y\|)$ in case $p = \infty$.

When X is a real Banach Lattice and $p=1$ this is the condition of an abstract L_1 space and when $p=\infty$ it is that of an abstract M space (see section 3).

Definition 1. A (real or complex) Banach lattice is said to be an *abstract L_p space* $(1 \leqslant p < \infty)$ if the norm is p additive on X.

We need to know that an abstract L_p space is order complete. To accomplish this we prove that for the real case, we have the following additional property: if $x, y \geqslant 0$, then $\|x+y\|^p \geqslant \|x\|^p + \|y\|^p$. Recall that we used this property in section 3 to observe that, in fact, the norm is order continuous in $L_p(\mu, \mathbb{R})$. The same argument will work here when we have established the above inequality. This is analogous to the equality proved for abstract M spaces (see lemmas 2 and 3 of section 3) and this inequality is also due to Bernau [40].

For the following lemmas and theorem we suppose that X is a real abstract L_p space and x, y are fixed elements of X. We shall use the fact that orderings on a vector space with respect to maximal cones (with respect to set theoretic inclusion) are total orderings.

Lemma 1. *The positive cone X^+ of X is the intersection of all the maximal cones containing it.*

Proof. If $x \notin X^+$, let C be a maximal cone containing $\{-x\} \cup X^+$. Then $x \notin C$. ∎

Lemma 2. *Let $u = x \vee y$ and $v = x \wedge y$. If $-1 < t_1 < \cdots < t_n \leqslant 0$ and $\delta = \min\{t_2 - t_1, t_3 - t_2, \ldots, t_n - t_{n-1}, t_1 + 1 - t_n\}$ and $k > 2/\delta$, then* $\sum_{i=1}^{n} \sum_{r=0}^{n} x \wedge k[((t_i + r + 1)u - nv)^+ \wedge (nv - (t_i + r)u)^+] \geqslant (n-1)x.$

Proof. Let $\{a_j\}$, $1 \leqslant j \leqslant n(n+1)$, denote the increasing sequence whose range is the set of all $t_i + r (1 \leqslant i \leqslant n, 0 \leqslant r \leqslant n)$. Then we want to show that

$$(*) \qquad \sum_{j=1}^{n(n+1)} x \wedge k[((a_i + 1)u - nv)^+ \wedge (nv - a_j u)^+] \geqslant (n-1)x.$$

By the preceeding lemma, we can assume that X is totally ordered.

Suppose $x \geq y$. Then $u = x$ and $v = y$. Let $\theta = \sup\{\alpha \in [0,n] : \alpha x \leq n y\}$ and let s be the largest integer so that $a_s \leq \theta$. Since $a_n = t_n \leq 0$ and $a_{n(n+1)} = t_n + n \leq n$, we have that $n \leq s \leq n(n+1)$. Moreover, $a_{j+1} - a_j \geq \delta$ implies that either $\{a_{s-n+2}, \ldots, a_s\} \subset (\theta - 1 + \delta/2,\ \theta - \delta/2)$ or $\{a_{s-n+1}, \ldots, a_{s-1}\} \subset (\theta - 1 + \delta/2, \theta - \delta/2)$.

Thus there is an integer r such that $n-1 \leq r \leq n(n+1)$ and $a_r < \theta - \delta/2 < \theta + \delta/2 < a_{r-n+2} + 1$. Hence for $r - n + 1 \leq j \leq r$, $k[((a_j+1)x - ny)^+ \wedge (n - a_j y)^+] \geq k(\delta/2)x > x$. It follows that $(n-1)$ terms of (*) are equal to x and hence (*) is valid.

Now suppose that $x < y$. Then $x = v$ and $y = u$. As in the above there are $(n-1)$ values of j such that $k[((a_j+1)y - nx)^+ \wedge (nx - a_j y)^+] > y > x$. ∎

Corollary. *For some i with* $1 \leq i \leq n$,

$$\left\| \sum_{r=0}^{n} x \wedge k[((a_i + r + 1)u - nv)^+ \wedge (nv - (a_i + r)u)^+ \right\| \geq (n(n-1)) \|x\| . \quad ∎$$

Lemma 3. *There is an* $a \in (-1,0)$ *and an integer* k *such that for* $z_r = ((a+r+1)u - nv)^+ \wedge (nv - (a+r)u)^+$, *we have* $\left\| \sum_{r=0}^{n} x \wedge k z_r \right\| \geq (n - 1/n)\|x\|$ *and* $\left\| \sum_{r=0}^{n} y \wedge k z_r \right\| \geq (n - 1/n)\|y\|$.

Proof. By lemma 2 and its corollary there are n distinct numbers $t_1, \ldots, t_n$ in $(-1,0)$ and integers $k_1, \ldots, k_n$ such that the corollary is valid for the pair (t_i, k_i), $i = 1, \ldots, n$. Applying the corollary again, there is a $t_i = a$ and an integer $k > \max\{k_1, \ldots, k_n\}$ such that the corollary is valid for y, a, k. The two inequalities follow.

Now fix a, k so that the inequalities in lemma 3 are valid and observe that $z_r \wedge z_s = 0$ for $r \neq s$. Let $w_r = ((a+r+1)x - nv)^+ \wedge (nv - (a+r)u)^+$ and $v_r = ((a+r+1)y - nv)^+ \wedge (nv - (a+r)u)^+$. Then $w_r \vee v_r = z_r$ and if $r < n$, $w_r \wedge v_r = 0$ and $w_r + v_r = w_r \vee v_r = z_r$. ∎

Lemma 4. *If* $0 \leq r \leq n$ *and* $m \geq 0$, *then* $m w_r \wedge u \geq m w_r \wedge x \geq m w_r \wedge (n + a/n)u$ *and* $m w_r \wedge ((a+r+1)/n)u \geq m w_r \wedge y \geq m w_r \wedge ((a + r/n))u$ *and similarly for* v_r *replacing* w_r *and* y *replacing* x.

Proof. As in lemma 2, we may assume that X is totally ordered. If $w_r \neq 0$, then $(a+r+1)x > nv > (a+r)u$.

If $v = x$, then $r = n$ and we are done. If $v = y$, then $u = x > (n + a/n)u$ and the inequalities are clear. ∎

Note that if α, β are real numbers with $\alpha \geq \beta \geq 0$, then $(\alpha - \beta)^p \geq (1 - (1/n)^{1/2})^p \alpha^p - n^{p/2} \beta^p$.

Lemma 5. *If the norm on X is p additive for some $1 \leqslant p < \infty$ and $k_1 > (k/1+a)$, then for $r=0, \dots, n$,*

$$\|n k_1 z_r \wedge x + n k_1 z_r \wedge y\|^p \geqslant \left(1 - \left(\frac{1}{n}\right)^{\frac{1}{2}}\right)^p [\|k z_r \wedge x\|^p + \|k z_r \wedge y\|^p] - 2 n^{p/2} \left\|k z_r \wedge \left(\frac{1}{n}\right) u\right\|^p.$$

Proof. We consider three cases.

Case 1. $0 < r < n$. Since $w_r \wedge v_r = 0$, we have $\|n k_1 z_r \wedge x + n k_1 z_r \wedge y\|^p = \|n k_1 w_r \wedge x + n k_1 w_r \wedge y\|^p + \|n k_1 v_r \wedge x + n k_1 v_r \wedge y\|^p$. Now, $w = n k_1 w_r \wedge x + n k_1 w r \wedge y$. By lemma 4, $w \geqslant n k_1 w_r \wedge (n + a/n) u + n k_1 w_r \wedge (r + a/n) u \geqslant ((n + a/n) + (r + a/n)) (n k_1 w_r \wedge u)$ and $\|w\|^p \geqslant [(n + a/n)^p + (r + a/n)^p] \|n k_1 w_r \wedge u\|^p$. By lemma 4 again, $(n + a/n)(n k_1 w_r \wedge u) \geqslant k w_r \wedge (n + a/n) u \geqslant k w_r \wedge x - (k w_r \wedge (n+1+a/n) u - k w_r \wedge (n + a/n) u) \geqslant k w_r \wedge x - k w_r \wedge (1/n) u$. Hence by the remark preceeding the lemma,

$$\begin{aligned}\left\|\left(\frac{n+a}{n}\right)(n k_1 w_r \wedge u)\right\|^p &\geqslant \left\|\left(k w_r \wedge x - k w_r \wedge \frac{1}{n} u\right)^+\right\|^p \\ &\geqslant \left(\|k w_r \wedge x\| - \left\|k w_r \wedge x \wedge \frac{1}{n} u\right\|\right)^p \\ &\geqslant \left(1 - \left(\frac{1}{n}\right)^{\frac{1}{2}}\right)^p \|k w_r \wedge x\|^p - n^{\frac{p}{2}} \left\|k w_r \wedge \frac{1}{n} u\right\|^p.\end{aligned}$$

Similarly, $\|(r + a/n)(n k_1 w_r \wedge u)\|^p \geqslant (1 - (1/n)^{1/2})^p \|k w_r \wedge y\|^p - n^{p/2} \|k w_r \wedge (1/n) u\|^p$ and adding gives

(1) $\|n k_1 w_r \wedge x + n k_1 w_r \wedge y\|^p \geqslant (1 - (1/n)^{1/2})^p [\|k w_r \wedge x\|^p + \|k w_r \wedge y\|^p] - 2 n^{p/2} \|k w_r \wedge (1/n) u\|^p$. By interchanging w_r and v_r and x and y we have

(2) $\|n k_1 v_r \wedge x + n k_1 v_r \wedge y\|^p \geqslant (1 - (1/n)^{1/2})^p [\|k v_r \wedge x\|^p + \|k v_r \wedge y\|^p] - 2 n^{p/2} \|k v_r \wedge (1/n) u\|^p$. Adding this with the last inequality gives the result.

Case 2. $r = 0$. The two inequalities above are still valid.

Case 3. $r = n$. We observe that $n k_1 z_n \wedge x \geqslant n k_1 z_n \wedge (n + a/n) u$ and $n k_1 z_n \wedge y \geqslant n k_1 z_n \wedge (n + a/n) u$. A proof similar to that of the proof of (1) above establishes the lemma. ∎

Theorem 1 (Bernau). *Let X be a Banach lattice in which the norm is p additive for some p with $1 \leqslant p < \infty$. Then $\|x + y\|^p \geqslant \|x\|^p + \|y\|^p$ for all $x, y \geqslant 0$ in X.*

Proof. By lemma 5 and the pairwise disjointness of the z_r's we have

$$\|x+y\|^p \geqslant \left\| \sum_{r=0}^{n} (n k_1 z_r \wedge x + n k_1 z_r \wedge y) \right\|^p = \sum_{n=0}^{n} \|n k_1 z_r \wedge x + n k_1 z_r \wedge y\|^p$$

$$\geqslant \left(1-\left(\frac{1}{n}\right)^{\frac{1}{2}}\right)^p \left[\sum_{r=0}^{n} (\|k z_r \wedge x\|^p + \|k z_r \wedge y\|^p) \right]$$

$$-2n^{\frac{p}{2}} \sum_{r=0}^{n} \left\| k z_r \wedge \frac{1}{n} u \right\|^p = \left(1-\left(\frac{1}{n}\right)^{\frac{1}{2}}\right)^p \left[\left\| \sum_{r=0}^{n} k z_r \wedge x \right\|^p \right.$$

$$\left. + \left\| \sum_{r=0}^{n} k z_r \wedge y \right\|^p \right] - 2n^{\frac{p}{2}} \left\| \sum_{r=0}^{n} k z_r \wedge \frac{1}{n} u \right\|^p .$$

Hence by choice of z_r and k,

$$\|x+y\|^p \geqslant \left(1-\left(\frac{1}{n}\right)^{\frac{1}{2}}\right)^p \left[\left\| \left(\frac{n-1}{n}\right) x \right\|^p + \left\| \left(\frac{n-1}{n}\right) y \right\|^p \right] - 2n^{\frac{p}{2}} \left\| \frac{1}{n} u \right\|^p$$

$$= \left(1-\left(\frac{1}{n}\right)^{\frac{1}{2}}\right)^p \left(1-\frac{1}{n}\right)^p (\|x\|^p + \|y\|^p) - 2n^{-\frac{p}{2}} \|u\|^p .$$

As $n \to \infty$, we obtain the result. ∎

We also give a theorem due to Ando which gives an alternate approach to the characterization of abstract L_p spaces.

Theorem 2 (Ando). *Let X be a real Banach lattice in which the norm is p additive for $1<p<\infty$. Then the norm in X^* is q additive where $1/p+1/q=1$.*

Proof. Let x^*, y^* be in X^* with $x^* \wedge y^* = 0$ and let $\varepsilon > 0$ be given. There is an $x \geqslant 0$ in X such that $\|x\| = 1$ and $\|x^* + y^*\| \leqslant (x^* + y^*)(x) + \varepsilon$. Since $0 = (x^* \wedge y^*)(x) = \inf\{x^*(y) + y^*(x-y) : 0 \leqslant y \leqslant x\}$, there is a $0 \leqslant y \leqslant x$ with $x^*(y) + y^*(x-y) < \varepsilon$. Thus $(x^*+y^*)(x) \leqslant x^*(x-y) + y^*(y) + \varepsilon \leqslant x^*((x-y)-(x-y)\wedge y) + y^*(y-(x-y)\wedge y) + 3\varepsilon \leqslant \|x^*\| \, \|(x-y) - (x-y)\wedge y\| + \|y^*\| \, \|y-(x-y)\wedge y\| + 3\varepsilon \leqslant (\|x^*\|^q + \|y^*\|^q)^{1/q} [\|(x-y) - (x-y)\wedge y\|^p + \|y-(x-y)\wedge y\|^p]^{1/p} + 3\varepsilon$. On the other hand, $[(x-y) - (x-y)\wedge[y-(x-y)\wedge y] = 0$. Hence, by assumption, $\|(x-y)-(x-y) \wedge y\|^p + \|y-(x-y)\wedge y\|^p = \|(x-y)-(x-y)\wedge y + y - (x-y)\wedge y\|^p \leqslant \|x\|^p = 1$. Thus, $\|x^* + y^*\|^q \leqslant \|x^*\|^q + \|y\|^q$.

Conversely, let $0 \leqslant x, y \in X$ be such that $\|x\| = \|y\| = 1$ and $\|x^*\|^q + \|y^*\|^q \leqslant x^*(x)^q + y^*(y)^q + \varepsilon$. Put $z = x^*(x)^{q-1} x$ and $w = y^*(y)^{q-1} y$. Then $\|x^*\|^q + \|y^*\|^q \leqslant x^*(z) + y^*(w) + \varepsilon$. Since $x^* \wedge y^* = 0$ as above, there are $0 \leqslant z_0 \leqslant z$ and $0 \leqslant w_0 \leqslant w$ such that $x^*(z-z_0) + y^*(z_0) < \varepsilon$ and $y^*(w-w_0) + x^*(w_0) < \varepsilon$. Therefore $x^*(z) + y^*(w) \leqslant x^*(z_0 - z_0 \wedge w_0) + y^*(w - z_0 \wedge w_0) + x^*(z-z_0) + y^*(z_0) + y^*(w-w_0) + x^*(w_0) \leqslant (x^*+y^*)(z_0 - z_0 \wedge w_0 + w_0 - z_0 \wedge w_0) + 2\varepsilon$.

Since $(z_0 - z_0 \wedge w_0) \wedge (w_0 - w_0 \wedge z_0) = 0$, we have $\|(z_0 - z_0 \wedge w_0) + (w_0 - w_0 \wedge z_0)\|^p = \|z_0 - z_0 \wedge w_0\|^p + \|w_0 - w_0 \wedge z_0\|^p \leqslant \|z\|^p + \|w\|^p = |x^*(x)|^{(q-1)p} + |y^*(y)|^{(q-1)p} \leqslant \|x^*\|^q + \|y^*\|^q$. Thus we have $\|x^*\|^q + \|y^*\|^q \leqslant x^*(z) + y^*(w) + \varepsilon \leqslant \|x^* + y^*\| (\|x^*\|^q + \|y^*\|^q)^{1/p} + 3\varepsilon$. By letting $\varepsilon \to 0$, we are done. ∎

We now wish to prove that a complex abstract L_p space is linearly isometric and lattice isomorphic to $L_p(\mu, \mathbb{C})$ for some measure μ. This result, which we shall call *Bohnenblust's theorem*, has been the subject of some confusion over the years. In [48] Bohnenblust proved the theorem for a real separable σ-complete abstract L_p space. For $p=1$ Kakutani proved the theorem in [147]. Although he stated and used the more restrictive condition that the norm is additive over the positive cone, the result follows from this and his duality theory for abstract M and abstract L_1 spaces as developed in [148] (see theorem 6 of section 3). In [217] Nakano proved Bohnenblust's theorem for a σ-complete Banach lattice (i.e., he removed the separability condition used by Bohnenblust). In [12] Ando stated and used the theorem in its most general form. His (unpublished) solution follows from theorem 2 above which he graciously supplied to the author. We leave it to the reader to deduce how theorem 2 implies Bohnenblust's theorem (see exercise 6). In [188] Bohnenblust's theorem is also used. However, the Banach lattice constructed in [188] is clearly order complete and the authors correctly appeal to Nakano's solution as adequate in their case. Various authors have proposed other axioms on the norm to obtain the result. In [116] Gordon proved Bohnenblust's theorem under the additional assumption that the inequality of theorem 1 above is valid. In a paper which contains many important and interesting results, Bretagnolle, Dacunha-Castelle, and Krivine [53] assumed the norm satisfied $\|x+y\|^p \geqslant \|x\|^p + \|y\|^p \geqslant \|x \vee y\|^p$ whenever $x, y \geqslant 0$, and Ng essentially duplicated this assumption in [298]. In [201] Marti proved it when X is weakly complete.

Theorem 3 (Bohnenblust, Nakano). *Let Y be a complex abstract L_p space. Then Y is linearly isometric and lattice isomorphic to $L_p(\mu, \mathbb{C})$ for some measure μ.*

Proof. Put $X = \mathrm{Re}(Y)$. We first assume that Y has a *weak order unit* u. That is, $u \geqslant 0$, $\|u\| = 1$ and if $|y| \wedge u = 0$, then $y = 0$.

The idea is to create a Boolean algebra which acts like the Boolean algebra of characteristic functions in an $L_p(\mu, \mathbb{C})$ space. This is accomplished as follows. Let $\mathscr{A} = \{x \in X : x \wedge (u - x) = 0\}$. It is easily verified that $\mathscr{A}$ is a Boolean algebra under the lattice supremum and infimum of X and complementation given by $u - x$. In fact, since X is order complete, so is $\mathscr{A}$. For, let $\{x_t\}$ be a collection of elements in $\mathscr{A}$. Then $x=\sup_t x_t$ exists in X. It clearly sufficies to show that x is in $\mathscr{A}$. Now $x\wedge(u-x)=2x\wedge u-x=\sup_t\{2x_t\wedge u\}-x=\sup_t\{x_t\}-x=x-x=0$. Moreover, by assumption the function $\| \|^p$ is additive on $\mathscr{A}$ and $\|u\|^p=1$. Hence there is a (unique) totally disconnected Hausdorff space T and a regular Borel measure μ on T such that $\mu(\phi(x))=\|x\|^p$ where ϕ is an isomorphism from $\mathscr{A}$ onto the Boolean algebra of closed and open sets in T. This gives us a unique linear isometry $L\colon L_p(\mu,\mathbb{C})\to Y$ which is given by $L\left(\sum_{i=1}^n a_i f_{A_i}\right)=\sum_{i=1}^n a_i\phi^{-1}(A_i)$ for all continuous simple functions on T. By theorem 8 of section 1 the range of L contains all $x\in X$ such that $|x|\leqslant nu$ for some integer n. If $y\in X$ and $y\geqslant 0$, $y=\sup y\wedge nu$. If $m>n$, we have $\|y\wedge mu-y\wedge nu\|^p=\|L^{-1}(y\wedge mu)-L^{-1}(y\wedge nu)\|^p\leqslant\|L^{-1}(y\wedge mu)\|^p-\|L^{-1}(y\wedge nu)\|^p\to 0\ (m,n\to\infty)$. We conclude that $\|y-y\wedge nu\|\to 0\ (n\to\infty)$ and hence L is onto. Since L is clearly a lattice isomorphism we are done.

The general case is handled by decomposition and Zorn's lemma. There is a maximal family $\{u_i\}_{i\in I}$ of positive norm one elements in X such that $u_i\wedge u_j=0$ for $i\neq j$. Let $Y_i=\{u_i\}^{\perp\perp}$. Then u_i is a weak order unit in Y_i and each Y_i is an order complete complex abstract L_p space. Thus there is a linear isometry and lattice isomorphism L_i of $L_p(\mu_i,\mathbb{C})$ onto Y_i for some measure μ_i. One can easily verify that Y is linearly isometric and lattice isomorphic to $\left(\oplus\sum_{i\in I} Y_i\right)_p$ which, in turn, is equivalent to $\left(\oplus\sum_{i\in I} L_p(\mu_i,\mathbb{C})\right)_p$. ∎

A corollary which is worth stating is the following.

Corollary. *Any abstract L_p space X is linearly isometric to*

$$\left[l_p(\Gamma,\mathbb{C})\oplus\left(\oplus\sum_{\alpha A} L_p([0,1]^{m_\alpha},\mathbb{C})\right)_p\right]_p$$

for some index set Γ (possibly $\Gamma=\emptyset$ so that $l_p(\Gamma,\mathbb{C})=\{0\}$) and some set of cardinal numbers $m_\alpha\geqslant\aleph_0$ (possibly $A=\emptyset$ so that the second term above is $\{0\}$).

Proof. Note that in the proof of the theorem we decomposed X into a direct sum of spaces each representable as $L_p(\mu_i,\mathbb{C})$ for μ_i a finite measure. By theorem 9 of section 14, each $L_p(\mu_i,\mathbb{C})$ can be represented in terms of $l_p(\Gamma_i,\mathbb{C})$ and $[\oplus\sum L_p([0,1]^{m_j},\mathbb{C})]_p$. ∎

We now wish to show that the only Banach lattices with the property that $\|x+y\|$ is a function of $\|x\|$ and $\|y\|$ whenever $x\wedge y=0$ are abstract M and abstract L_p spaces. To accomplish this we need another technical result due to Bohnenblust [48].

Theorem 4 (Bohnenblust). *Let $f: R^+ \times R^+ \to R^+$ be a continuous function. Then there is a p with $0 < p \leqslant \infty$ such that $f(\xi,\eta) = (\xi^p + \eta^p)^{1/p}$ (if $p < \infty$) and $f(\xi,\eta) = \max(\xi,\eta)$ (if $p = \infty$) if and only if the following conditions hold.*

(1) $f(\tau\xi, \tau\eta) = \tau f(\xi,\eta)$ *for all* $\tau, \xi, \eta \geqslant 0$
(2) $f(\xi,\eta) \leqslant f(\xi',\eta')$ *when* $0 \leqslant \xi \leqslant \xi'$ *and* $0 \leqslant \eta \leqslant \eta'$
(3) $f(\xi,\eta) = f(\eta,\xi)$ *for* $\xi,\eta \geqslant 0$
(4) $f(0,1) = 1$
(5) $f(\xi, f(\eta,\tau)) = f(f(\xi,\eta),\tau)$ *for all* $\xi, \eta, \tau \geqslant 0$.

Moreover, $f(t, 1-t) \leqslant 1$ when $0 \leqslant t \leqslant 1$ if and only if $p \geqslant 1$.

Proof. When $p = \infty$, $\max(\xi,\eta)$ replaces $(\xi^p + \eta^p)^{1/p}$ in the interpretation. Clearly if f is of this form, then conditions 1—5 hold.

Let $\{a_n\}$ be defined by $a_1 = 1$, $a_n = f(1, a_n - 1)$ for $n > 1$.

A. $a_{n+m} = f(a_n, a_m)$ for all n, m.

This is equivalent to $a_n = f(a_m, a_{n-m})$ for $n > 1$ and $m = 1, \ldots, n-1$. For $m = 1$ it is true and the general case follows by induction since $a_n = f(a_{m-1}, a_{n-m+1}) = f(a_{m-1}, f(1, a_{n-m})) = f(f(1, a_{m-1}), a_{n-m})$ $= f(a_m, a_{n-m})$.

B. $a_n a_m = a_{nm}$ for all n, m.

This is clear for $m = 1$ and if it is true for $m-1$, then $a_n a_m = a_n f(1, a_{m-1}) = f(a_n, a_{nm-n}) = a_{nm}$.

Moreover, $a_2 \geqslant 1$, and the a_n's are monotone nondecreasing.

Case 1. $a_2 = 1$. By (B) $a_{2^n} = 1$ and thus $a_n = 1$ for all n. Furthermore for $0 \leqslant \xi \leqslant 1$, $1 = f(1,0) \leqslant f(1,\xi) \leqslant f(1,1) = 1$ and $f(1,\xi) = 1$. Hence $f(\xi,\eta) = \max(\xi,\eta)$ and $p = \infty$.

Case 2. $a_n > 1$ for $n > 1$. In this case $a_n = n^{1/p}$ for some $0 < p < \infty$.

Let $m, n > 1$ and k determine h so that $m^h \leqslant n^k < m^{h+1}$. By (B), $a_m^h = a_{mn}$ and similarly for k and $h+1$. Hence $h \log a_m \leqslant k \log a_n \leqslant (h+1) \log a_m$. Thus $\dfrac{\log a_m}{\log m} = \dfrac{\log a_n}{\log n}$. If $\dfrac{1}{p} = \dfrac{a_2}{\log 2}$, then $a_n = n^{1/p}$.

C. $f(1, r^{1/p}) = (1+r)^{1/p}$ for any rational $r > 0$. Let $r = m/n$. Then $f(1, m^{1/p}/n^{1/p}) = n^{-1/p} f(n^{1/p}, m^{1/p}) = n^{-1/p} f(a_n, a_m) = n^{-1/p}(n+m)^{1/p}$ $= (1+r)^{1/p}$.

Thus the theorem is established. ∎

Theorem 5 (Bohnenblust). *Let X be a real Banach lattice of dimension $\geqslant 3$. The the following are equivalent.*

(1) *For each* $x = x_1 + x_2$, $y = y_1 + y_2$ *with* $|x_1| \wedge |x_2| = 0 = |y_1| \wedge |y_2|$ *in X, if* $\|x_1\| = \|y_1\|$ *and* $\|x_2\| = \|y_2\|$, *then* $\|x\| = \|y\|$.

(2) *There is a function* $g:\mathbb{R}^+\times\mathbb{R}^+\to\mathbb{R}^+$ *such that for each* x,y *in* X *with* $|x|\wedge|y|=0$, $g(\|x\|,\|y\|)=\|x+y\|$.

(3) X *is an abstract* M *or* L_p *space* $(1\leqslant p<\infty)$.

Proof. Clearly (3) implies (2) and (2) implies (1). Thus we need only show that (1) implies (3).

For any two positive elements x,y of norm 1 with $x\wedge y=0$ let $f(\xi,\eta)=\|\xi x+\eta y\|$. Condition (1) shows that f is independent of the choice of x and y. It remains to show that f satisfies the conditions of theorem 4.

Clearly $f(\tau\xi,t\eta)=\tau f(\xi,\eta)$ for all $\xi,\eta,\tau\geqslant 0$. Let $0\leqslant\xi\leqslant\xi'$ and $0\leqslant\eta\leqslant\eta'$. Then by the monotonicity of the norm, $f(\xi,\eta)\leqslant f(\xi',\eta')$.

Clearly $f(\xi,\eta)=f(\eta,\xi)$ by condition (1). Condition (1) clearly holds for any three mutually disjoint elements instead of two. Thus $f(\xi,f(\eta,\tau))=f(f(\xi,\eta),\tau)$ for all $\xi,\eta,\tau\geqslant 0$. Clearly $f(1,0)=1$.

Thus by theorem 2, $f(\xi,\eta)=(\xi^p+\eta^p)^{1/p}$ for some $0<p<\infty$ or $f(\xi,\eta)=\max(\xi,\eta)$. Since the norm is convex, $p\geqslant 1$. ∎

Exercises. 1. Show that if (T,Σ,μ) is a σ finite measure space, then $L_p(T,\Sigma,\mu,\mathbb{C})=L_p(T',\Sigma',\mu',\mathbb{C})$ for some finite measure space (T',Σ',μ').

2. Let $n\leqslant m$ be cardinal numbers. Show that $L_p([0,1]^n,\mathbb{C})$ is linearly isometric to a subspace of $L_p([0,1]^m,\mathbb{C})$. Prove that if $\aleph_0\leqslant n,m$ and $L_p([0,1]^n,\mathbb{C})$ is linearly isometric to a subspace of $L_p([0,1]^m,\mathbb{C})$, then $n\leqslant m$.

3. Let $\{m_\alpha\}$ be a collection of cardinal numbers. Show that $\left(\oplus\sum_\alpha L_p([0,1]^{m_\alpha},\mathbb{C})\right)_p=L_p(\mu,\mathbb{C})$ for some measure μ.

4. Let X be a real normed linear lattice for which the norm is p additive for some $1\leqslant p\leqslant\infty$. Show that the norm is also p additive in the completion of X.

5. Let Γ be an infinite set and $m=\operatorname{card}\Gamma$. Show that $l_p(\Gamma,\mathbb{C})$ is linearly isometric to a subspace of $L_p([0,1]^m,\mathbb{C})$.

6. Prove Bohnenblust's theorem using theorem 2 instead of theorem 1.

7. Prove that the norm in an abstract L_p space is order continuous by using theorem 1. Show that the operator defined in the proof of theorem 3 is indeed a linear isometry which preserves the lattice structures involved.

8. Prove that in any (purely nonatomic) measure algebra there is a separable σ-subalgebra containing the unit. Give the transfinite induction step in Maharam's theorem (theorem 8 of section 14).

9. Let Y be a complex vector space and suppose that X is a real linear subspace of Y such that X is a normed linear lattice and the norm is p additive on X for some $1\leqslant p<\infty$. Furthermore suppose that $Y=X\oplus iX$. Show that there is a norm and modulus on Y such that the completion of Y in this norm is a complex L_p space.

10. Let Y be a complex Banach lattice and put $X = \operatorname{Re}(Y)$. Suppose that there is a linear isometry and lattice isomorphism L of X into $L_p(\mu, \mathbb{R})$ for some measure μ. Show that $\hat{L}$ defined by $\hat{L}(x+iy) = L(x) + iL(y)$ has the property that $\hat{L}|x| = |\hat{L}x|$ for all x in Y and, thus, $\hat{L}$ is a linear isometry.

11. Let X be a real vector lattice. Show that if M is an ideal in X, then X/M is a vector lattice where $(x+M) \wedge (y+M) \equiv x \wedge y + M$. Let ρ be a seminorm on X such that $|y| \leqslant |x|$ implies that $\rho(y) \leqslant \rho(x)$. Show that if ρ is p additive, then so is the norm on X/M determined by ρ with respect to the above vector lattice structure and where M is the kernel of ρ.

Chapter 6

Characterizations of Abstract M and L_p Spaces

In section 16 we prove that the only real Banach lattices which have the property that every closed sublattice is the range of a positive contractive projection are the lattices $L_p(\mu, \mathbb{R})$ and $c_0(\Gamma, \mathbb{R})$ for some measure μ and some index set Γ. We also prove a similar joint characterization of abstract M and L_p spaces in terms of simultaneous linear extensions.

In section 17 we study contractive projections on $L_p(\mu, \mathbb{C})$ and show that a subspace of $L_p(\mu, \mathbb{C})$ is the range of a contractive projection (not necessarily positive) if and only if it is linearly isometric to $L_p(\nu, \mathbb{C})$ for some measure ν (the same is true for the real case also). We then use this to characterize $L_p(\mu, \mathbb{C})$ as a Banach space without assuming a priori a lattice structure.

Section 18 is devoted to a study of abstract L_1 spaces including characterizations of spaces $l_1(\Gamma, \mathbb{R})$ for Γ an arbitrary index set.

§ 16. Positive Contractive Projections in Abstract M and L_p Spaces

We shall demonstrate the existence (and uniqueness for $1 \leqslant p < \infty$) of positive contractive projections onto sublattices of real Banach lattices which have the property that the norm is both order continuous and p additive for some $1 \leqslant p \leqslant \infty$. In turn, these two conditions characterize the spaces $L_p(\mu, \mathbb{R})$ and $c_0(\Gamma, \mathbb{R})$. We also show that closed sublattices of Banach lattices in which the norm is p additive for some $1 \leqslant p \leqslant \infty$ admit positive simultaneous linear extensions and that this property gives a joint characterization of abstract M and L_p spaces.

All results in this section are for real spaces only.

To accomplish these joint characterizations we use theorem 5 of section 15.

We first give a characterization of $c_0(\Gamma, \mathbb{R})$ due to Bohnenblust [48] (who proved it for the separable case).

Theorem 1 (Bohnenblust). *Let X be an abstract M space. Then X is linearly isometric and order isomorphic to $c_0(\Gamma, \mathbb{R})$ for some index set Γ if and only if the norm is order continuous in X.*

Proof. It is first shown that for each $x>0$, there is an atom y with $0<y\leqslant x$. Suppose $x>0$ with $\|x\|=1$ is not an atom. Then there is a component y of x which is different from x and 0. Since $y\wedge(x-y)=0$ and X is an abstract M space, $\|y\|=1$ or $\|x-y\|=1$. Let $Y=\{y:0\leqslant y\leqslant x, y\neq 0, x, \text{ and } \|y\|=1\}$. If $\{y_\lambda\}$ is a maximal chain in Y directed downwards by the order of X, then $y=\inf y_\lambda$ exists and since the norm is order continuous, $\|y\|=1$. Suppose y is not an atom. Then there is a $z\neq 0$, y with $z\wedge(y-z)=0$ and $0\leqslant z\leqslant y$ with $\|z\|=\|y\|=1$. Thus $\{y_\lambda, z\}$ is a chain larger than the maximal chain $\{y_\lambda\}$. Hence y is an atom and every positive normalized x dominates an atom.

Let Γ denote the set of all positive normalized atoms in X. Clearly if x and y are distinct elements of Γ, then $x\wedge y=0$. Furthermore, $\Gamma^{\perp\perp}=X$ since, in fact, $\Gamma^\perp=0$. Moreover, $\Gamma^{\perp\perp}$ is the norm closure of the span of Γ since the norm is order continuous. The operator A defined by $Af=\sum_{x\in\Gamma} f(x)x$ for all $f\in c_0(\Gamma,\mathbb{R})$ with only finitely many nonzero values is clearly a linear isometry. Since the set of all such functions f is norm dense in $c_0(\Gamma,\mathbb{R})$ and the range of A is the span of Γ, A can be extended to a lattice preserving linear isometry of $c_0(\Gamma,\mathbb{R})$ onto X.

The proof of fact that the norm is order continuous in $c_0(\Gamma,\mathbb{R})$ is left to the reader. ∎

Definition 1. If X and Y are Banach spaces with $X\subset Y$, a contractive linear operator $L: X^*\to Y^*$ is said to be a *simultaneous linear extension* if $(Lx^*)(x)=x^*(x)$ for all $x^*\in X^*$ and $x\in X$. Clearly such an operator is a linear isometry.

We shall also need the following theorem due to Ando [12].

Theorem 2 (Ando). *Let X be an abstract M space and Y a closed vector sublattice of X, then Y^* admits a positive simultaneous linear extension into X^*.*

Proof. Let $Y^\perp$ be the annihilator of Y in X^*. Then Y^* is isometrically isomorphic to $X^*/Y^\perp$ and Y^{**} can be considered as the closed subspace $Y^{\perp\perp}$ of X^{**}.

Take $y^*\in Y^*$. Since Y^* is a vector lattice $y^*=(y^*)^+-(y^*)^-$. Extend $(y^*)^+$ to X dominated by the sublinear functional $p(x)=\|(y^*)^+\|\,\|x^+\|$. Call this extension x^* and let z^* be the corresponding extension of $(y^*)^-$. Then x^*-z^* extends y^* and in X^*, $|x^*-z^*|\leqslant x^*+z^*$. Hence $|x^*-z^*|y=|y^*|y$ for all y in Y.

Suppose $x^{**}\in Y^{\perp\perp}$, as an element of X^{**} we compute $|x^{**}|$ by the formula,

$$|x^{**}|(x^*)=\sup\{x^{**}(w^*):|w^*|\leqslant x^*\}$$

for $0 \leqslant x^* \in X^*$. Since x^{**} annihilates $Y^\perp$ our argument above shows that we need only take the supremum over $w^* \in X^*$ such that $|w^*| \mid Y = |w^* \mid Y|$. With the second absolute value computed for $w^* \mid Y$ as an element of Y^*. It follows that $|x^{**}|(x^*) = |x^{**}|(x^* \mid Y)$ where the second absolute value is computed for x^{**} as an element of Y^{**}. We conclude that Y^{**} is a closed vector sublattice of X^{**}.

By theorem 7 of section 3, Y^{**} is an abstract M space with a strong unit u defined by $u(x^* \mid Y) = \|(x^* \mid Y)^+\| - \|(x^* \mid Y)^-\|$.

By Hahn-Banach extension type arguments we extend the identity map on Y^{**} to a linear map Q of X^{**} onto Y^{**} such that $Q(x^*) \leqslant \|x^{**+}\| u$ $(x^{**} \in X^{**})$. Clearly Q is a positive contractive projection. Define $L: Y^* \to X^*$ by $L(x^*)x = (QJx)x^*$ $(x \in X, X^* \in Y^*)$ where J is the canonical embedding of X into X^{**}. If $x \in Y$, $Jx \in Y^{\perp\perp} = Y^{**}$ and $(Lx^*)x = (Jx)x^* = x^*(x)$. Thus L is a simultaneous linear extension of Y^* into X^*. Positivity of L follows from positivity of Q. ∎

We now come to the central theorem of this section (also due to Ando [12]). Its proof is carried out in the theorems and lemmas which occupy the rest of the section.

Theorem 3 (Ando). *Let X be a real Banach lattice of dimension at least* 3. *Then*

(1) *The norm is order continuous and p additive for some $1 \leqslant p \leqslant \infty$ if and only if each closed sublattice of X is the range of a positive contractive projection on X.*

(2) *The norm is p additive for some $1 \leqslant p \leqslant \infty$ if and only if each closed sublattice of X admits a positive simultaneous linear extension.*

Thus (1) is a joint characterization of $c_0(\Gamma, \mathbb{R})$ and $L_p(\mu, \mathbb{R})$ and (2) is a joint characterization of $L_p(\mu, \mathbb{R})$ and abstract M spaces.

We shall need the following lemma in the course of the proof.

Lemma 1. *Let X be a Banach lattice. If $\{x_n\}$ is a sequence of mutually disjoint positive elements of X, then for any $y \geqslant 0$, $\|y\| = \lim \|y - y \wedge x_n\|$.*

Proof. Let $\varepsilon > 0$ be given and let $K = \{n: \|y - y \wedge x_n\| \leqslant \|y\| - \varepsilon\}$. Suppose that $k_1, \ldots, k_j$ are distinct elements of K and observe that $(y - y \wedge x_{k_1}) + \cdots + (y - y \wedge x_{k_j}) = jy - \max\{y \wedge x_{k_r}: r = 1, \ldots, j\} \geqslant (j-1)y$. Hence $(j-1)\|y\| \leqslant \sum_{r=1}^{j} \|y - y \wedge x_{k_r}\| \leqslant j(\|y\| - \varepsilon)$. This yields $j\varepsilon \leqslant \|y\|$. Thus K is a finite set and we are done. ∎

One of the key steps in proving the above theorem is now given. This theorem is due to Ando [12]. (His proof is incorrect; in particular it requires that $\|x + y\| > \max\{\|x\|, \|y\|\}$ whenever x and y are positive disjoint elements, which is not the case in an abstract M space. However, the main idea in the proof is still that of [12].)

Theorem 4. *Let X be a Banach lattice with dimension $X \geqslant 3$. If every two dimensional sublattice of X is the range of a positive contractive projection on X, then the norm is p additive on X for some $1 \leqslant p \leqslant \infty$.*

It is first established that the norm is p additive in three dimensional sublattices of X by showing the norm of the sum of two disjoint elements is a function of their norms. The general proof then follows from this. The following observation is useful in the proof.

Let x, y be two positive nonzero elements in X with $\|x\| < 1$. Then there is a unique $\xi > 0$ such that $\|x + \xi y\| = 1$. For, let $f(\xi) = \|x + \xi y\|$ for all $\xi \geqslant 0$. Then $f(0) < 1$ and $\lim_{\xi \to \infty} f(\xi) = \infty$ so that there is a ξ with $f(\xi) = 1$. Now suppose $0 < \eta < \xi$ and $f(\eta) = f(\xi) = 1$. Let $x^* \in X^*$ with $\|x^*\| = 1$ and $x^*(x + \eta y) = \|x + \eta y\|$. Then $\|x + \eta y\| \geqslant x^*(x + \xi y)$ $= x^*(x + \eta y) + x^*(\xi - \eta) y = \|x + y\| + x^*(\xi - \eta) y$ and thus $x^*(y) \leqslant 0$. But $\|x\| \geqslant x^*(x) \geqslant x^*(x + \eta y) = \|x + \eta y\| = 1$ which is a contradiction since $\|x\| < 1$. Thus ξ is unique.

If x, y, z are positive nonzero elements in X with $\|x\| < 1$ and $\rho = \sup\{\xi \geqslant 0 : \|x + \xi y\| \leqslant 1\}$, then for $0 \leqslant \xi < \rho$ there is a unique $g(\xi) > 0$ such that $\|x + \xi y + g(\xi) z\| = 1$. Moreover, g is concave, decreasing, and continuous.

Proof of the theorem. Let x_1, x_2, x_3 satisfy $x_i \wedge x_j = 0$ if $i \neq j$ and $\|x_i\| = 1$. Then for $0 \leqslant \xi < 1$ and $0 \leqslant \theta \leqslant \pi/2$ there is a unique positive $r = r(\xi, \theta)$ such that $\|\xi x_1 + r(\xi, \theta)(\cos\theta\, x_2 + \sin\theta\, x_3)\| = 1$. Let $K = \operatorname{span}\{x_1, x_2, x_3\}$ and $H = \operatorname{span}\{x_1, \cos\theta\, x_2 + \sin\theta\, x_3\}$. Then K and H are closed sublattices of X. Let P be a positive contractive projection of X onto H. Since $\dim K = 3$ and $\dim H = 2$ there is a nonzero $z \in K$ such that $Pz = 0$. Since $Px_1 = x_1 \neq 0$, it follows that z is in the span of x_2 and x_3. Since P is positive we can obtain z in the form $\cos\phi\, x_2 + \sin\phi\, x_3$ with $\pi/2 \leqslant \phi \leqslant \pi$ and $\phi = \phi(\theta)$.

Consider the curve K_ξ defined in the plane using polar coordinates by $r = r(\xi, \theta)$ with $0 \leqslant \theta \leqslant \pi/2$. The region in the first quadrant bounded by K_ξ and the two axes is clearly convex. Observe that $\|\xi x_1 + r(\xi, \theta)$ $(\cos\theta\, x_2 + \sin\theta\, x_3) + \alpha z\| \geqslant \|P(\xi x_1 + r(\xi, \theta)(\cos\theta\, x_2 + \sin\theta\, x_3) + \alpha z)\|$ $= \|(\xi x_1 + r(\xi, \theta)(\cos\theta\, x_2 + \sin\theta\, x_3)\|$. Hence for each $0 \leqslant \xi < 1$, the line through the point $(r(\xi, \theta), \theta)$ with inclination ϕ to the horizontal axis is a line of support for K_ξ. It follows that if any one of the curves K_ξ has a tangent at the point θ, then they all do and all the tangents are parallel. (There is a tangent if and only if there is a unique line of support.)

If we fix ξ, convexity shows that K_ξ has a tangent except perhaps on a countable set of points. This can also be seen by observing that

$\theta \to \phi(\theta)$ determines an increasing function of θ for any choice of support line at θ (by convexity). Hence $\phi(\theta)$ is uniquely determined except perhaps for countably many θ. Since the gradient of a curve $r = r(\theta)$ is given by $(r' \sin\theta + r\cos\theta)/(r'\cos\theta - r\sin\theta)$, we conclude that $\frac{1}{r}\frac{\partial r}{\partial \theta}$ exists for all but countably many θ and is independent of ξ. It follows that $g(\xi,\theta) = r(\xi,\theta)/r(0,\theta)$ satisfies $\frac{\partial g}{\partial \theta} = 0$ on $[0,\pi/2]$ except perhaps at a countable set of points. From the mean value theorem (see [288]) we conclude that $g(\xi,\theta)$ is independent of θ. Now observe that $g(\xi,\theta) = \|r(\xi,\theta)(\cos\theta\, x_2 + \sin\theta\, x_3)\|$. Hence there exists a decreasing real valued function F on $[0,1)$ such that if $0 \leqslant \xi < 1$, $\|\xi x_1 + \eta x_2 + \zeta x_3\| = 1$ if and only if $\|\eta x_2 + \zeta x_3\| = F(\xi)$.

Consider $0 \leqslant \xi < 1$. We have $\dfrac{1}{\|\xi x_1 + x_2\|} = F\left(\dfrac{\xi}{\|\xi x_1 + x_2\|}\right)$ and $\|\xi x_1 + x_2\| = \|\xi x_1 + x_2\| \left\| \dfrac{\xi}{\|\xi x_1 + x_2\|} x_1 + \dfrac{1}{\|\xi x_1 + x_2\|} x_3 \right\| = \|\xi x_1 + x_3\|$. By continuity, $\|x_1 + x_2\| = \|x_1 + x_3\|$. Arguing similarly with x_2 and then x_3 in place of x_1, we have $\|x_1 + \eta x_2\| = \|\eta x_2 + x_3\|$ for $0 \leqslant \eta \leqslant 1$ and $\|x_1 + \zeta x_3\| = \|x_2 + \zeta x_3\|$ for $0 \leqslant \zeta \leqslant 1$.

Now define $G(\alpha,\beta) = \|\alpha x_1 + \beta x_2\|$. We conclude that $\|\alpha x_i + \beta x_j\| = G(\alpha,\beta)$ for $x_i \neq x_j$, $\alpha,\beta \geqslant 0$.

Suppose $\alpha,\beta,\gamma \geqslant 0$. We claim that $\|\alpha x_1 + \beta x_2 + \gamma x_3\| = G(\alpha, G(\beta,\gamma)) = G(\beta, G(\gamma,\beta)) = G(\gamma, G(\alpha,\beta))$. For, choose $\lambda > 1$ such that $\|\alpha x_1 + \lambda(\beta x_2 + \gamma x_3)\| > \alpha$ and let λ_0 be the infimum of all such λ's. We have

$$\frac{\lambda \|\beta x_2 + \gamma x_3\|}{\|\alpha x_1 + \lambda(\beta x_2 + \gamma x_3)\|} = F\left(\frac{\alpha}{\|\alpha x_1 + \lambda(\beta x_2 + \gamma x_3)\|}\right)$$

and hence $\|\alpha x_1 + \lambda(\beta x_2 + \gamma x_3)\| = \|\alpha x_1 + \lambda G(\beta,\gamma) x_2\| = G(\alpha, \lambda G(\beta,\gamma))$. By continuity we conclude that $\|\alpha x_1 + \lambda_0(\beta x_2 + \gamma x_3)\| = G(\alpha, \lambda_0 G(\beta,\gamma))$. If $\lambda_0 = 1$, we have $\|\alpha x_1 + \beta x_2 + \gamma x_3\| = G(\alpha, G(\beta,\gamma))$. If $\lambda_0 > 1$, we have $\alpha = \|\alpha x_1 + \lambda_0(\beta x_2 + \gamma x_3)\| = G(\alpha, \lambda_0 G(\beta,\gamma)) \geqslant G(\alpha, G(\beta,\gamma)) \geqslant g(\alpha, 0) = \alpha$. Thus $\|\alpha x_1 + \beta x_2 + \gamma x_3)\| = G(\alpha, \lambda_0 G(\beta,\gamma))$ as claimed. The other two equalities follow similarly. It follows that if $x, y \in K$ and $x \wedge y = 0$, then $\|x + y\|$ is a function of $\|x\|$ and $\|y\|$. By theorem 5 of section 15 the norm is p additive on K for some $1 \leqslant p \leqslant \infty$.

If X is finite dimensional we can choose a basis $\{x_1, \dots, x_n\}$ for X consisting of mutually disjoint positive elements. By our arguments above we have that the norm is p_i additive on the span of $\{x_i, x_{i+1}, x_{i+2}\}$ for some $1 \leqslant p_i \leqslant \infty$. Since these three dimensional sublattices intersect consecutively in two dimensional sublattices we conclude that all the p_i's are equal say to p and the norm is p additive on X.

Suppose that X is infinite dimensional. Then there is an infinite pairwise disjoint sequence $\{x_n\}$ in X consisting of nonzero positive elements. As in the finite dimensional case we see that there is a single p such that $1 \leqslant p \leqslant \infty$ and the norm is p additive on the span of any three of the elements x_n and, hence, on the span of any set of four mutually disjoint positive elements a, b, c, d such that $a \in \{x_i\}^{\perp\perp}$ and $b \in \{x_j\}^{\perp\perp}$ with $i \neq j$.

Now suppose $u \wedge v = 0$. Because X is Archimedean we can replace each x_n by a (large) scalar multiple of itself and assume that $x_n \nleqslant u+v$ for any n. Put $w_n = x_{2n-1} + x_{2n}$ and observe that the sequence $\{w_n\}$ is also pairwise disjoint so by lemma 1 $\|y\| = \lim \|y - y \wedge w_n\|$ for any $y \geqslant 0$. For each n, the four elements $u - u \wedge w_n$, $v - v \wedge w_n$, $x_{2n-1} - (u+v) \wedge x_{2n-1}$, $x_{2n} - (u+v) \wedge x_{2n}$ are mutually disjoint. Since the last two are positive and in $\{x_{2n-1}\}^{\perp\perp}$ and $x_{2n}^{\perp\perp}$ respectively, we conclude that $\|u+v - (u+v) \wedge w_n\| = \|u - u \wedge w_n + v - v \wedge w_n\| = (\|u - u \wedge w_n\|^p + \|v - v \wedge w_n\|^p)^{1/p}$ (interpret correctly for $p = \infty$). Letting $n \to \infty$ we conclude that the norm is p additive on X. ∎

Theorem 5. *Let X be a Banach lattice with dimension $\geqslant 3$.*

(1) *If each closed sublattice of X is the range of positive contractive projection, then there is a p such that the norm is p additive and order continuous.*

(2) *If each closed sublattice of X admits a positive simultaneous linear extension, then there is a p such that the norm is p additive.*

Proof. (1) is a consequence of theorem 4 and theorem 4 of section 3, (2) is a consequence of theorem 4 and the fact that if Y is a reflexive sublattice of X and Y admits a positive simultaneous linear extension, then it admits a contractive positive projection (in particular, when Y is finite dimensional this is true). ∎

We have already shown that a closed sublattice of an abstract M space admits a positive simultaneous linear extension. Clearly if there is a positive contractive projection P of a Banach lattice X onto a closed sublattice Y, then $Ly^* = y^* \circ P$ for $y^* \in Y^*$ defines a positive simultaneous linear extension of Y^* into X^*. Thus when we show that if X is a Banach lattice in which the norm is order continuous and p additive for some $1 \leqslant p \leqslant \infty$, then each closed sublattice is the range of a positive contractive projection, it will follow that the converses of (1) and (2) in theorem 5 are valid. A proof of these using the full force of the representation as $L_p(\mu, \mathbb{R})$, which obtain the projection in terms of conditional expectations, will be given in the next section, we give a proof here which uses only order continuity of the norm and p additivity. The exercises at the end of this section contain yet another approach.

Theorem 6. *Let X be a Banach lattice in which the norm is order continuous and p additive for some $1 \leqslant p \leqslant \infty$. Then each closed sublattice of X is the range of a positive contractive projection.*

Clearly if the norm is p additive for $1 \leqslant p < \infty$, then the norm is already order continuous. Thus the assumption is only necessary in the case $p = \infty$. The proof is accomplished in a series of lemmas below. Throughout X denotes a Banach lattice under the hypotheses of the theorem and Y is a closed sublattice of X. Let $Y_1 = \{x \in X : |x| \leqslant y$ for some $y \in Y\}$. Then Y_1 is norm dense in $Y^{\perp\perp}$ and since $Y^{\perp\perp}$ is a band in X, it suffices to show that there is a positive contractive projection of Y_1 onto Y. If $y \in Y$, recall that $[y]$ denotes the band projection of X onto $y^{\perp\perp}$. Moreover, for $x \geqslant 0$, $[y](x) = \sup n y \wedge x = \lim n y \wedge x$ and thus for $y \in Y$, $[y](Y) \subset Y$.

Let $\mathscr{D}$ denote the family of all maximal sets of pairwise disjoint positive nonzero elements of Y. Then $\mathscr{D}$ is directed by $D_1 \geqslant D_2$ if and only if for each $y \in D_1$ there is a $z \in D_2$ with $y \leqslant z$.

Let $x \in Y_1$, $x > 0$ and choose $y \in Y$ such that $x \leqslant y$ and put $f(x,z,y) = \dfrac{\|[z]x\|}{\|[z]y\|}$ for all $z \in Y$, $z > 0$, and $[z](y) \neq 0$, put $f(x,z,y) = 0$ if $[z](y) = 0$.

For $D \in \mathscr{D}$ let $x(D,y) = \sup\{f(x,z,y)[z](y) : z \in D\}$ and $q(x,y) = \limsup_{\mathscr{D}} x(D,y)$. By the hypothesis on X all the suprema and infima involved exist in X and $x(D,y)$, $q(x,y)$ are in Y. Moreover, since each D is maximal, $\|x(D,y)\| = \|x\|$ and since $f(x,z,y) \leqslant 1$, $q(x,y) \leqslant y$.

In the following lemma x is as above.

Lemma 2. *If u, v are in Y and $u \wedge v \geqslant x$, then $q(x,u) = q(x,v)$.*

Proof. Let n be a positive integer and put $g_r = \left[\left(\dfrac{r+1}{n} u - v\right)^+\right] u - \left[\left(\dfrac{r}{n} u - v\right)^+\right] u$ for $r = 0, 1, 2, \ldots$. Then $g_r \in Y$ and $g_r \wedge g_s = 0$ for $r \neq s$ and $\dfrac{r}{n}[g_r]u \leqslant [g_r]v \leqslant \dfrac{r+1}{n}[g_r]u$. This inequality is valid for g_r replaced by any g with $0 < g \leqslant g_r$. Hence, if $0 < g \leqslant g_r$ and $r \geqslant \sqrt{n}$, $f(x,g,v) \leqslant \left(\dfrac{n}{r}\right) f(x,g,u)$ and $f(x,g,v)[g]v \leqslant \dfrac{n}{r} f(x,g,u)(r+1)/n[g]u = f(x,g,u)\left(1 + \dfrac{1}{r}\right)[g]u \leqslant (f(x,g,u) + 1/\sqrt{n})[g]u$. If $r \leqslant \sqrt{n}$, then $f(x,g,v)[g]v \leqslant [g]v \leqslant \dfrac{(r+1)}{n}[g]u \leqslant (2/\sqrt{n})[g]u$. Hence for all r and $0 \leqslant g \leqslant g_r$, $f(x,g,v)[g]v \leqslant f(x,g,u)[g]u + (2/\sqrt{n})[g]n$. If $D \supset \{g_r : r = 0, 1, 2, \ldots\}$ and $D_1 \geqslant D$, then $x(D_1, v) \leqslant x(D_1, u) = (2/\sqrt{n})u$ and hence

$q(x,v) \leqslant q(x,u) + (2/\sqrt{n})u$. If follows that $q(x,v) \leqslant q(x,u)$ and interchanging u and v gives the equality. ∎

Corollary. *If* $x \in Y$ *and* $x > 0$ *and* $u \in Y$ *with* $u \geqslant x$, *then* $q(x,u) = q(x,x) = x$. ∎

For any $x \in Y_1$, let $q(x)$ denote the unique element of Y defined by $q(x^+, u) = q(x)$ for any $u \in Y_1$ with $u \geqslant x^+$.

Lemma 3. *For* x, y *in* Y, *we have*

(i) $q(x+y) \leqslant q(x) + q(y)$,

(ii) $q(ax) = a\,q(x)$ *for* $a \geqslant 0$,

(iii) $\|q(x)\| \leqslant \|x\|$.

Proof. Parts (i) and (ii) are clear. To prove (iii) it suffices to consider $x \geqslant 0$. Let $u \in Y$ with $u \geqslant x$ and let n be a positive integer. For $r = 1, \ldots, n+1$ let $P_r = \left\{ g \in Y : 0 < g \text{ and } 1 - \frac{r}{n} < f(x, g', u) \leqslant 1 - \frac{r-1}{n} \right.$ for all $g' \in Y$ with $\left. 0 < g' \leqslant g \right\}$. It is easy to see that $P_r \perp P_s$ for $r \neq s$, and $u^\perp \subset P_{n+1}$. Moreover, if $0 < g \in P_1^\perp \cap \cdots \cap P_r^\perp \cap Y$, then $f(x,g,u) \leqslant 1 - \frac{r}{n}$ $(r = 1, 2, \ldots, n)$ and in particular $P_{n+1} = P_1^\perp \cap \cdots \cap P_n^\perp \cap (Y^+ \setminus \{0\})$. For, assume this is valid for $r-1$ and take $h \in P_1^\perp \cap \cdots \cap P_r^\perp (Y^+ \setminus \{0\})$. By hypothesis $f(x,g,u) \leqslant 1 - \frac{r-1}{n}$ for $0 < g \leqslant h$. Let S be a maximal pairwise dijoint subset of Y^+ such that $0 \leqslant s \leqslant h$ and $f(x,s,u) \leqslant 1 - \frac{r}{n}$ for all $s \in S$. If $S^{\perp\perp} \neq \{h\}^{\perp\perp}$, we can find a $g_1 \in Y^+$ such that $0 < g_1 \leqslant h$ and $g_1 \in S^\perp$. Now $g_1 \notin P_r$ so there is a $g \in Y$ with $0 < g \leqslant g_1$ and $f(x,g,u) \notin \left(1 - \frac{r}{n}, 1 - \frac{r-1}{n}\right]$. It follows that $f(x,g,u) \leqslant 1 - \frac{r}{n}$ contradicting the maximality of S. Thus $S^{\perp\perp} = \{h\}^{\perp\perp}$ and $[h]x = \sup\{[s](x) : s \in S\}$.

Now $\|[s]x\| = \|f(x,s,u)[s]u\| \leqslant \left(1 - \frac{r}{n}\right)\|[h]u\|$ for all $s \in S$ and hence $\|[h]x\| \leqslant \left(1 - \frac{r}{n}\right)\|\sup_{s \in S}[s]u\| = \left(1 - \frac{r}{n}\right)\|[h]u\|$ and $f(x,h,u) \leqslant 1 - \frac{r}{n}$ as required. Let D_0 be a maximal pairwise disjoint subset of $P_1 \cup \cdots \cup P_n \cup P_{n+1}$. By the argument above, $D_0 \in \mathscr{D}$. Suppose $D \geqslant D_0$. If $h \in D$, then there is a $g \in D_0$ with $0 < h \leqslant g$ and there is an r such that $g \in P_r$. Hence $f(x,h,u)$ and $f(x,g,u)$ are both in $\left(1 - \frac{r}{n}, 1 - \frac{r-1}{n}\right]$

and $f(x,h,u) \leqslant f(x,g,u) + \frac{1}{n}$. Thus $x(D,u) = \sup\{f(x,h,u)[h]u : h \in D\}$ $= \sup_{g \in D_0}(\sup\{f(x,h,u)[h]u : h \in D,\, h \leqslant g\}) \leqslant \sup_{g \in D_0}\left(\sup\left\{\left(f(x,g,u) + \frac{1}{n}\right)[g]u : g \in D_0\right\}\right) = x(D_0,u) + \frac{1}{n}u$. Hence $\|q(x)\| = \|q(x,u)\| \leqslant \left\|x(D_0,u) + \frac{1}{n}u\right\|$ $\leqslant \|x(D_0,u)\| + \frac{1}{n}\|u\| = \|x\| + \frac{1}{n}\|u\|$. Letting $n \to \infty$ we deduce that $\|q(x)\| \leqslant \|x\|$ as required. ∎

Lemma 4. *There is a linear map $P: Y_1 \to Y_1$ such that $Py = y$ for $y \in Y$ and $Px \leqslant q(x)$ for $x \in Y_1$.*

Proof. Note that Y is order complete, $q(y) = y^+ \geqslant y$ for $y \in Y$ and q is subadditive and positive homogeneous. A Hahn-Banach type argument allows us to obtain P by extending the identity map on Y to a linear map of Y_1 into Y such that $Px \leqslant q(x)\,(x \in Y)$ (see theorem 6 of section 1). ∎

Proof of the theorem. We show that the map P above is positive and contractive. Let $x \in Y_1$ with $x \geqslant 0$. Then $p(-x) \leqslant q(-x) = q((-x)^+) = 0$ so $P(x) \geqslant 0$. For any $x \in Y_1$, $\|Px\| = \|Px^+ - Px^-\| = \|\,|Px^+ - Px^-|\,\|$ $\leqslant \|Px^+ + Px^-\| = \|P|x|\,\| \leqslant \|q(|x|)\| \leqslant \|\,|x|\,\| = \|x\|$. ∎

We now investigate the uniquess of contractive projections in abstract L_p spaces. In particular, we show that if X is an abstract L_p space for some $1 \leqslant p < \infty$ and Y is a closed sublattice of X, then there is a unique positive contractive projection P of X onto Y such that $PY^\perp = 0$.

Lemma 5. *Suppose $1 \leqslant p < \infty$, then there is a constant M (depending on p only), such that if n is a positive integer, $a_0, \ldots, a_n$ are positive numbers, $\sum_{i=0}^{n} a_i^p = 1$, and $0 \leqslant k \leqslant 1/2$, then*

$$\left[\sum_{i=0}^{n}\left(1 + \frac{k(i+1)}{n}\right)^p a_i^p\right]^{1/p} + \left[\sum_{i=0}^{n}\left(1 - \frac{ki}{n}\right)^p a_i^p\right]^{1/p} \leqslant 2 + \frac{k}{n} + Mk^2.$$

Proof. Let $F(k,n)$ denote the left hand side of the above inequality. For $0 \leqslant k \leqslant 1/2$ and $i = 0, 1, \ldots, n$, $1 - ki/n \neq 0$. Thus $F(k,n)$ is, for each n an infinitely differentiable function of k. Elementary computations show that $F(0,n) = 2$, $\partial F(0,n)/\partial k = 1/n$ and $\partial^2 F(k,n)/\partial k^2$ is bounded, uniformly in n, for $0 \leqslant k \leqslant 1/2$. The result now follows. ∎

In the next lemma, M is as in lemma 5.

Lemma 6. *Let X be an abstract L_p space for some $1 \leqslant p < \infty$. If x,u are in X, $0 \leqslant x \leqslant u$, $\|u\| = 1$ and $0 \leqslant k \leqslant 1/2$, then $\|u + kx\| + \|u - kx\|$ $\leqslant 2 + (2M)k^2$.*

Proof. This is clear for $k=0$. If $k>0$ choose $n>1/k$ and put $g_i=\left[\left(\frac{(i+1)}{n}u-x\right)^+\right]u-\left[\left(\frac{i}{n}u-x\right)^+\right]u$ for $i=0,\ldots,n$. Note that $g_i\wedge g_j=0$ for $i\neq j$ and $\sum_{i=0}^{n}[g_i]u=u$. Furthermore, $[g_i](u+kx)\leqslant\left(1+\frac{k(i+1)}{n}\right)[g_i]u$, and $[g_i](u-kx)\leqslant\left(1-\frac{ki}{n}\right)[g_i]u$. If $a_i=\|[g_i]u\|$, then $\sum_{i=0}^{n}a_i^p=1$ and by lemma 5, $\|u+kx\|+\|u-kx\|$
$\leqslant\left[\sum_{i=0}^{n}\left(1+\frac{k(i+1)}{n}\right)^p a_i^p\right]^{1/p}+\left[\sum_{i=0}^{n}\left(1-\frac{ki}{n}\right)^p a_i^p\right]^{1/p}\leqslant 2+Mk\left(k+\frac{1}{n}\right)$
$\leqslant 2+2Mk^2$. ∎

Lemma 7. *Let X be a Banach lattice in which the norm is order continuous and $Y\subset X$ a closed sublattice. If P is a positive contractive projection of X onto Y such that $PY^\perp=0$ and B is a band projection such that $BY\subset Y$, then $BP=PB$.*

Proof. Since $Bx\leqslant x\,(x\geqslant 0)$ we have $BPBx\leqslant BPx$ and $BPBx\leqslant PBx\,(x\geqslant 0)$. If $u\in Y$ and $0\leqslant x\leqslant u$ we also have $BPB(u-x)\leqslant BP(u-x)$. Since $BPu=Bu=PBu=BPBu$ we conclude that $BPx=BPBx$, and similarly $PBx=BPBx$. This gives $Pbx=BPx\,(x\in Y_1)$ and, by continuity $PBx=BPx\,(x\in\overline{Y}_1=Y^{\perp\perp})$. Since $BY^\perp\subset Y^\perp$, $PBx=0=BPx\,(x\in Y^\perp)$. Thus $PB=BP$ as required. ∎

Let q be the function defined as in lemma 3.

Lemma 8. *Let X be an abstract L_p space for some $1\leqslant p<\infty$ and Y a closed linear sublattice of X. If P is a positive contractive projection of X onto Y such that $PY^\perp=0$, then $Px\leqslant q(x)$ for all $x\in Y$.*

Proof. By lemma 7, $P[x]=[x]P$ for all $x\in Y$. Hence if $u\in Y$, and $0\leqslant x\leqslant u$, $Px(D,u)=\sup\left\{\frac{\|[z]Px\|}{\|[z]u\|}[z]u\colon z\in D\right\}=\sup\left\{\frac{\|P[z]x\|}{\|[z]u\|}[z]u\colon z\in D\right\}$ $\leqslant\sup\left\{\frac{\|[z]x\|}{\|[z]u\|}[z]u\colon z\in D\right\}\leqslant x(D,u)$ for $D\in\mathscr{D}$. Hence $Px=q(Px,u)\leqslant q(x,u)=q(x)$. For a general $x\in Y_1$, $Px\leqslant q(x+)=q(x)$. ∎

Theorem 7. *Let X be an abstract L_p space for some $1\leqslant p<\infty$ and Y a closed linear sublattice of X. Then the positive contractive projection P of X onto Y such that $PY^\perp=0$ is unique.*

Proof. Let Q be a positive contractive projection of Y_1 onto Y such that $QY^\perp=0$. Then $Qx\leqslant q(x)$ for all $x\in Y_1$. Hence $-v-q(-v-x)\leqslant Q(x)\leqslant -u+q(u+x)$ for all $u,v\in Y$. Thus it is sufficient to show that $\inf\{v-u+q(-v-x)+q(u+x)\colon u,v\in Y\}=0$.

Fix $u \in Y$ such that $u \geqslant x$ and assume $x \geqslant 0$. Then we will show that $\inf\{-2\lambda u + q(\lambda u - x) + q(\lambda u + x) : \lambda \geqslant 1\} = 0$. For, let $\lambda > 2$, and let $k = 1/\lambda$. As in the proof of Lemma 3 (iii) there is a $D \in \mathscr{D}$ such that $q(u \pm kx) \leqslant (u \pm kx)(D, 2u) + 2k^2 u$. By lemma λ, applied to $\|[z](u+kx)\| + \|[z](u-kx)\|$ for each $z \in D$, we have

$$\begin{aligned} q(u+kx) + q(u-kx) - 2u - 2k^2 u &\leqslant (u+kx)(D,2u) + (u-kx)(D,2u) \\ &= \sup\left\{\left(\frac{(\|[z](u+kx)\| + \|[z](u-kx)\|) - 2\|[z]u\|}{\|[z]2u\|}\right)[z]2u : z \in D\right\} \\ &\leqslant \sup\left\{\left(\frac{(2M)\|[z]u\|\,k^2}{\|[z]2u\|}\right)[z]2u : z \in D\right\} \\ &= (2M)k^2 u. \end{aligned}$$

Thus $q(\lambda u + x) + q(\lambda u - x) - 2\lambda u \leqslant (2M)\lambda k^2 u + \lambda 2k^2 u = [(2M+1)/\lambda]u$. Since $\inf\{(1/\lambda)u : \lambda \geqslant 1\} = 0$, we are done. ∎

A similar characterization of Hilbert space is due to Kakutani (real case, see [149]) and Bohnenblust (complex case, see [49]).

Theorem (Kakutani, Bohnenblust). *A (real or complex) Banach space of dimension at least 3 is linearly isometric to a Hilbert space if and only if each closed subspace is the range of a contractive projection.*

The isomorphic version was recently accomplished by Lindenstrauss and Tzafriri in [192].

Theorem (Lindenstrauss-Tzafriri). *A Banach space is linearly isomorphic to a Hilbert space if and only if each closed subspace is complemented.*

An isomorphic version of Ando's theorem (theorem 3) was given by Tzafriri in [273] and improved by Lindenstrauss and Tzafriri in [192].

Theorem. *Let X be a σ complete Banach lattice such that each closed sublattice of X is complemented. Then X is isomorphic to either $c_0(\Gamma, \mathbb{R})$ or $L_p(\mu, \mathbb{R})$.*

There are several results which jointly characterize the classical sequence spaces $c_0(\mathbb{N}, \mathbb{R})$ and $l_p(\mathbb{N}, \mathbb{R})$ in terms of basis theory. We briefly mention them here. They also are related to theorem 5 of section 15 (see exercise 1).

A Schauder basis for a real Banach space X is said to be *unconditional* if and only if for each x the series $\sum_{n=1}^{\infty} x_n^*(x) x_n$ converges unconditionally to x (in norm). The basis $\{x_n\}$ is said to be *perfectly homogeneous* if every normalized block basis with respect to $\{x_n\}$ is equivalent to $\{x_n\}$.

The following theorem is found in [283].

Theorem (Zippin). *A Banach space X has a perfectly homogeneous Schauder basis $\{x_n\}$ if and only if it is isometric to $l_p(\mathbb{N},\mathbb{R})$ or $c_0(\mathbb{N},\mathbb{R})$ in such a way that $\{x_n\}$ is equivalent to the unit vector basis.* ∎

In [194] the following theorem is proved.

Theorem (Lindenstrauss-Zippin). *The only Banach spaces with an unconditional basis which have the property that all normalized unconditional bases are equivalent are $l_p(\mathbb{N},\mathbb{R})$ and $c_0(\mathbb{N},\mathbb{R})$.* ∎

Finally, in [192] the following theorem is proved.

Theorem (Lindenstrauss-Tzafriri). *The only unconditional Schauder bases $\{x_n\}$ which have the property that the closed linear span of any block basis with respect to a permutation of $\{x_n\}$ is complemented are the unit vector bases for $l_p(\mathbb{N},\mathbb{R})$ and $c_0(\mathbb{N},\mathbb{R})$.* ∎

Exercises. 1. Let X be a Banach space with a normalized unconditional Schauder basis $\{x_n\}$ and suppose that $\{x_n^*\}$ is the biorthogonal sequence to $\{x_n\}$. Suppose that $\left\|\sum_{n=1}^{\infty} x_n^*(x)x_n\right\| = \left\|\sum_{n=1}^{\infty} \pm x_n^*(x)x_n\right\|$ for all $x\in X$. Show that X is a Banach lattice with respect to the positive cone $C=\{x: x_n^*(x)\geqslant 0$ for all $n\}$. Prove that if every normalized block basis $\{y_n\}$ with to $\{x_n\}$, $\left\|\sum_{n=1}^{\infty} a_n x_n\right\| = \left\|\sum_{n=1}^{\infty} a_n y_n\right\|$, X is linearly isometric to $l_p(\mathbb{N},\mathbb{R})$ or $c_0(\mathbb{N},\mathbb{R})$ and $\{x_n\}$ is equivalent to the unit vector basis.

The following exercises concern alternate proofs to theorems 6 and 7. By $[Y^{\perp\perp}]$ we mean the band projection onto $Y^{\perp\perp}$.

2. Let $1<p<\infty$ and suppose that X_1, X_2 are closed vector sublattices of $L_p(\mu,\mathbb{R})$ (for some measure μ) and $X_1^{\perp}\supset X_2$. Suppose $P_i: L_p(\mu,\mathbb{R})\to X_i$ is a positive contractive projection for $i=1,2$.

(i) $P_i[X_i^{\perp\perp}]$ is a positive contractive projection on X_i for $i=1,2$.

(ii) $P_1[X_1^{\perp\perp}]+P_2[X_2^{\perp\perp}]$ is a positive contractive projection of $L_p(\mu,\mathbb{R})$ onto X_1+X_2.

3. If $0<x\in L_p(\mu,\mathbb{R})$ and $\|x\|=1$, then there is exactly one positive contractive projection of $L_p(\mu,\mathbb{R})$ onto $\operatorname{span}(x)$ and it is given by $Py=(\int y x^{p/q})x$ for all $y\in L_p(\mu,\mathbb{R})$ $\left(\text{where } \frac{1}{p}+\frac{1}{q}=1\right)$.

4. If X is a closed vector sublattice of $L_p(\mu,\mathbb{R})$ and $P: L_p(\mu,\mathbb{R})\to X$ is a positive contractive projection such that $P(X^{\perp})=0$, then $PB=BP$ for any band projection B such that $B(X)\subset X$.

5. If X is a finite dimensional vector sublattice of $L_p(\mu,\mathbb{R})$, then there is a unique positive contractive projection $P: L_p(\mu,\mathbb{R})\to X$ such that $P(X^{\perp})=0$.

6. Use 2 through 5 to show that for each closed vector sublattice of $L_p(\mu,\mathbb{R})$ $(1<p<\infty)$ there is a unique positive contractive projection P of $L_p(\mu,\mathbb{R})$ onto X such that $P(X^\perp)=0$.

7. Let Γ a nonempty index set. Give a direct proof that each closed vector sublattice of $c_0(\Gamma,\mathbb{R})$ is the range of a positive contractive projection on $c_0(\Gamma,\mathbb{R})$.

8. Give a direct proof that each closed sublattice of $L_1(\mu,\mathbb{R})$ is the range of a positive contractive projection.

§ 17. Contractive Projections in Abstract L_p Spaces

In this section we wish to prove that a closed linear subspace of $L_p(\mu,\mathbb{C})$ is linearly isometric to $L_p(v,\mathbb{C})$ for some measure v if and only if it is the range of a contractive projection in $L_p(\mu,\mathbb{C})$. We then use this to give a Banach space characterization of $L_p(\mu,\mathbb{C})$ in terms of its finite dimensional subspaces.

All of the results of this section are also valid for the real case. We state and prove all results in the complex case only. The proofs in the real case follows along the same lines. The results on contractive projections in this section are basically due to Douglas [90] who proved them for the case $p=1$ and μ is finite, Ando [13] who proved them for $1<p<\infty$ by showing how to reduce this case to the case $p=1$ (and μ is finite), and Tzafriri [272] who proved that the range of a contractive projection in an arbitrary real abstract L_p space is again an abstract L_p space. The development here follows Bernau and Lacey [42].

We first prove two technical lemmas. The notation for band projections used throughout is the following. If M is a set in X, J_M denotes the band projection onto $M^{\perp\perp}$. If $M=\{f\}$, we denote J_M by J_f.

The first lemma is due to Ando [13].

Lemma 1 (Ando). *Suppose $0<p<\infty$ and let M be a closed subspace of $L_p(T,\Sigma,\mu,\mathbb{C})$. If $\{f_n\}$ is a sequence in M, then there exists $f\in M$ such that $S(f)=\bigcup_{n=1}^{\infty} S(f_n)$. In particular if μ is finite or M is separable there exists $f\in M$ such that $J_f=J_{M^{\perp\perp}}$; that is, f is a function in M of maximum support.*

Proof. If $f,g\in L_p(\mu,\mathbb{C})$ and α is a scalar, the zero sets $\{t\in T\colon (f+\alpha g)(t)=0\}$ have disjoint intersection with $S(f)\cup S(g)$ for differing values of α. Since $S(f)\cup S(g)$ is σ finite, $\mu(S(f)\cup S(g)\setminus S(f+\alpha g))=0$ except, perhaps for countably many values of α.

Assume, as we may, that $\int|f_n|^p=1$ for all n. We define, inductively, two sequences $\{\alpha_n\},\{\varepsilon_n\}$ of positive real numbers such that, if we write

$g_n=\alpha_1 f_1+\cdots+\alpha_n f_n$, $A_n=\{t\in T: |g_n(t)|\leqslant\varepsilon_n\}$, and $B_n=\{t\in T: |\alpha_{n+1} f_{n+1}(t)| \geqslant \varepsilon_n/2\}$, then

(i) $\alpha_{n+1}<2^{-n/p}$ and $\varepsilon_{n+1}<\varepsilon_n/2$;

(ii) $\mu(S(g_n)\cup S(f_{n+1})\setminus S(g_{n+1}))=0$;

(iii) $\int_{A_n\cup B_n}|f_i|^p\,d\mu<2^{-n}$ $(i=1,2,\ldots,n)$.

Start with $\alpha_1=1$. Suppose $\alpha_1,\ldots,\alpha_n$; $\varepsilon_1,\ldots,\varepsilon_{n-1}$ have been chosen. Note that $\mu(S(f_i)\setminus S(g_n))=0$ $(i=1,\ldots,n)$ so if $C_\varepsilon=\{t\in T: |g_n(t)|\leqslant\varepsilon\}$, $\int_{C_\varepsilon}|f_i|^p\,d\mu\to 0$ $(\varepsilon\to 0)$ for $i=1,\ldots,n$. Also if $D_\eta=\{t\in T: |f_{n+1}(t)\geqslant\eta\}$, $\int_{D_\eta}|f_i|^p\,d\mu\to 0$ $(\eta\to\infty)$ for $i=1,\ldots,n$. Thus we choose ε_n such that $0<\varepsilon_n<\varepsilon_{n-1}/2$, and $\int_{A_n}|f_i|^p\,d\mu<2^{-n-1}$ $(i=1,2,\ldots,n)$; then choose η such that $\int_{D_\eta}|f_i|^p\,d\mu<2^{-n-1}$ $(i=1,2,\ldots,n)$, and α_{n+1} such that $0<\alpha_{n+1}<2^{-n/p}$, (ii) is satisfied, and $\alpha_{n+1}\eta<\varepsilon_n/2$. Since $B_n\subset D_\eta$ we also have (iii) satisfied.

By (i) $\{g_n\}$ converges in $L_p(\mu,\mathbb{C})$ to an element $f\in M$, and $S(f)\subset\bigcup S(f_n)$. Let $E=\limsup\,(A_n\cup B_n)=\bigcap_{n=1}^{\infty}\bigcap_{k=n}^{\infty}(A_n\cup B_n)$. Fix i and let $N>i$, then, by (iii)

$$\begin{aligned}\int_E|f_i|^p\,d\mu&\leqslant\int_{\bigcup_N^{\infty}(A_n\cup B_n)}|f_i|^p\,d\mu\\&\leqslant\sum_N^{\infty}\int_{A_n\cup B_n}|f_i|^p\,d\mu\\&\leqslant\sum_N^{\infty}2^{-n}\\&=2^{1-N}\to 0\quad(N\to\infty).\end{aligned}$$

Thus $\mu(E\cap S(f_i))=0$ for all i and $\mu(E\cap\bigcup S(f_n))=0$. We complete our proof by showing that $T\setminus E\subset S(f)$. If $t\in T\setminus E$ choose the smallest integer n such that $t\notin\bigcup_{k=n}^{\infty}(A_k\cup B_k)$, then $|g_n(t)|>\varepsilon_n$ and $|\alpha_k f_k(t)| <\varepsilon_{k-1}/2<\varepsilon_n/2^{k-n}$ $(k\geqslant n+1)$. Hence

$$\begin{aligned}|g_k(t)&\geqslant|g_n(t)|-|\alpha_{n+1}f_{n+1}(t)|-\cdots-|\alpha_k f_k(t)|\\&>|g_n(t)-\varepsilon_n(2^{-1}+\cdots+2^{-(k-n)})\\&>|g_n(t)|-\varepsilon_n\qquad(k>n).\end{aligned}$$

Thus $|f(t)|=\lim|g_k(t)|\geqslant|g_n(t)|-\varepsilon_n>0$, and we are done. ∎

The second lemma is due to Tzafriri [272].

Lemma 2 (Tzafriri). *Let M be a separable subspace of $L_p(T,\Sigma,\mu,\mathbb{C})$ $(p\geqslant 1)$ and L a bounded linear operator on $L_p(\mu,\mathbb{C})$. Then there is a σ finite* *set $T_0 \in \Sigma$ and a σ subring Σ_0 of Σ such that Σ_0 consists of subsets of T_0 and $L_p(T_0, \Sigma_0, \mu, \mathbb{C})$ is separable, L-invariant and contains M.*

Proof. The subspace $M + LM$ is separable, L-invariant and generates a separable vector sublattice M_1 of $L_p(\mu, \mathbb{C})$. We construct a sequence of separable vector sublattices M_n such that $M_n + LM_n \subset M_{n+1}$. Then $\overline{\bigcup M_n}$ is a separable L-invariant closed vector sublattice of $L_p(\mu, \mathbb{C})$. Writing $K_1 = \overline{\bigcup M_n}$ we have K_1 closed under all band projections J_x with $x \in K_1$. Let $\Sigma_1 = \{S(x): x \in K_1\}$ then Σ_1 is a σ subring of Σ and if $x, y \in K_1$ with $x \in y^{\perp\perp}$ then x/y is Σ_1 measurable. If f_n is dense in K_1, $f = \Sigma 2^{-n} \|f_n\|^{-1} |f_n| \in K_1$ and $\mu(S(x) \setminus S(f)) = 0$ $(x \in K_1)$. Consider $L_p(S(f), \Sigma_1, \mu, \mathbb{C})$. It is easy to see that this is the closure of the vector sublattice spanned by K_1 and the functions $f_{f^{-1}(0,\alpha]}$ with α rational. Thus, writing $T_1 = S(f)$ we have

$$K_1 \subset L_p(T_1, \Sigma_1, \mu, \mathbb{C})$$

with $L_p(T_1, \Sigma_1, \mu, \mathbb{C})$ separable. Continuing inductively, we obtain a sequence $T_1 \subset T_2 \subset \cdots \subset T_n \subset \cdots$ of σ finite subsets of T and sequence $\Sigma_1 \subset \Sigma_2 \subset \cdots \subset \Sigma_n \subset \cdots$ of σ subrings of Σ, such that each Σ_n consists of subsets of T_n, $L_p(T_n, \Sigma_n, \mu, \mathbb{C}) + LL_p(T_n, \Sigma_n, \mu, \mathbb{C}) \subset L_p(T_{n+1}, \Sigma_{n+1}, \mu, \mathbb{C})$ and each $L_p(T_n, \Sigma_n, \mu, \mathbb{C})$ is separable.

Let $K_0 = \overline{\bigcup_{n=1}^{\infty} L_p(T_n, \Sigma_n, \mu, \mathbb{C})}$ then K_0 is a separable L-invariant closed vector sublattice of $L_p(T, \Sigma, \mu, \mathbb{C})$. Define $\Sigma_0 = \{S(f): f \in K_0\}$ and find, as for K_1, $f \in K_0$ such that $\mu(S(x) \setminus S(f)) = 0$ $(x \in K_0)$. It is routine to show that $K_0 = L_p(S(f), \Sigma_0, \mu, \mathbb{C})$. This proves our lemma with $T_0 = S(f)$. ∎

Lemma 3. *If M is a subspace of $L_p(T, \Sigma, \mu, \mathbb{C})$, $h \in L_p(\mu, \mathbb{C})$. Then there is a sequence $\{f_n\}$ in M such that $\lim f_{S(f_n)} h = J_M(h)$.*

Proof. Choose a sequence $\{f_n\}$ in M such that

$$\|f_{S(f_n)} h\|_p \to \sup\{\|f_{S(f)} h\|_p : f \in M\}.$$

By using lemma 1, this lemma can be strengthened, in case M is closed, to say that for each $h \in L_p(\mu, \mathbb{C})$ there exists $f \in M$ such that $J_M(h) = J_f h = f_{S(f)} h$. ∎

We shall denote the *range* of an operator L by $R(L)$. The following lemma is basically due to Douglas [90].

Lemma 4. *Let P be a contractive projection on $L_1(X,\Sigma,\mu,\mathbb{C})$ and suppose $f\in R(P)$; then*

(i) $PJ_f=J_fPJ_f$;

(ii) $P(h\operatorname{sgn} f)=|P(h\operatorname{sgn} f)|\operatorname{sgn} f \quad (0\leqslant h\in L_1(\mu,\mathbb{C}))$;

(iii) $\|P(h\operatorname{sgn} f)\|=\|J_f h\| \quad (0\leqslant h\in L_1(\mu,\mathbb{C}))$.

Proof. Suppose $0\leqslant h\leqslant |f|$, then

$$\begin{aligned}\|f\|-\|h\operatorname{sgn} f\| &= \|f-h\operatorname{sgn} f\|\\ &\geqslant \|P(f-h\operatorname{sgn} f)\|\\ &= \|f-P(h\operatorname{sgn} f)\|\\ &\geqslant \|f\|-\|P(h\operatorname{sgn} f)\|\\ &\geqslant \|f\|-\|h\operatorname{sgn} f\|.\end{aligned}$$

This gives equality throughout so (iii) is valid for $0\leqslant h\leqslant |f|$. In addition we have $0\leqslant |f-P(h\operatorname{sgn} f)|=|f|-|P(h\operatorname{sgn} f)|$ μ almost everywhere, and (ii) also follows for $0\leqslant h\leqslant |f|$. We extend immediately to $h\in L_1(\mu,\mathbb{C})$ such that $0\leqslant h\leqslant n|f|$ for some n, and since linear combinations of such h are dense in $f^{\perp\perp}$ (theorem 8 of section 1) we have (ii) and (iii) for $0\leqslant h\in f^{\perp\perp}$. If $h\in L_1(\mu,\mathbb{C})$ and $h\geqslant 0$, $(J_f h)\operatorname{sgn} f=h\operatorname{sgn} f$ so (ii) and (iii) are proved.

For (i) take $g\in L_1(\mu,\mathbb{C})$ and put $h=(\operatorname{Re}(g\operatorname{sgn}\bar{f}))^+$, by (ii) $Ph\in f^{\perp\perp}$ so $Ph=J_fPh$. We conclude easily that $P(J_f g)=P((g\operatorname{sgn}\bar{f})\operatorname{sgn} f)$ $=J_fPJ_f g$ and (i) is proved. ∎

Suppose $1<p<\infty$; then identify the dual of $L_p(T,\Sigma,\mu,\mathbb{C})$ with $L_q(T,\Sigma,\mu,\mathbb{C})$ in the usual way $(1/p+1/q=1)$. Let P be a contractive projection on $L_p(\mu,\mathbb{C})$. The conjugate operator P^* is defined uniquely on $L_q(\mu,\mathbb{C})$ by the equation

$$\int Pf\cdot g\,d\mu=\int f\cdot P^*g\,d\mu \qquad (f\in L_p(\mu,\mathbb{C}),\ g\in L_q(\mu,\mathbb{C}))\,.$$

Clearly P^* is a contractive projection on $L_q(\mu,\mathbb{C})$.

The following lemma is due to Ando.

Lemma 5. *Suppose $1<p<\infty$ and let P be a contractive projection on $L_p(T,\Sigma,\mu,\mathbb{C})$, then $f\in R(P)$ if and only if $|f|^{p-1}\operatorname{sgn}\bar{f}\in R(P^*)$.*

Proof. Suppose $f\in R(P)$; by Holder's inequality

$$\begin{aligned}\|f\|_p^p &= \int |f|^p\,d\mu=\int Pf\cdot|f|^{p-1}\operatorname{sgn}\bar{f}\,d\mu\\ &= \int f\cdot P^*(|f|^{p-1}\operatorname{sgn}\bar{f})\,d\mu\\ &\leqslant \|f\|_p\|P^*(|f|^{p-1}\operatorname{sgn}\bar{f})\|_q\\ &\leqslant \|f\|_p\left\||f|^{p-1}\operatorname{sgn}\bar{f}\right\|_q\\ &= \|f\|_p\|f\|_p^{p/q}\\ &= \|f\|_p^p.\end{aligned}$$

The conditions for equality in Holder's inequality lead to $P^*(|f|^{p-1}\operatorname{sgn}\bar{f}) = |f|^{p-1}\operatorname{sgn}\bar{f}$ as required. This proves necessity. Sufficiency follows dually. ∎

We next generalize an argument contained in Ando's Theorem 1 (see [13]).

Lemma 6. *Suppose* $1<p<\infty$, $p\neq 2$; *and let* P *be a contractive projection on* $L_p(T,\Sigma,\mu,\mathbb{C})$; *if* $f\in R(P)$ *then,*

(i) $|f|\operatorname{sgn} g\in R(P)$ $\quad (g\in R(P))$,

(ii) $PJ_f=J_fP$,

(iii) $P(h\operatorname{sgn} f)=|P(h\operatorname{sgn} f)|\operatorname{sgn} f$ $\quad (0\leqslant h\in L_p(\mu,\mathbb{C}))$.

Proof. (i) Suppose first that $p>2$, let $\lambda\in\mathbb{R}$, $0\leqslant|\lambda|<1$, and let $g\in R(P)$. By Lemma 5, $g_\lambda=\lambda^{-1}(|f+\lambda g|^{p-1}\operatorname{sgn}\overline{(f+\lambda g)}-|f|^{p-1}\operatorname{sgn}\bar{f})\in R(P^*)$. Since $p>2$,

$$\begin{aligned} g_\lambda &= \lambda^{-1}[(|f+\lambda|^{p-2}-|f|^{p-2})(\overline{f+\lambda g})+|f|^{p-2}\cdot\lambda g] \\ &= \lambda^{-1}[(|f+\lambda g|^{p-2}-|f|^{p-2})\overline{(f+\lambda g)}]+|f|^{p-2}g\,. \end{aligned}$$

Recall, that for real λ and complex w,z, $\left.\dfrac{d}{d\lambda}|w+\lambda z|\right|_\lambda=\operatorname{Re} z\operatorname{sgn}\overline{w+\lambda z}$, provided $w+\lambda z\neq 0$. It follows that as $\lambda\to 0$,

$$g_\lambda\to(p-2)|f|^{p-3}\operatorname{Re}(g\operatorname{sgn}\bar{f})\cdot\bar{f}+|f|^{p-2}\bar{g}$$

at all points of T where $f\neq 0$.

If $2|\lambda g|<|f|$ we have $|f|/2<|f+\lambda\theta g|<2|f|$ if $0<\theta<1$; and, by the mean value theorem there exists θ, $0<\theta<1$, such that

$$\begin{aligned} |g_\lambda| &\leqslant (p-2)|f+\theta\lambda g|^{p-3}\,|\operatorname{Re}(g\operatorname{sgn}\overline{(f+\theta\lambda g)})|\,|f+\lambda g|+|f|^{p-2}|g| \\ &\leqslant 2(p-2)|f|\,2^{p-3}|f|^{p-3}|g|+|f|^{p-2}|g| \\ &= ((p-2)2^p+1)|f|^{p-2}|g|\in L_q(\mu,\mathbb{C})\,. \end{aligned}$$

If $2|\lambda g|\geqslant|f|$, then $|f+\lambda g|\leqslant 3|\lambda g|$ and

$$\begin{aligned} |g_\lambda| &\leqslant \lambda^{-1}[(3|\lambda g|)^{p-2}+(2|\lambda g|)^{p-1}] \\ &= (3^{p-1}+2^{p-1})|g|^{p-1}|\lambda|^{p-2} \\ &\leqslant (2^{p-1}+3^{p-1})|g|^{p-1}\in L_q(\mu,\mathbb{C})\,. \end{aligned}$$

Thus we have that $g_\lambda\to 0$ $(\lambda\to 0)$ if $f=0$. This shows that g_λ converges to

$$g_0=(p-2)|f|^{p-2}\operatorname{sgn}\bar{f}\operatorname{Re}(g\operatorname{sgn}\bar{f})+|f|^{p-2}(g)\,,$$

pointwise almost everywhere on T and that the convergence is dominated by an element of $L_q(\mu,\mathbb{C})$. Hence $\|g_\lambda-g_0\|_q\to 0$ and $g_0\in R(P^*)$ because $R(P^*)$ is closed.

By the same argument, applied to $-ig$, we have using $\operatorname{Re}(-iz)=\operatorname{Im} z$,

$$k_0=(p-2)|f|^{p-2}\operatorname{sgn}\bar{f}\operatorname{Im}(g\operatorname{sgn}\bar{f})+i|f|^{p-2}g\in R(P^*).$$

Now,

$$\begin{aligned} g_0-ik_0&=(p-2)|f|^{p-2}\operatorname{sgn}\bar{f}(g\operatorname{sgn}\bar{f})+2|f|^{p-2}\bar{g}\\ &=(p-2)|f|^{p-2}\operatorname{sgn}\bar{f}\,g\operatorname{sgn}f+2|f|^{p-2}\bar{g}\\ &=p|f|^{p-2}\bar{g}\in R(P^*). \end{aligned}$$

(Note that this last is valid in the real case too.)

Using Lemma 5 again, we conclude that $|f|^{p-2}g^{q-1}\operatorname{sgn}\overline{|f|^{p-2}\bar{g}}=|f|^{1-(q-1)}|g|^{q-1}\operatorname{sgn}g\in R(P)$. Set

$$k_n=|f|^{1-(q-1)^n}|g|^{(q-1)^n}\operatorname{sgn}g \qquad (n=1,2,\ldots).$$

We have just shown that $k_1\in R(P)$ and the same method, applied inductively, gives $k_n\in R(P)$ for all n. Since $0<q-1<1$,

$$|k_n|\leqslant\max\{|f|,|g|\}\leqslant|f|+|g|\in L_p(\mu,\mathbb{C}),$$

so $\{k_n\}$ is dominated in $L_p(\mu,\mathbb{C})$. Since $k_n\to|f|\operatorname{sgn}g$ μ almost everywhere on T, we have $\|k_n-|f|\operatorname{sgn}g\|_p\to 0$ and since $R(P)$ is closed, $|f|\operatorname{sgn}g\in R(P)$ which proves (i) for $p>2$.

Suppose $1<p<2$; as we have already stated P^* is a contractive projection on $L_q(\mu,\mathbb{C})$, and $q>2$. By Lemma 5, $f_1=|f|^{p-1}\operatorname{sgn}\bar{f}$ and $g_1=|g|^{p-1}\operatorname{sgn}\bar{g}$ are in $R(P^*)$. By our proof above $|f_1|\operatorname{sgn}g_1=|f|^{p-1}\operatorname{sgn}\bar{g}\in R(P^*)$, and, by Lemma 5 again, $|f|\operatorname{sgn}g\in R(P)$.

This completes the proof of (i).

For (ii) we have by (i), that $|f|\operatorname{sgn}Pk\in R(P)$ $(k\in L_p(\mu,\mathbb{C}))$. By (i) again,

$$J_f Pk=|Pk|\operatorname{sgn}(|f|\operatorname{sgn}Pk)\in R(P).$$

Thus $J_fP=PJ_fP$. Further, since P^* is a contractive projection on $L_q(\mu,\mathbb{C})$, and $|f|^{p-1}\operatorname{sgn}\bar{f}\in R(P^*)$ we have $J_gP^*=P^*J_gP^*$ with $g=|f|^{p-1}\operatorname{sgn}\bar{f}$. In addition $J_g=J_f^*$, since J_g and J_f are each multiplication by the same characteristic function. We conclude

$$J_fP=PJ_fP=(P^*J_f^*P^*)^*=(P^*J_gP^*)^*=(J_gP^*)^*=PJ_f,$$

which is (ii).

(iii) The proof is like the proof of Lemma 4 (ii). Suppose $0\leqslant h\leqslant|f|$. By (i), $|f|\operatorname{sgn}P(h\operatorname{sgn}f)\in R(P)$, so by Lemma 5, $|f|^{p-1}\operatorname{sgn}\overline{P(h\operatorname{sgn}f)}\in R(P^*)$. Hence,

$$\begin{aligned} \int|P(h\operatorname{sgn}f)|\,|f|^{p-1}d\mu&=\int P(h\operatorname{sgn}f)\cdot|f|^{p-1}\operatorname{sgn}\overline{P(h\operatorname{sgn}f)}\,d\mu\\ &=\int h\operatorname{sgn}f\cdot|f|^{p-1}\operatorname{sgn}\overline{P(h\operatorname{sgn}f)}\,d\mu\\ &\leqslant\int h|f|^{p-1}d\mu. \end{aligned}$$

Also $0 \leqslant |f - h \operatorname{sgn} f| = |f| - h \leqslant |f|$. Hence,

$$\begin{aligned} \|f\|_p^p &= \int |P(|f| \operatorname{sgn} f)| \, |f|^{p-1} d\mu \\ &= \int |P(h \operatorname{sgn} f) + P((|f| - h) \operatorname{sgn} f)| \, |f|^{p-1} d\mu \\ &\leqslant \int |P(h \operatorname{sgn} f)| \, |f|^{p-1} d\mu + \int |P((|f| - h) \operatorname{sgn} f)| \, |f|^{p-1} d\mu \\ &\leqslant \int h |f|^{p-1} d\mu + \int (|f| - h) |f|^{p-1} d\mu \\ &= \|f\|_p^p . \end{aligned}$$

We have equality at each stage and hence, (μ almost everywhere),

$$|f| = P(|f| \operatorname{sgn} f) = |P(h \operatorname{sgn} f)| + |f - P(h \operatorname{sgn} f)| .$$

This proves (iii) for $0 \leqslant h \leqslant |f|$. The extension to $0 \leqslant h \in L_p(\mu, \mathbb{C})$ is the same as in the proof of Lemma 4 (ii) and (iii) so we are done. ∎

We now develop the notion of a conditional expectation operator. First we need the necessary σ subring.

Lemma 7. *Suppose $1 \leqslant p < \infty$, $p \neq 2$, and let P be a contractive projection on $L_p(T, \Sigma, \mu, \mathbb{C})$. Define Σ_0 to be the set of supports of all functions whose equivalence classes are in $R(P)$; then*

(i) $PJ_g f = J_g f \quad (f, g \in R(P))$;

(ii) *Σ_0 is a σ subring of Σ.*

Proof. (i) By lemma 6 (ii), (i) is valid if $p \neq 1$. We give a proof that uses only the identity $J_g P J_g = P J_g$ valid for $1 \leqslant p < \infty$, $p \neq 2$ (Lemma 4 (i) or 6 (ii) weakened). Since $f - J_g f \in g^{\perp}$ and $J_g f - P J_g f \in g^{\perp\perp}$, we have

$$\begin{aligned} \|P(f - J_g f)\|^p &= \|f - P J_g f\|^p \\ &= \|f - J_g f\|^p + \|J_g f - P J_g f\|^p \\ &\geqslant \|P(f - J_g f)\|^p + \|J_g f - P J_g f\|^p . \end{aligned}$$

Thus $P J_g f = J_g f$ which is (i).

(ii) By (i), $S(f) \setminus S(g) = S(f - J_g f) = S(P(f - J_g f)) \in \Sigma_0$. Thus Σ_0 is closed under differences. If $\{f_n\}$ is a sequence of nonzero elements in $R(P)$ such that $S(f_n) \cap S(f_m) = \emptyset \, (m \neq n)$ then $f = \Sigma 2^{-n} \|f_n\|^{-1} f_n \in R(P)$ and $S(f) = \bigcup S(f_n)$. This proves (ii). ∎

Corollary. *Let P be a contractive projection on $L_p(T, \Sigma, \mu, \mathbb{C})$. $(1 \leqslant p < \infty, p \neq 2)$. If $h \in R(P)^{\perp\perp}$, there exists $f \in R(P)$ such that $h \in f^{\perp\perp}$.*

Proof. By Lemma 3 there is a sequence $\{f_n\}$ in $R(P)$ such that $h = \lim f_{S(f_n)} h$. Choose $f \in R(P)$ such that $S(f) = \bigcup S(f_n)$, then $h \in f^{\perp\perp}$. ∎

Observe now that if $f \in L_p(\mu, \mathbb{C})$ the measure $|f|^p d\mu$ restricted to any subring, Σ_0, of Σ, is finite. By the Radon-Nikodym theorem we may define the *conditional expectation operator*, $\mathscr{E}_f = \mathscr{E}(\Sigma_0, |f|^p)$, for the measure $|f|^p d\mu$ relative to Σ_0. $\mathscr{E}_f$ is uniquely determined by the equation

$$\int_A h|f|^p d\mu = \int_A (\mathscr{E}_f h)|f|^p d\mu \qquad (A \in \Sigma_0)$$

for $h \in L_1(T, \Sigma, |f|^p d\mu, \mathbb{C})$, and the condition that $\mathscr{E}_f h$ is Σ_0 measurable.

Lemma 8. *Suppose* $1 \leqslant p < \infty$, $p \neq 2$; *let* P *be a contractive projection on* $L_p(T, \Sigma, \mu, \mathbb{C})$ *and let* Σ_0 *be the* σ *subring of* Σ, *consisting of supports of functions in* $R(P)$. *If* $M_f = f^{-1} J_f R(P) = \{f^{-1} J_f g : g \in R(P)\}$, *then* $M_f = L_p(S(f), \Sigma_0 | S(f), |f|^p d\mu, \mathbb{C})$ *where* $\Sigma_0 | S(f) = \{A \in \Sigma_0 : A \subset S(f)\}$ *and we make the obvious identification of functions on* $S(f)$ *and functions on* T *which vanish off* $S(f)$. *In addition the map* $h \to f^{-1} h$ *is an isometric isomorphism between* $J_f R(P)$ *and* $L_p(S(f), \Sigma_0 | S(f), |f|^p d\mu, \mathbb{C})$.

Proof. Observe that $|f|^p d\mu$ is finite on $S(f)$, and that the isometry claim is obviously true. If $A \in \Sigma_0 | S(f)$ then $A = S(g)$ for some $g \in R(P)$. By lemmas 4 and 7 (if $p=1$) or 6 (if $p>1$) we have $J_g f = P J_g f$ so that $f_A = f^{-1} J_g f \in M_f$. Let h be a simple function with respect to $\Sigma_0 | S(f)$. Then $h \in M_f$ and $hf \in R(P)$. In addition

$$\int_{S(f)} |h|^p \cdot |f|^p d\mu = \int_T |hf|^p d\mu .$$

We conclude that

$$M_f \supset L_p(S(f), \Sigma_0 | S(f), |f|^p d\mu, \mathbb{C})$$

Conversely, let $h \in M_f$, then $h \in L_p(S(f), \Sigma | S(f), |f|^p d\mu, \mathbb{C})$ and it is enough to show that h is Σ_0 measurable. Let $g = (\operatorname{Re} h)^+$, then $gf \in L_p(T, \Sigma, \mu, C)$. By Lemma 4(ii) and Lemma 6(iii)

$$P(gf) = P(|gf| \operatorname{sgn} f) = |P(|gf| \operatorname{sgn} f)| \operatorname{sgn} f$$

so $f^{-1} P(gf) = |f|^{-1} |P(|gf| \operatorname{sgn} f)| \in M_f$. It follows that

$$\operatorname{Re} h = f^{-1} P((\operatorname{Re} h)^+ f) - f^{-1} P((\operatorname{Re} h)^- f) \in M_f .$$

Since each of these functions is positive it is sufficient to consider $0 \leqslant h \in M_f$. Suppose $\alpha > 0$ and put $k = h \vee \alpha f_{S(f)}$. Arguing as above, we have $f^{-1} P(kf) \geqslant h$ and $f^{-1} P(kf) \geqslant \alpha f_{S(f)}$ so that $f^{-1} P(kf) \geqslant k \geqslant 0$. Since P is contractive we have

$$\|kf\|^p \geqslant \|P(kf)\|^p = \|P(kf) - kf + kf\|^p \geqslant \|P(kf) - kf\|^p + \|kf\|^p .$$

This gives $P(kf) = kf$, so that $k \in M_f$. This shows, incidentally, that M_f is a lattice. For our purpose, however, we have

$$\begin{aligned} \{t \in S(f) : h(t) > \alpha\} &= \{t \in S(f) : (k - \alpha f_{S(f)})(t) \neq 0\} \\ &= S(kf - \alpha f) \in \Sigma_0 . \end{aligned}$$

Thus M_f consists of Σ_0 measurable functions and we are done. ∎

Theorem 1. *Suppose* $1 \leqslant p < \infty$, $p \neq 2$ *and that* P *is a contractive projection on* $L_p(T, \Sigma, \mu, \mathbb{C})$. *If* $f \in R(P)$ *and* $h \in f^{\perp\perp}$ *then*

$$Ph = f\,\mathscr{E}(\Sigma_0, |f|^p)(hf^{-1}).$$

Proof. Since $f^{-1}Ph \in M_f$ we know $f^{-1}Ph$ is Σ_0 measurable. Thus we have only to show

$$\int_A f^{-1}Ph|f|^p d\mu = \int_A hf^{-1} \cdot |f|^p d\mu \qquad (A \in \Sigma_0).$$

Choose $g \in R(P)$ such that $A = S(g)$. By Lemma 4 (i), $u = J_g f \in R(P)$.

Suppose $p = 1$ and $0 \leqslant k \in L_1(\mu, \mathbb{C})$. By Lemma 6 (ii) and (iii), $\int_A k \operatorname{sgn} f \cdot f^{-1}|f|\,d\mu = \int_{A \cap S(f)} k\,d\mu = \|J_u k\| = \|P(k \operatorname{sgn} u)\| = \| |P(k \operatorname{sgn} J_g f)| \times \operatorname{sgn} J_g f \| = \int_A f^{-1} P(J_g k \operatorname{sgn} f) \cdot |f|\,d\mu$. Putting $v = f - u = f - J_g f \in R(P)$, we have, by Lemma 4 (i),

$$P(k \operatorname{sgn} f) = J_u P(J_u k \operatorname{sgn} f) + J_v P(J_v k \operatorname{sgn} f).$$

Hence

$$\int_A f^{-1} P(J_g k \operatorname{sgn} f) \cdot |f|\,d\mu = \int_A f^{-1} P(k \operatorname{sgn} f) \cdot |f|\,d\mu.$$

We conclude that

$$\int_A hf^{-1} \cdot |f|\,d\mu = \int_A f^{-1} P(hf^{-1}) \cdot |f|\,d\mu$$

for all $h \in f^{\perp\perp}$ and all $A \in \Sigma_0$ so we are finished for $p = 1$.

If $p > 1$ we have $PJ_g = J_g P$ by Lemma (6) (ii) and $|f|^{p-1} \operatorname{sgn} \bar{f} \in R(P^*)$ by Lemma 5. Hence,

$$\begin{aligned}
\int_A hf^{-1} \cdot |f|^p d\mu &= \int_T J_g h \cdot |f|^{p-1} \operatorname{sgn} \bar{f}\,d\mu \\
&= \int_T J_g h \cdot P(|f|^{p-1} \operatorname{sgn} \bar{f})\,d\mu \\
&= \int_T PJ_g h \cdot |f|^{p-1} \operatorname{sgn} \bar{f}\,d\mu \\
&= \int_T J_g Ph \cdot f^{-1}|f|^p d\mu \\
&= \int_A f^{-1} Ph \cdot |f|^p d\mu \qquad (A \in \Sigma_0).
\end{aligned}$$

Thus

$$Ph = f^{-1}\mathscr{E}(\Sigma_0, |f|^p)(hf^{-1}) \qquad (h \in f^{\perp\perp})$$

as claimed. ∎

Theorem 2. *Suppose* $1 \leqslant p < \infty$, $p \neq 2$, *let* P *be a contractive projection on* $L_p(T, \Sigma, \mu, \mathbb{C})$ *and let* J *be the band projection on* $R(P)^{\perp\perp}$*; then* PJ *is the unique contractive projection on* L_p *which satisfies* $R(PJ) = R(P)$ *and* $PJ\,R(P)^{\perp} = \{0\}$. *If* $p \neq 1$, $P = PJ$ *so* P *is uniquely determined by its range. If* $p = 1$, *and* A *is a linear contraction on* $L_1(\mu, \mathbb{C})$ *which satisfies* $PA = A$ *and* $AJ = 0$, *then* $PJ + A$ *is a contractive projection on* $L_1(\mu, \mathbb{C})$ *with the same range as* P.

Proof. Let Q be a contractive projection on $L_p(\mu, \mathbb{C})$ such that $R(Q) = R(P)$ and $Q\,R(P)^{\perp} = \{0\}$. Then $Q = QJ$ and if $h \in L_p(\mu, \mathbb{C})$ there exists, by the corollary to lemma 7, $f \in R(P) = R(Q)$ such that $Jh = J_f h$. By Theorem 1 $Qh = QJh = f^{-1}\mathscr{E}(\Sigma_0, |f|^p)(Jh \cdot f^{-1}) = PJh$. Thus $Q = PJ$. (It is clear that PJ satisfies the stated conditions.)

If $p \neq 1$ take h, f as above and put $u = Ph - PJh = Ph - PJ_f h = Ph - J_f Ph$, by lemma 6 (ii). Since band projections commute and $u \in R(P) \cap f^{\perp}$, $J_u h = J_u J h = J_u J_f h = 0$. By lemma 6 (ii) again,

$$u = J_u u = J_u Ph - J_u PJ_f h = PJ_u h - J_u J_f Ph = 0 - 0 = 0.$$

Hence $P = PJ$ as required.

If $p = 1$, $PA = A$ and $AJ = 0$, we have $AP = AJP = 0$ and $A^2 = APA = 0$. Also $(PJ + A)^2 = PJPJ + PJA + APJ + A^2 = PPJ + PJPA$ $0 + 0 = PJ + A$. Thus $PJ + A$ is a projection. Observe that

$$R(PJ + A) = R(PJ + PA) \subset R(P) = R(PJP + AP) = R((PJ + A)P \subset R(PJ + A)$$

It remains to show that if A is contractive, $PJ + A$ is contractive. If $h \in L_1(\mu, \mathbb{C})$,

$$\begin{aligned} \|(PJ + A)h\|_1 &= \|PJh + A(h - Jh)\|_1 \\ &\leqslant \|PJh\|_1 + \|A(h - Jh)\|_1 \\ &\leqslant \|Jh\|_1 + \|h - Jh\|_1 \\ &= \|Jh + h - Jh\|_1 \\ &= \|h\|_1 . \quad \blacksquare \end{aligned}$$

We are now ready to prove the equivalences of various conditions on a subspace of $L_p(\mu, \mathbb{C})$ so that it is the range of a contractive projection.

Let $\mathscr{S}(T, \Sigma)$ denote the set of Σ measurable functions h such that $S(h)$ is σ finite. By a *multiplication operator* on $\mathscr{S}(T, \Sigma)$ we mean a map $h \to kh$ defined for functions h in some subset of $\mathscr{S}(T, \Sigma)$ and some fixed Σ measurable function k. If k satisfies $|k| = 1$ on $S(k)$ we will call k a *unitary multiplication.*

A multiplication operator on $\mathscr{S}(T, \Sigma)$ preserves equality almost everywhere and hence induces a multiplication operator on each $L_p(T, \Sigma, \mu, \mathbb{C})$ into $\mathscr{S}(T, \Sigma)$ modulo null functions $(1 \leqslant p < \infty)$. Further, k_1 and k_2 will induce the same such multiplication operator on $L_p(\mu, \mathbb{C})$ if k_1 and k_2 agree locally almost everywhere.

Suppose that $\mathscr{K}$ is a set of Σ measurable functions such that if $k_1, k_2 \in \mathscr{K}$ and $k_1 \neq k_2$, $\mu(S(k_1) \cap S(k_2)) = 0$. If $f \in \mathscr{S}(T, \Sigma)$ then, because $S(f)$ has σ finite measure, $S(f)$ meets at most countably many $S(k)$, with $k \in \mathscr{K}$, in a set of positive measure. Enumerate these as $\{k_n\}$, then there is a unique set $N \in \Sigma$ such that, $N \subset S(f)$ and each $t \in S(f) \setminus N$ lies in at most one set $S(k_n)$. (In fact $N = \bigcup_{1 \leqslant n < m < \infty} (S(k_n) \cap S(k_m))$.) On $S(f) \setminus N$ the series $\sum_{n=1}^{\infty} f(t) k_n(t)$ has at most one nonzero term. Thus $\mathscr{K}$

determines a map $U_{\mathscr{K}}:\mathscr{S}(T,\Sigma)\to\mathscr{S}(T,\Sigma)$ by taking, for f as above, $U_{\mathscr{K}}f(t)=\sum_{n=1}^{\infty} f(t)k_n(t)$ for $t\in S(f)\setminus N$ and $U_{\mathscr{K}}f(t)=0$ elsewhere. We call $U_{\mathscr{K}}$ the *direct sum* of the (disjoint) multiplication operators induced by the elements of $\mathscr{K}$. If $U_{\mathscr{K}}$ maps $L_p(\mu,\mathbb{C})$ to $L_p(\mu,\mathbb{C})$ $(1\leqslant p<\infty)$ it is not hard to check that the net of finite sums of the multiplication operators in $\mathscr{K}$ is strongly convergent to $U_{\mathscr{K}}$.

We can now state the main theorem. The equivalence of (i) and (ii) generalizes theorem 4 of [13] and extends theorem 6 of [272].

Theorem 3. *Suppose $1\leqslant p<\infty$ and $p\neq 2$ and let M be a subspace of $L_p(T,\Sigma,\mu,\mathbb{C})$. Then the following conditions on M are equivalent.*

(i) *M is the range of a contractive projection on $L_p(\mu,\mathbb{C})$.*

(ii) *There is a measure space (Ω,Ξ,λ) such that M is isometrically isomorphic to $L_p(\Omega,\Xi,\lambda,\mathbb{C})$.*

(iii) *There is a direct sum of unitary multiplication operators $U:L_p(T,\Sigma,\mu,\mathbb{C})\to L_p(T,\Sigma,\mu,\mathbb{C})$ such that U is an isometry and UM is a closed vector sublattice of $L_p(T,\Sigma,\mu,\mathbb{C})$.*

Furthermore, in (ii) *we can always choose $\Omega=T$, Ξ a σ subring of Σ, λ absolutely continuous with respect to μ, and the isometry a direct sum of multiplication operators.*

If μ is σ finite, then the direct sum of multiplication operators can be taken to be ordinary multiplications.

Proof. Assume (i). By Zorn's Lemma there is a maximal subset $\mathscr{K}$ of M consisting of functions $f\in M$, such that $\mu(S(f_1)\cap S(f_2))=0$ if $f_1\neq f_2$. If $g\in M$, $S(g)$ is σ finite and there is a countable subset $\{f_n\}$ of $\mathscr{K}$ such that if $f\in\mathscr{K}\setminus\{f_n\}$, $\mu(S(f)\cap S(g))=0$. By Lemma 7, Σ_0 is a σ ring so, there exists $h\in M$ such that $S(h)=S(g)\setminus\bigcup S(f_n)$ and by maximality of $\mathscr{K}$, $h=0$. Define a measure λ on Σ_0 by $\lambda A=\sum_{f\in\mathscr{K}}\int_A|f|^p d\mu$. This definition is meaningful since A has σ finite μ measure and at most countably many of the integrals are nonzero. For $f\in\mathscr{K}$ define f^{-1} by

$$f^{-1}(t)=\begin{cases}1/f(t) & t\in S(f),\\ 0 & t\notin S(f),\end{cases}$$

and let V be the direct sum of the multiplications f^{-1} $(f\in\mathscr{K})$. By Lemma 9 $J_f M\to f^{-1}M$ is an isometric isomorphism of $J_f M$ with $L_p(S(f),\Sigma_0|S(f),|f|^p\mu,\mathbb{C})$. It is routine to check that V is an isometric isomorphism of M with $L_p(T,\Sigma_0,\lambda,\mathbb{C})$. ($M$ is the direct sum of its subspaces $J_f M$ $(f\in\mathscr{K})$ and similarly for the L_p-spaces.)

If μ is σ finite $\mathscr{K}$ will be countable, say $\mathscr{K}=\{f_n\}$ and we can find $f\in M$ such that $S(f)=\bigcup S(f_n)$. Then Σ_0 consists entirely of subsets of $S(f)$ and sets of measure zero so that $M_f = L_p(T, \Sigma_0, |f|^p d\mu, \mathbb{C})$, $J_f M = M$, and V can be multiplication by f^{-1}.

Assume (ii) and let $L: L_p(\Omega, \Xi, \lambda, \mathbb{C}) \to L_p(T, \Sigma, \mu, \mathbb{C})$ be a linear isometry with range M. Suppose $a, b \in L_p(\Omega, \Xi, \lambda, C)$ and $|a| \wedge |b| = 0$, we claim that $|La| \wedge |Lb| = 0$.

(At this point we need the fact that $|a| \wedge |b| = 0$ if and only if $\|a+b\|^p + \|a-b\|^p = 2(\|a\|^p + \|b\|^p)$.) This is due to Lamperti [177] and the reader is referred to [253] for a proof of it. Now, since L is a linear isometry, $\|La+Lb\|^p + \|La-Lb\|^p = 2(\|La\|^p + \|Lb\|^p)$ so that $|La| \wedge |Lb| = 0$.

Take a maximal subset of Ξ consisting of sets of nonzero finite λ measure which intersect pairwise in sets of λ measure zero and let $\mathscr{K}$ be the corresponding set of characteristic functions. Let $a \in \mathscr{K}$ and suppose $B \in \Xi$ and $B \subset S(a)$. Write $b = f_B$, then $L(a-b)$, Lb are disjoint in M so we have $Lb = |Lb| \operatorname{sgn} Ta$. This extends to positive simple functions b in $a^{\perp\perp}$ and then to all positive $b \in a^{\perp\perp}$. Define $U: L_p(T, \Sigma, \mu, \mathbb{C}) \to L_p(T, \Sigma, \mu, \mathbb{C})$ to be the direct sum of the unitary multiplications $\operatorname{sgn} \overline{La}$ $(a \in \mathscr{K})$. It is easy to see that UL is positive and hence $UM = UL[L_p(\Omega, \Xi, \lambda, \mathbb{C})]$ is a closed vector sublattice of $L_p(T, \Sigma, \mu, \mathbb{C})$ (compare the proof in Lemma 9 where we showed that functions in M_f were Σ_0 measurable) since U is an isometry.

Assume (iii) and let Σ_0 be the set of supports of functions (whose equivalence classes are) in M. Then Σ_0 is a σ subring of Σ. (If $\{f_n\}$ is a sequence in M, $S(f_n) = S(Uf_n) = S(|Uf_n|)$ so $\bigcup S(f_n) = S(U^{-1} \Sigma 2^{-n} \|f_n\|^{-1} |Uf_n|)$. If $f, g \in M$, $J_g = J_{Ug}$;

$$J_g |Uf| = \lim |Uf| \wedge n |Ug| \in UM$$

and $S(f) \setminus S(g) = S(U^{-1}(|Uf| - J_g |Uf|))$.) Let $f, g \in UM$ and suppose f is real, $g \geqslant 0$ and $f \in g^{\perp\perp}$, then $\{t \in T: (f/g)(t) > \alpha\} = S((f - \alpha g)^+) \in \Sigma_0$. Thus f/g is Σ_0 measurable. This extends to all $f \in UM \cap g^{\perp\perp}$ and hence $J_f g / J_g f$ is Σ_0 measurable if $f, g \in UM$ and $g \geqslant 0$. This now extends to all $f, g \in UM$ and, since $U^{-1} J_f g / U^{-1} J_g f = J_f g / J_g f$ we have f/g is Σ_0 measurable for $f, g \in M$ and $f \in g^{\perp\perp}$. It follows that M is the set of all elements in $L_p(T, \Sigma, \mu, \mathbb{C})$ which can be written in the form hf with h, Σ_0 measurable and $f \in M$. (If $h = f_{S(g)}$ with $g \in M$, $hf = J_g f = U^{-1} J_{Ug} U f \in U^{-1}(UM) = M$.)

Let J be the band projection on $M^{\perp\perp}$, let $h \in L_p(T, \Sigma, \mu, \mathbb{C})$ choose $f \in M$ such that $Jh = J_f h$, (such an f exists by the argument used in the corollary to lemma 7) and define

$$Ph = f \mathscr{E}(\Sigma_0, |f|^p)(hf^{-1}).$$

Then $Ph \in M$ and this definition is independent of the choice of f in M such that $h \in f^{\perp\perp}$. To see this suppose $g \in M$ and $h \in g^{\perp\perp}$. Then

h is zero outside $S(f)\cap S(g)\in\Sigma_0$ and so is $\mathscr{E}(\Sigma_0,|f|^p)(hf^{-1})$. Let $B=S(f)\cap S(g)$, then $f_1=f_B\in M$ and

$$\int_A hf^{-1}|f|^p d\mu=\int_{A\cap B} hf^{-1}|f|^p d\mu=\int_A hf_1^{-1}|f_1|^p d\mu \qquad (A\in\Sigma_0),$$

so that $f\mathscr{E}(\Sigma_0,|f|^p)(hf^{-1})=f_1\mathscr{E}(\Sigma_0,|f_1|^p)(hf^{-1})$. Thus we may assume $S(f)=S(g)$. Now

$$g^{-1}f\mathscr{E}(\Sigma_0,|f|^p)(hf^{-1})\in L_1(X,\Sigma_0,|g|^p d\mu,\mathbb{C}),$$

so we have, for $A\in\Sigma_0$,

$$\int_A g^{-1}f\mathscr{E}(\Sigma_0,|f|^p)(hf^{-1})|g|^p d\mu=\int_A g^{-1}f|f^{-1}g|^p\mathscr{E}(\Sigma_0,|f|^p)(hf^{-1})|f|^p d\mu.$$

Because $g^{-1}f$ and $f^{-1}g$ are Σ_0 measurable and the integrals are finite, the second integral is

$$\int_A g^{-1}f|f^{-1}g|^p hf^{-1}|f|^p d\mu=\int_A hg^{-1}|g|^p d\mu.$$

Thus

$$f\mathscr{E}(\Sigma_0,|f|^p)(hf^{-1})=g\mathscr{E}(\Sigma_0,|g|^p)(hg^{-1})$$

and our definition of Ph is unambiguous. If $h_1,h_2\in L_p(\mu,\mathbb{C})$ we can take $f\in M$ such that $Jh_1=J_f h_1$ and $Jh_2=J_f h_2$. Thus P is linear. Since $f^{-1}Ph=\mathscr{E}(\Sigma_0,|f|^p)(hf^{-1})$ we see $P^2=P$. Finally, if $p>1$, write $u=\mathscr{E}(\Sigma_0,|f|^p)(hf^{-1})$, we have

$$\|Ph\|_p^p=\int|u|^{p-1}\operatorname{sgn}\bar{u}\cdot\mathscr{E}(\Sigma_0,|f|^p)(hf^{-1})|f|^p d\mu.$$

Since u is Σ_0-measurable, this is

$$\begin{aligned}\int|u|^{p-1}\operatorname{sgn}\bar{u}\cdot hf^{-1}|f|^p d\mu &\leqslant \left(\int\left||u|^{p-1}\operatorname{sgn}\bar{u}\right|^q|f|^p d\mu\right)^{1/q}\left(\int|hf^{-1}|^p|f|^p d\mu\right)^{1/p}\\ &=\left(\int|Ph|^p\right)^{1/q}\|h\|_p\\ &=\|Ph\|_p^{p/q}\|h\|_p.\end{aligned}$$

(We used Hölder's inequality for the measure $|f|^p d\mu$, q for the conjugate index to p.) We conclude $\|Ph\|_p\leqslant\|h\|_p$ if $p>1$ and the proof if $p=1$ is trivial.

Since $Ph=h$ $(h\in M)$ we have shown that M is the range of the contractive projection P. ∎

Corollary. *If M is a closed vector sublattice of $L_p(\mu,\mathbb{C})$ $(1\leqslant p<\infty, p\neq 2)$, then M is the range of a positive contractive projection.*

Proof. Clearly M satisfies condition (iii) with $U=I$. In the definition of Ph we may always choose a positive $f\in M$ such that $h\in f^{\perp\perp}$. Positivity of P follows from positivity of conditional expectation. ∎

We now wish to prove a Banach space characterization of $L_p(\mu,\mathbb{C})$. This will require the notion of an $\mathscr{L}_p$ space which is due to Lindenstrauss and Pełczyński [188].

If X and Y are isomorphic Banach spaces, then we put $d(X,Y) = \inf\{\|L\|\,\|L^{-1}\| : L$ is a linear isomorphism of X onto $Y\}$. Clearly $d(X,Y)\geqslant 1$ and if X and Y are linearly isometric, then $d(X,Y)=1$ (it is not true that $d(X,Y)=1$ implies that X and Y are linearly isometric, see exercise 1).

Definition 1. Let $1\leqslant p\leqslant\infty$ and $1\leqslant\lambda$. We say that X is a $\mathscr{L}_{p,\lambda}$ *space* if for each finite dimensional subspace Y of X, then is a finite dimensional subspace Z of X which contains Y and we have that $d(Z, l_p(n,\mathbb{C}))\leqslant\lambda$, where n is the Hamel dimension of Z.

We say that X is an $\mathscr{L}_p$ space if it is an $\mathscr{L}_{p,\lambda}$ space for some λ.

The main theorems we wish to prove are that for $1<p<\infty$ a $\mathscr{L}_p$ space is linearly isomorphic to a complemented subspace of $L_p(\mu,\mathbb{C})$ for some measure μ and that a $\mathscr{L}_1$ space has the property that its dual is a $\mathscr{P}(\mathbb{C})$ space (see section 11). We use these to prove that a Banach space is a $\mathscr{L}_{p,\lambda}$ space for all $\lambda>1$ if and only if it is linearly isometric to $L_p(\mu,\mathbb{C})$ for some measure μ. The reader should check [188] and [190] for many important isomorphic results concerning these spaces such as a complemented subspace of a $\mathscr{L}_p$ space is isomorphic to either a $\mathscr{L}_p$ or a $\mathscr{L}_2$ space ($\mathscr{L}_2$ is isomorphic to a Hilbert space) and if X is a $\mathscr{L}_p$ space, then X^* is a $\mathscr{L}_q$ space where $1/p+1/q=1$.

Theorem 4. *Let X be a Banach space and $1\leqslant p<\infty$ and $\lambda\geqslant 1$. If for each finite dimensional space Y of X there is a finite dimensional space $Z\subset l_p(\mathbb{N},\mathbb{C})$ such that $d(Y,Z)\leqslant\lambda$, then there is an abstract L_p space V such that $d(X,X_0)\leqslant\lambda$ for some $X_0\subset V$.*

Proof. Let U^* be the closed unit ball of X^* and $B(U^*,\mathbb{C})$ denote the Banach space of all bounded complex valued functions on U^*. For each $x\in X$, let $f_x = Jx|_{U^*}$ where J is the natural map of X into X^{**}. Clearly $\|f_x\| = \|x\|$ for all $x\in X$. Let Y be a finite dimensional subspace of X. By hypothesis there is a bounded linear operator $L_Y\colon Y\to l_p(\mathbb{N},\mathbb{C})$ such that $\lambda^{-1}\|x\|\leqslant\|L_Y(x)\|\leqslant\|x\|$ for all $x\in Y$. If P_m denotes the natural projection of $l_p(\mathbb{N},\mathbb{C})$ onto $\operatorname{span}(e_1,\ldots,e_m)$, where e_i is the unit vector basis in $l_p(\mathbb{N},\mathbb{C})$, and n is the Hamel dimension of Y, then there is an m such that $\|P_m L_Y x\|\geqslant\left(1-\frac{1}{n}\right)\|L_Y(x)\|$ for all $x\in Y$. Thus $\tilde{L}_Y = P_m L_Y$ is a bounded linear operator from Y into $l_p(m,\mathbb{C})$ such that $\lambda^{-1}\left(1-\frac{1}{n}\right)\|x\| \leqslant\|\tilde{L}_Y x\|\leqslant\|x\|$ for all $x\in Y$. Let q be such that $\frac{1}{p}+\frac{1}{q}=1$ and $\xi_1,\ldots,\xi_m$ the usual basis for $l_q(m,\mathbb{C}) = l_p(m,\mathbb{C})^*$. Let x_i^* be a norm preserving extension of $\tilde{L}_Y^*\xi_i$ for $i=1,\ldots,m$ and ϕ_Y be defined by $\phi_Y(f) = \sum_{i=1}^{m} f(x_i)$

for $f \in B(U^*, \mathbb{C})$. Then ϕ_Y is a positive linear functional and each $x \in Y$, $\phi_Y |f_x|^p = \sum_{i=1}^{m} |L_Y \xi_i(x)|^p = \sum_{i=1}^{m} |\xi_i(L_Y x)|^p = \|L_Y x\|^p$ and hence $\lambda^{-1}\left(1 - \frac{1}{n}\right)\|x\| \leqslant (\phi_Y |f_x|^p)^{1/p} \leqslant \|x\|$ for $x \in Y$. Let $\mathbb{C}_\infty$ denote the Alexandroff one point compactification of the complex numbers and let T be the product of card $B(U^*, \mathbb{C})$ copies of $\mathbb{C}_\infty$. For each finite dimensional subspace Y in X let $\Pi_Y \in T$ be defined by $\Pi_Y(f) = \phi_Y(f)$ for $f \in B(U^*, \mathbb{C})$. Clearly the net $\{\Pi_Y : Y$ a finite dimensional subspace of $X\}$ has a limit point, say Π, in T. Let $Z = \{f \in B(U^*, \mathbb{C}) : \Pi(|f|^p) < \infty\}$. Then Z is a linear subspace and a sublattice of $B(U^*, \mathbb{C})$ and clearly if $f \in Z$, $g \in B(U^*, \mathbb{C})$, and $|g| \leqslant |f|$, then $g \in Z$. Moreover, $p(f) = (\Pi |f|^p)^{1/p}$ defines a seminorm on Z which has the property that if $|f| \wedge |g| = 0$ on U^*, then $p(f+g)^p = p(f)^p + p(g)^p$. Thus the completion of $Z/p^{-1}(0)$ is an abstract L_p space. We denote this space by V and leave the details of the verification that V is indeed an abstract L_p space to the reader.

Moreover, for $x \in X$, $f_x \in Z$ and $\lambda^{-1}\|x\| \leqslant p(f_x) \leqslant \|x\|$. The operator $L : X \to V$ defined by $L(x) = f_x + p^{-1}(0)$ satisfies $\lambda^{-1}\|x\| \leqslant \|L(x)\| \leqslant \|x\|$ for all $x \in X$ and we are done. ∎

We have a stronger result in the case $1 < p < \infty$.

Theorem 5. *Let X be an $\mathscr{L}_p$ space for $1 < p < \infty$. Then there is an abstract L_p space V and a complemented subspace Z of V isomorphic to X.*

Proof. Let $U^*, B(U^*, \mathbb{C})$ and f_x have the same meaning as in theorem 4. Consider the finite dimensional subspaces Y of X where n is the Hamel dimension of Y and $d(Y, l_p(n, \mathbb{C})) \leqslant \lambda$. Since X is an $\mathscr{L}_{p,\lambda}$ space, these spaces are directed under inclusion. For each such Y let $L_Y : Y \to l_p(n, \mathbb{C})$ satisfy $\lambda^{-1}\|x\| \leqslant \|L_Y x\| \leqslant \|x\|$ for all $x \in Y$ and let ϕ_Y be defined relative to L_Y as in theorem 4.

By using only the Y's stated above we obtain by the construction in theorem 4 a space Z and the abstract L_p space V. That is, Z consists of all $f \in B(U^*, \mathbb{C})$ for which $\pi(f) = \lim_Y \phi_Y |f|^p < \infty$ and V is the completion of $Z/\pi^{-1}(0)$. For each such Y let $P_Y : B(U^*, \mathbb{C}) \to Y$ be defined by $P_Y(f) = L_Y^{-1}\left(\sum_{i=1}^{n} f(L_Y^* \xi_i)\eta_i\right)$ where $L_Y^* \xi_i$ denotes a norm preserving extension of $L_Y^*(\xi_i)$ and (ξ_i), (η_i) are the usual bases in $l_q(n, \mathbb{C})$, $l_p(n, \mathbb{C})$ respectively and $n =$ dimension of Y. Then for $x \in Y$, $P_Y f_x = L_Y^{-1}\left(\sum_{i=1}^{n} L_Y^* \xi_i(x)\eta_i\right) = L_Y^{-1}\left(\sum_{i=1}^{n} \xi_i(L_Y x)\eta_i\right) = L_Y^{-1}(L_Y x) = x$. Also since $\|L_Y^{-1}\| \leqslant \lambda$ and $\left\|\sum_{i=1}^{n} f(L_Y^* \xi_i)\eta_i\right\|^p = \phi_Y |f|^p$, $\|P_Y f\| \leqslant \lambda(\phi_Y |f|^p)^{1/p}$ for $f \in B(U^*, \mathbb{C})$. Hence

$\limsup_{Y} \|P_Y f\| < \infty$ if $f \in Z$. Since $1 < p < \infty$, X is reflexive and there is a subset of the P_Y's which converges in the weak operator topology to some bounded linear operator P, that is, $Pf = \lim_{Y'} P_{Y'} f$ in the weak topology for all $f \in Z$ where $\lim_{Y'}$ denotes the limit over a subnet. Since $P_Y f_x = x$ for all $x \in Y$, $Pf_x = x$ for all $x \in X$ and since $\|P_Y\| \leqslant \lambda$ for all Y, $\|P\| \leqslant \lambda$. Passing from Z to V we get an operator $\tilde{P}: V \to X$ with $\|\tilde{P}\| \leqslant \lambda$ and $\tilde{P}Lx = x$ for all $x \in X$ where L is the operator at the end of the proof of theorem 4 namely, $Lx = fx + q^{-1}(0)$ for $x \in X$. Hence $L\tilde{P}$ is a projection of V onto LX which is isomorphic to X. ∎

Corollary 1. *If X is an $\mathscr{L}_{p,\lambda}$ space for all $\lambda > 1$, then X is linearly isometric to a complemented subspace Z of an abstract L_p space V in such a way that Z is the range of a contractive projection on V.*

Proof. In the proofs of theorems 4 and 5 we can replace the nets of finite dimensional subspaces by nets of elements (Y, λ) where in theorem 4 where $d(Y, l_p(n, \mathbb{C})) \leqslant \lambda$. The ordering is given by $(Y_1, \lambda_1) \leqslant (Y_2, \lambda_2)$ if and only if $Y_1 \subset Y_2$ and $\lambda_1 \geqslant \lambda_2$ (we can assume $1 < \lambda < 2$). In particular, the operator L constructed in theorem 4 via the above ordering has the property that $\lambda^{-1}\|x\| \leqslant \|Lx\| \leqslant \|x\|$ for all $1 < \lambda < 2$ and all $x \in X$, that is, L is a linear isometry.

Similarly, by replacing Y with (Y, λ) $(1 < \lambda < 2)$ in the proof of theorem 5 we obtain an operator $P: V \to X$ with $\|P\| \leqslant \lambda$ for all $1 < \lambda < 2$, that is, $\|P\| \leqslant 1$ and PL is a contractive projection of V onto $L(X)$. So by theorem 3, $L(X)$ is an abstract L_p space. ∎

For $p = 1$ we have the following result.

Corollary 2. *If X is an $\mathscr{L}_1$ space, then X^* is a $\mathscr{P}(\mathbb{C})$ space. In particular, if X is an $\mathscr{L}_{1,\lambda}$ space for all $\lambda > 1$, then X^* is a $\mathscr{P}_1$ space and X is linearly isometric to an abstract L_1 space.*

Proof. The reason for restricting p to $1 < p < \infty$ in the above theorem and corollary was to guarantee the existence of subnets which converge in the weak operator topology.

We can avoid this difficulty by embedding X into X^{**} by the natural map J and using the weak* topology on X^{**}. Using this in the proof of theorem 5 we obtain an abstract L_1 space V, an operator $P: V \to X^{**}$, an operator $L: X \to V$ such that $PL = J$. Let Q be the natural map of X^* into X^{**}. Then $(PL)^* Q = J^* Q$ is the identity on X^* and $P^* Q L^*$ is a projection of V^* onto $P^* Q(X^*)$ which is isomorphic under $P^* Q$ to X^*. Since V^* is a $\mathscr{P}(\mathbb{C})$ space, X^* is a $\mathscr{P}(\mathbb{C})$ space.

If X is an $\mathscr{L}_{1,\lambda}$ space for all $\lambda > 1$, then similar to the proof of corollary 1, we can conclude that the projection $P^* Q L^*$ is contractive and P^*Q is an isometric embedding of X^* onto the range of P^*QL^*. Hence X^* is a $\mathscr{P}_1(\mathbb{C})$ space and by theorem 3 X is linearly isometric to an abstract L_1 space. ∎

We shall need the following theorem before proceeding. In it we shall use the facts that all finite dimensional Banach spaces are $\mathscr{P}_\lambda(\mathbb{C})$ spaces (for some $\lambda \geqslant 1$) and that any two n dimensional spaces are linearly isomorphic.

Theorem 6. *Let X be a Banach space and suppose that $\{X_\gamma\}_{\gamma \in \Gamma}$ is an upwards directed family of finite dimensional subspaces of X whose union is dense in X. Then for any $\lambda > 1$ and any finite dimensional subspace Y of X, there is a finite dimensional subspace Z of X containing Y and such that $d(Z, X_\gamma) \leqslant \lambda$ for some γ.*

Proof. Let $y_1, \ldots, y_n$ be a Hamel basis for Y and let $K > 0$ be such that Y is a $\mathscr{P}_K(\mathbb{C})$ space and such that $\sum_{i=1}^n |a_i| \leqslant K \left\| \sum_{i=1}^n a_i y_i \right\|$ for all scalars $a_1, \ldots, a_n$. Let $\varepsilon > 0$ be such that if $\|x_i - y_i\| < \varepsilon$, then $x_1, \ldots, x_n$ is linearly independent. Since the X_γ's are upwards directed, there is a γ such that $\|x_i - y_i\| < \varepsilon$ for some $x_i \in X_\gamma$ $(i = 1, \ldots, n)$. It is easily seen that there is a projection P of X_γ onto $\operatorname{span}(x_1, \ldots, x_n)$ such that $\|P\| \leqslant K \dfrac{1 + K\varepsilon}{1 - K\varepsilon}$. Now, let $x_{n+1}, \ldots, x_{n+p}$ be the completion of $x_1, \ldots, x_n$ to a Hamel basis for X_γ such that $x_{n+1}, \ldots, x_{n+p}$ are chosen in the kernel of P and let $Z = \operatorname{span}(y_1, \ldots, y_n, x_{n+1}, \ldots, x_{n+p})$. Then $Y \subset Z$ and the operator $L: Z \to X_\gamma$ defined by $L\left(\sum_{i=1}^n a_i y_i + \sum_{j=1}^p b_j x_{n+j} \right) = \sum_{i=1}^n a_i x_i + \sum_{j=1}^p b_j x_{n+j}$ is a linear isomorphism of Z onto X_γ such that $\|L\| \leqslant 1 + \varepsilon K^2 (1 + K\varepsilon)$ and $\|L^{-1}\| \leqslant 1/(1 - \varepsilon K^2 (1 + K\varepsilon))$. Thus for ε small enough, we get the conclusion. ∎

The following theorem now follows readily from the results of this section.

Theorem 7. *Let X be a complex Banach space. Then the following are equivalent.*

(1) *X is linearly isometric to $L_p(T, \Sigma, \mu, \mathbb{C})$ for some measure space (T, Σ, μ).*

(2) *X is an $\mathscr{L}_{p,\lambda}$ space for all $\lambda > 1$,*

(3) *there is a upwards directed family $\{X_\gamma\}$ of finite dimensional subspaces of X whose union is dense in X and such that X_γ is linearly isometric to $l_p(n_\gamma, \mathbb{C})$ where n_γ is the Hamel dimension of X_γ.*

Proof. By corollaries 1 and 2 of theorem 5, theorem 3, and theorem 6 of section 11 we have that (2) implies (1). The fact that (3) implies (2) is immediate from theorem 6. By considering the span of characteristic functions of finite collections of pairwise disjoint sets of finite measure and using the fact that simple functions are dense in $L_p(\mu, \mathbb{C})$, we can easily see that (1) implies (3). ∎

The equivalence of (1) and (3) was first established by Zippin in [283]. The equivalence of (1) and (2) is basically due to Lindenstrauss and Pełczyński [183] (who observed it for μ a finite measure for $1<p<\infty$ and for a general measure when $p=1$). It was completely proved by Tzafriri in [272].

Exercises. 1. Renorm $c_0(\mathbb{N}, \mathbb{R})$ in such a way that the new space X has distance 1 from $c_0(N, R)$, that is, $d(X, c_0(\mathbb{N}, \mathbb{R}))=1$, but so that they are not linearly isometric.

For problems 2, 3, and 4 we assume that $1 \leqslant p<\infty$ and $p \neq 2$. If X is a Banach space, a b_p projection on X is a projection P such that $\|x\|^p=\|Px\|^p+\|x-Px\|^p$. The class $\mathscr{B}_p$ is the of all b_p projections on X. These results are due to Cunningham [74] and Sullivan [268].

2. Show that $\mathscr{B}_1$ is a complete Boolean algebra under the operations $P_1 \vee P_2=P_1+P_2-P_1 P_2$, $P_1 \wedge P_2=P_1 P_2$ and $P'=I-P$ for P, P_1, P_2 in $\mathscr{B}_1$.

3. Suppose $1<p<\infty$ and $p \neq 2$. We say the norm on a Banach space X satisfies *Clarkson's inequalities* if for all x, y in X

$$\|x+y\|^p+\|x-y\|^p \leqslant 2\left[\|x\|^p+\|y\|^p\right] \quad \text{for } 1<p<2,$$
$$\|x+y\|^p+\|x-y\|^p \geqslant 2\left[\|x\|^p+\|y\|^p\right] \quad \text{for } 2<p<\infty.$$

Show that if X satisfies Clarkson's inequalities for some $1<p<\infty$, $p \neq 2$, then

(i) if P, Q are in $\mathscr{B}_p=PQ=QP$, then PQ is in $\mathscr{B}_p$,
(ii) if P, Q are in $\mathscr{B}_p$, then $PQ=QP$,
(iii) $\mathscr{B}_p$ is a complete Boolean algebra under the operations given in 2.

4. Suppose $1 \leqslant p<\infty$ and $p \neq 2$ and if $1<p<2$, suppose X is a Banach space satisfying Clarkson's inequalities. Prove that if for each $x \in X$, the closed linear span of $\{Px: P \in \mathscr{B}_p\}$ is the range of some element of $\mathscr{B}_p$, then X is linearly isometric to $L_p(\mu, \mathbb{R})$ for some measure μ. (Hint, first consider the case where there is a $u \in X$ such that X is the closed linear span of $\{Pu: P \in \mathscr{B}_p\}$.)

5. Prove that the space V constructed in theorem 4 is a complex abstract L_p space. (See exercises 10 and 11 of section 15.)

6. An alternate approach to the embedding in theorem 4 is the following. Let $\tilde{X}=\operatorname{Re} Z$ where Z is as in the proof of theorem 4. Consider $\tilde{X}/M$ where M is the kernel of the seminorm p in theorem 4 re-

stricted to X. Prove that the completion of $\tilde{X}/M$ is a real L_p space (see exercise 9 of section 15) and use this to obtain the embedding.

7. Let $L: l_p(n,\mathbb{C}) \to l_p(m,\mathbb{C})$ be a linear isometry where n and m are positive integers. Show that for $1 \leqslant p < \infty \{L(e_i^n)\}$ is a block basis with respect to some rearrangement of $\{e_1^m, \ldots, e_m^m\}$ when $p \neq 2$.

8. Let μ be a measure and n be a positive integer. Give a direct proof of the fact that if L is a linear isometry of $l_p(n,\mathbb{C}) \to L_p(\mu,\mathbb{C})$, then the range of L is complemented with a projection of norm one.

9. Let X be a closed linear subspace of $L_p(\mu,\mathbb{C})$. Prove that X is linearly isometric to an abstract L_p space if and only if $|f| \operatorname{sgn} g$ is in X whenever f and g are for $1 \leqslant p < \infty$ and $p \neq 2$.

10. Let X be an $\mathscr{L}_{p,\lambda}$ for all $\lambda > 1$. Prove direct from the definition that X satisfies Clarkson's inequalities.

11. Prove that every $\mathscr{L}_2$ space is linearly isomorphic to a Hilbert space. (Hint due to Rosenthal: Consider the family of all finite dimensional subspaces of the space X. For each such space F there is a norm $\| \ \|_F$ on F which is a Hilbert space norm. Consider the product over all x in X of the intervals $[0, \|x\|]$ where X is a $\mathscr{L}_{2,\lambda}$ space. The net defined by $f_F(x) = \|x\|_F$ if x is in F and $f_F(x) = 0$ otherwise can be used to obtain a Hilbert space norm on X equivalent to the original norm since $\| \ \|_F$ can be chosen so that $\|x\| \leqslant \|x\|_F \leqslant \lambda \|x\|$ for all x in F.)

§ 18. Geometric Properties of Abstract L_1 Spaces and some Dual Abstract L_1 Spaces

In this section we present some geometric characterizations of abstract L_1 spaces. These are given in terms of the geometry of the closed unit ball of the space and, in particular, of the facial structure of the closed unit ball.

The second part of the section is concerned with certain dual abstract L_1 spaces, in particular, the dual of $C(T,\mathbb{R})$ for T certain specific compact Hausdorff spaces.

If X is a real linear space and K is a nonempty convex set in X, a nonempty convex subset $F \subset K$ is said to be a *face* of K if whenever $x \in F$ and $x = ay + (1-a)z$ with $0 \leqslant a \leqslant 1$ and y, z in K, it follows that y, z are in F. A face F' of K is said to be *complementary* to a face F of K if $F \cap F' = \emptyset$ and for each $x \in K \setminus (F \cup F')$ there are unique a, y, z with $0 < a < 1$, $y \in F$, $z \in F'$ and $x = ay + (1-a)z$.

We will be working with partially ordered Banach spaces (X, C) such that C is 1 generating and the norm is additive on C. Thus (X^*, C^*) has a strong order unit e (given by $e(x) = \|x_1\| - \|x_2\|$ where $x = x_1 - x_2$ with $x_1, x_2 \in C$) and order unit norm (see theorem 7 of section 2). It follows that C^{**} is 1 generating in X^{**} (see theorem 3 of section 2) and since e is the strong order unit and X^* has the strong order unit norm, the norm is additive on X^{**}.

An easy test as to when X can be so ordered is found below. The details of the proof are left to the reader.

Lemma 1. *Let X be a Banach space and $S=\{x\in X:\|x\|=1\}$. Then X has a closed proper cone C such that the norm on C is additive and C is 1 generating if and only if there is a maximal face $F\subset S$ such that $\mathrm{co}(-F\cup F)=b(X)$.* ∎

By combining this with the translate property for lattice cones (see theorem 2 of section 1) we get the following characterization of abstract L_1 spaces.

Theorem 1. *Let X be a Banach space. Then X is linearly isometric to an abstract L_1 space if and only if there is a maximal face $F\subset S$ such that $b(X)=\mathrm{co}(-F\cup F)$ and the cone $C=\{ax:a\geqslant 0, x\in F\}$ has the property that the intersection of any two translates of C is a translate of C.* ∎

We need a theorem which says that the particular cone C we select on X does not matter in the classification of X as a Banach lattice. It's proof is left to the reader.

Theorem 2. *Let X be an abstract L_1 space and F_1, F_2 two maximal faces in S. Then there is a linear isometry of X onto X which carries C_1 onto C_2, where C_i is the cone generated by F_i for $i=1,2$.* ∎

We now give a characterization of abstract L_1 spaces X in terms of the facial structure of $S=\{x\in X:\|x\|=1\}$. We assume for the lemma and theorem below that (X,C) is a partially ordered Banach space, the norm is additive on C, C is 1 generating, e is the strong order unit of X^* relative to C, $K=\{x\in C:\|x\|=1\}$, $L=\{x^{**}\in X^{**}:x^{**}(e)=1=\|x^*\|\}$ $=\{x^{**}\in C^{**}:x^{**}(e)=1\}$, and $J:X\to X^{**}$ is the natural map.

We first note that if W is a convex set in X, F is a face of W, then F has at most one complementary face. For, suppose G and H are complementary faces of W with respect to F. If $x\in G$ and $x\notin H$, then $x=ay+(1-a)z$ where $0<a<1$, $y\in F$ and $z\in H$. Since G is a face, y, $z\in G$ which is a contradiction to $F\cap G=\emptyset$. Thus $G\subset H$ and a similar argument shows that $H\subset G$. (Note that this argument shows that if F has a complementary face F', then $F'\supset G$ for all faces G with $G\cap F=\emptyset$.)

For a set $A\subset X^{**}$ we denote its weak* closure by $w^*(A)$.

Lemma 2. *If $F\subset K$ is a complemented face, then*

(i) *F is norm closed and*

(ii) *$w^*(JF)$ is a complemented face in L.*

Proof. Let f be defined on K by $f(x)=\begin{cases}1 & \text{if } x\in F,\\ 0 & \text{if } x\in F,\\ a & \text{if } x\in K\setminus(F\cup F') \text{ and}\end{cases}$ $x=ay+(1-a)z$ where $0<a<1$, $y\in F$, $z\in F'$. Clearly f is affine on K and has a unique linear extension x^* to all of X. Since the closed unit ball of X is the convex hull of $-K\cup K$, x^* is continuous. Clearly $F=f^{-1}(1)$ and F is norm closed. By the second basic separation theorem it is easy to see that $\mathrm{co}(w^*(JF)\cup w^*(JF'))=w^*(JK)=L$. Suppose $x^{**}\in L\ (w^*(JF)\cup w^*(JF'))$. Then $x^{**}=a_i y_i^{**}+(1-a_i)z_i^{**}$ for $0<a_i<1$, $y_i^{**}\in w^*(JF)$, $z_i^{**}\in w^*(JF')$ for $i=1,2$. Then $x^{**}(x^*)=a_1=a_2=a$. For each $y^*\in X^*$, $y^*=y_1^*+y_2^*$ in a unique manner where $y_1^*|F'=0$ and $y_2^*|F=0$. Thus $x^{**}(y_1^*)=ay_1^{**}(y_1^*)=ay_2^{**}(y_1^*)$ and $x^{**}(y_2^*)=(1-a)z_1^{**}(y_2^*)$ $=(1-a)z_2^{**}(y_2^*)$. Hence $y_1^{**}=y_2^{**}$ and $z_1^{**}=z_2^{**}$. By the uniqueness of complemented faces it follows that $w^*(JF)=\{x^{**}\in L: x^{**}(x^*)=1\}$ and $w^*(JF')=\{x^{**}\in L: x^{**}(x^*)=0\}$. ∎

Corollary. *Let F be a complemented face in K. Then $w^*(JF)\cap w^*(\mathrm{ext}\,L)$ is a weak* closed and open set in L.*

Proof. The map $x^{**}\to x^{**}(x^*)$ takes only the values 0 and 1 on $w^*(\mathrm{ext}\,L)$. ∎

Theorem 3. *The space (X,C) is an abstract L_1 space if and only if*

(a) *every norm closed face of K is complemented in K, and*

(b) *for each $x^*\in X^*$ and $a\in\mathbb{R}$ such that $K^+=\{x\in K: x^*(x)>a\}$ and $K^-=\{x\in K: x^*(x)<a\}$ are both nonempty and there are complementary faces F, F' with $F\subset\{x\in K: x^*(x)\geqslant a\}$ and $F'\subset\{x\in K: x^*(x)\leqslant a\}$.*

Proof. Suppose X has properties (a) and (b) above. Let $T=w^*(\mathrm{ext}\,L)$ and A be the set of all affine functions on K associated with the pair of complementary faces (F,F') in K (as in lemma 2). As noted in the above corollary, for $f\in A$, $\hat{f}$ defined by $\hat{f}(x^{**})=x^{**}(x^*)$ for $x^{**}\in T$, where x^* is the unique linear extension of f, has the property that $\hat{f}$ takes only the values 0 and 1. Let $f_1, f_2\in A$, $F_i=f_i^{-1}(1)$, and $F=F_1\cup F_2$. Then F is a closed face in K and hence by (a) has a complementary face F' in K. The function $f\in A$ associated with (F,F') is such that $\hat{f}_1\hat{f}_2=\hat{f}$. That is, the linear span, B, of $A=\{\hat{f}: f\in A\}$ is a subalgebra of $C(T,\mathbb{R})$. Now, clearly B contains the constant functions and $B\subset\hat{X}^*\subset C(T,\mathbb{R})$ where $\hat{X}^*$ is the set of restrictions $\hat{x}^*$ of the elements of X^* to T. Moreover, $x^*\to\hat{x}^*$ is a linear isometric embedding of X into $C(T,\mathbb{R})$. Thus it remains only to show that $\hat{A}$ separates the points of T to conclude from the Stone-Weierstrass theorem that $\hat{X}^*=C(T,\mathbb{R})$, that is, X^* is order isometric to $C(T,\mathbb{R})$. Hence by theorem 7 of section 3, X is an abstract L_1 space in the ordering induced by C.

Suppose x^{**} and y^{**} are two distinct elements of T and $x^*\in X^*$ is such that $x^{**}(x^*)\neq y^{**}(x^*)$. Thus without loss of generality there is an $a\in\mathbb{R}$ with $x^{**}(x^*)<a<y^{**}(x^*)$. Let $K^+=\{x\in K: x^*(x)>a\}$ and $K^-=\{x\in K: x^*(x)<a\}$. If $K^-=\emptyset$, then since JK is weak* dense in L, $z^{**}(x^*)\geqslant a$ for all $z^{**}\in L$ which contradicts $x^{**}(x^*)<a$. Hence $K^-\neq\emptyset$ and, similarly, $K^+\neq\emptyset$. Thus by (b) there are complementary faces F and F' of K such that $F'\subset\{x\in K: x^*(x)\geqslant a\}$ and $F\subset\{x\in K: x^*(x)\leqslant a\}$. Let g be the affine function associated with (F',F) and y^* its unique linear extension. Then $w^*(JF')=\{z^{**}\in L: z^{**}(y^*)=1\}$ and $w^*(JF)=\{z^{**}\in L: z^{**}(y^*)=0\}$. Since $\operatorname{ext} L\subset w^*(JF)\cup w^*(JF')$, it follows that $x^{**}\in w^*(JF)$ and $y^{**}\in w^*(JF')$ so that $x^{**}(y^*)\neq y^{**}(y^*)$ and $\hat{A}$ separates the points of T.

Now suppose that (X,C) is an abstract L_1 space. If $F\subset K$ is a nonempty closed face of K and $D=\{ax: a\geqslant 0, x\in F\}$, then D is a closed subcone of C and $Y=D-D$ is a closed solid subspace of X. Let P be the band projection of X onto Y and $F'=\{x\in K: Px=0\}$. Then F' is a face of K complementary to F.

Let $x^*\in X^*$ and $a\in\mathbb{R}$ such that K^+ and K^- are both nonempty. Recall that $X^*=C(T,\mathbb{R})$ where T is the set of extreme points of $L=\{x^{**}\in X^{**}: x^{**}(e)=1=\|x^{**}\|\}$. Moreover, by theorem 11 of section 11, T is extremally disconnected. Hence $U=\{x^{**}\in T: x^{**}(x^*)>a\}$ and $V=\{x^{**}\in T: x^{**}(x^*)<a\}$ are open and nonempty and $\bar{U}\cap\bar{V}\neq\emptyset$. Since the natural embedding of X into X^{**} carries X onto the normal measures on T (see theorem 10 of section 11), $F=\{v\in N(T,\mathbb{R})^+: v(U)=1=\|v\|\}$ and $F'=\{v\in N(T,\mathbb{R})^+: v(V)=1=\|v\|\}$ are closed complementary faces of K and $F\subset\{x\in K: {}^*(x)\geqslant a\}$, $F'\subset\{x\in K: x^*(x)\leqslant a\}$. ∎

A famous unsolved problem in the isomorphic theory of L_1 spaces is the following.

Problem. Give a characterization of those spaces $L_1(\mu,\mathbb{R})$ which are linearly isometric (or linearly isomorphic) to X^* for some Banach space X.

In [180] Lewis and Stegall proved that the only infinite dimensional separable $\mathscr{L}_1$ space which is isomorphic to a dual space is $l_1(\mathbb{N},\mathbb{R})$. Rosenthal [245] considers $C(T,\mathbb{R})^*$ for some special compact Hausdorff spaces T and gives a conjecture as to the structure of dual abstract L_1 spaces. We shall present some results on $l_1(\Gamma,\mathbb{R})$ and a sample of Rosenthal's results.

The following lemma is easy to verify and its proof is left to the reader.

Lemma 3. *Let $X\subset l_1(\Gamma,\mathbb{R})$ be an infinite dimensional separable subspace. Then $X\subset Y\subset l_1(\Gamma,\mathbb{R})$ where Y is linearly isometric to $l_1(\mathbb{N},\mathbb{R})$.* ∎

Lemma 4. *Weak and norm convergence of sequences coincide in* $l_1(\Gamma, \mathbb{R})$.

Proof. By Lemma 3 it suffices to prove this for $l_1(\mathbb{N}, \mathbb{R})$.

Let $x_n = \{a_m^n\} \in l_1(N, R)$ and $x_n \to 0$ in the weak topology. Then for any $\{b_m\} \in l_\infty(\mathbb{N}, \mathbb{R})$ $\sum_{m=1}^{\infty} b_m a_m^n \to 0$. In particular, $a_m^n \to 0$ for each $m = 1, 2, 3, \ldots$.

We want $\lim_{n\to\infty} \sum_{m=1}^{\infty} |a_m^n| \to 0$. Suppose $\limsup \sum_{m=1}^{\infty} |a_m^n| > \varepsilon > 0$.

1. Let n_1 be the smallest integer such that $\sum_{m=1}^{\infty} |a_m^{n_1}| > \varepsilon$.

2. Let r_1 be the smallest integer such that $\sum_{m=1}^{r_1} |a_m^{n_1}| > \varepsilon/2$ and $\sum_{m=r_1+1}^{\infty} |a_m^{n_1}| < \varepsilon/5$.

3. n_k is the smallest integer $> n_{k-1}$ such that $\sum_{m=1}^{\infty} |a_m^{n_k}| > \varepsilon$ and $\sum_{m=1}^{r_{k-1}} |a_m^{n_k}| < \varepsilon/5$.

4. r_k is the smallest integer $> r_{k-1}$ such that $\sum_{m=r_{k-1}+1}^{r_k} |a_m^{n_k}| > \varepsilon/2$ and $\sum_{m=r_k+1}^{\infty} |a_m^{n_k}| < \varepsilon/5$.

$$b_m = \begin{cases} \operatorname{sgn} a_m^{n_1}, & 1 \leqslant m \leqslant r_1, \\ \operatorname{sgn} a_m^{n_{k+1}}, & r_k < m \leqslant r_{k+1}. \end{cases}$$

Then $|b_m| = 1$ for $m = 1, 2, \ldots$. Moreover, $\lim_{k\to\infty} \sum_{m=1}^{\infty} b_m a_m^{n_k} = 0$. But,

$$\left| \sum_{m=1}^{\infty} m a_m^{n_k} \right| \geqslant \sum_{m=r_{k-1}+1}^{n_k} |a_m^{n_k}| - \sum_{m=1}^{r_{k-1}} |a_m^{n_k}| - \sum_{m=r_k+1}^{\infty} |a_m^{n_k}| \geqslant \varepsilon/2 - \varepsilon/5 - \varepsilon/5 = \varepsilon/10$$

which is a contradiction. ∎

The following theorem summarizes the main Banach space properties of $l_1(\Gamma, \mathbb{R})$.

Theorem 4. *Let X be an abstract L_1 space. Then the following statements are equivalent:*

(i) *X is linearly isometric to $l_1(\Gamma, \mathbb{R})$ for some index set Γ,*

(ii) *X contains no subspace isomorphic to $L_1([0,1]^n, \mathbb{R})$ for any cardinal $n > 0$,*

(iii) *X contains no subspace isomorphic to $L_1([0,1], \mathbb{R})$,*

(iv) *weak and norm convergence coincide in X,*

(v) *every separable subspace of X is contained in a subspace of X isomorphic to $l_1(\mathbb{N}, \mathbb{R})$,*

(vi) *X^* is linearly isometric to $l_\infty(\Gamma, \mathbb{R})$ for some index set Γ.*

Proof. That (i) implies (iv) implies (ii) implies (iii) is clear. Also, (iii) implies (i) by the corollary to theorem 3 of section 15. Also, (i) implies (v) and (v) implies (iii). Clearly (i) implies (vi) and (vi) implies (i) by theorem 11 of section 11. ∎

The next simplest model of a dual L_1 space is to consider $C(T,\mathbb{R})^*$ for T a compact Hausdorff space. This includes, of course, $l_1(\Gamma,\mathbb{R})$ since by theorem $C(T,\mathbb{R})^* = l_1(\Gamma,\mathbb{R})$ where $\Gamma = T$ if and only if T is dispersed. However, in general, $C(T,\mathbb{R})^*$ is linearly isometric and lattice isomorphic to $\left[l_1(\Gamma,\mathbb{R}) \oplus \left(\oplus \sum_{i\in I} L_1([0,1]^{m_i},\mathbb{R})\right)_1\right]_1$ where $\Gamma = T$ and $m_i \geqslant \aleph_0$ for each $i \in I$ (note: $I = \emptyset$ if and only if T is dispersed). The results in this section are concerned with the cardinality of I and the set of possible values for m_i for certain compact Hausdorff spaces T. The general analysis of $C(T,\mathbb{R})^*$ in terms of this decomposition is incomplete.

Some terminology is needed. A compact Hausdorff space T is said to be *measure separable* if for each regular Borel measure μ on T, $L_1(\mu,\mathbb{R})$ is separable. Clearly any dispersed space or any compact metric space is measure separable. By the *free join* of a family of compact Hausdorff spaces we mean the Alexandroff one point compactification of the disjoint union of this family with the disjoint union topology. Clearly the free join of measure separable spaces is again measure separable. In particular, the free join of a dispersed space of cardinal $2^{\aleph_0}$ and $[0,1]$ is an example of a non metrizable measure separable space of cardinality $2^{\aleph_0}$. By theorem 5 of section 14 any purely non atomic measure μ on a measure separable compact Hausdorff space T has the property that $L_1(\mu,\mathbb{R})$ is linearly isometric and lattice isomorphic to $L_1([0,1],\mathbb{R})$. Hence in such a case, $C(T,\mathbb{R})^*$ has a simple structure.

Theorem 5. *If T is a non dispersed measure separable compact Hausdorff space, then* $C(T,\mathbb{R})^* = \left[l_1(\Gamma,\mathbb{R}) \oplus \left(\oplus \sum_{i\in I} L_1([0,1],\mathbb{R})\right)_1\right]_1$, *where* $\Gamma = T$ *and* $\operatorname{card} I \geqslant \aleph_0$.

Proof. The first part is clear. Since T is not dispersed, there is a continuous map h of T onto $[0,1]$ so $C([0,1],\mathbb{R})$ is lattice isomorphic and linearly isometric to the sublattice $Y = \{f \circ h : f \in C([0,1],\mathbb{R})\}$ of $C(T,\mathbb{R})$. Thus by theorem 2 of section 16 there is a positive simultaneous linear extension $A: Y^* \to C(T,\mathbb{R})^*$. Now there is a pairwise disjoint family $\{v\}_{j\, j\in J}$ of positive normalized purely nonatomic measures on $[0,1]$ with $\operatorname{card} J = 2^{\aleph_0}$. Moreover, $A(v_j) \wedge A(v_k) = 0$ for $j \neq k$ and $A(L_1(v_j,\mathbb{R})) \subset L_1(A v_j,\mathbb{R})$ for all $j \in J$ and since $L_1(v_j,\mathbb{R}) = L_1([0,1],\mathbb{R})$, $L_1(A v_j,\mathbb{R})$ is not purely atomic. Thus there is a family of pairwise disjoint positive purely nonatomic measures on T whose cardinal is $2^{\aleph_0}$ and it follows that $\operatorname{card} I \geqslant 2^{\aleph_0}$. ∎

Corollary. *If T is a non dispersed compact Hausdorff space of cardinal $2^{\aleph_0}$ and for each purely nonatomic measure μ on T, the support $S(\mu)$ of μ is metrizable, then* $C(T,\mathbb{R})^* = \left[l_1(\mathbb{R},\mathbb{R}) \oplus \left(\sum_{2^{\aleph_0}} \oplus L_1([0,1],\mathbb{R})\right)_1\right]_1$.

Proof. From the above it is only necessary to show that $\operatorname{card} C(T,\mathbb{R})^* = 2^{\aleph_0}$.

Let μ be a positive purely nonatomic measure on T and $S = S(\mu)$. Then $\mu \in C(S,\mathbb{R})^*$ and by the fact that S is metrizable and by the Krein-Milman theorem there is a sequence of purely atomic measures on S which converge to μ in the weak* topology. Thus the set of purely non-atomic measures on T has cardinal $(2^{\aleph_0})^{\aleph_0} = 2^{\aleph_0}$ and $\operatorname{card} C(T,\mathbb{R})^* = 2^{\aleph_0}$. ∎

We note that the above proof in theorem 5 actually shows that if T is a non dispersed compact Hausdorff space, then there is a family of cardinality $2^{\aleph_0}$ of pairwise disjoint positive purely nonatomic measures on T.

We develop some notation that will be used in the next theorems. For a cardinal $m \geqslant \aleph_0$ let $C_m = \{n : \aleph_0 \leqslant n \leqslant m\}$. Let G denote the unit circle in the plane and G^m denote m products of G with the product topology. Then G^m is a compact Hausdorff topological group in the product group structure associated with the natural circle group structure on G. Let I be a set with cardinal m. Since $m + m = m$, there are disjoint subsets I_1, I_2 of I such that $m = \operatorname{card} I_1 = \operatorname{card} I_2$ and $I = I_1 \cup I_2$. Let G_J denote the set of all functions from J to G (i.e. J products of G) and for $x \in G^{I_2}$ let $H_x = \{y \in G^I : y_i = x_i \text{ for all } i \in I_2\}$. Then H_x is homeomorphic to G^m and $H_x \cap H_y = \emptyset$ for all $x \neq y$. Now, $\operatorname{card} C_m \leqslant m \leqslant 2^m$ and $2^m 2^m = 2^m$ implies that $\mathscr{F} = \{H_x : x \in G^2\} = \bigcup_{n \in C_m} \mathscr{F}_n$ where $\mathscr{F}_n \cap \mathscr{F}_p = \emptyset$ for $n \neq p$ and $\operatorname{card} \mathscr{F}_n = 2^m$.

Lemma 5. *Let $\aleph_0 \leqslant n$. If $L_1([0,1]^n,\mathbb{R})$ is isomorphic to a subspace of $C(T,\mathbb{R})^*$, then $n \leqslant \dim C(T,\mathbb{R})$.*

Proof. By theorem 12 of section 14 there is a Hilbert space H with $\dim H = n$ and H is isomorphic to a subspace of $L_1([0,1]^n,\mathbb{R})$. Since H is reflexive, there is a bounded linear operator of $C(T,\mathbb{R})$ and H^*. Thus $n = \dim H = \dim H^* \leqslant \dim C(T,\mathbb{R})$. ∎

Theorem 6. *Let T be a compact Hausdorff space with weight $m \geqslant \aleph_0$. Suppose for each $n \in C_m$ there is a family $\mathscr{F}_n$ of closed sets in T such that*

(1) $\operatorname{card} \mathscr{F}_n = 2^m$,

(2) *for each $F \in \mathscr{F}_n$ there is a positive normalized regular Borel measure μ_F on F such that* $L_1(\mu_F,\mathbb{R}) = L_1([0,1]^n,\mathbb{R})$,

(3) *distinct members of* $\mathscr{F} = \bigcup_{n\in C_m} \mathscr{F}_n$ *are disjoint. Then* $C(T,\mathbb{R})^*$ *is linearly isometric and order isomorphic to*

$$X = \left[l_1(2^m,\mathbb{R}) \oplus \left(\oplus \sum_{n\in C_n} \left(\oplus \sum_{2^m} L_1([0,1]^n,\mathbb{R}) \right)_1 \right)_1 \right]_1 .$$

In particular, $C(G^m,\mathbb{R})^* = C(D^m,\mathbb{R})^* = C([0,1]^m,\mathbb{R}) = X$.

Proof. By Zorn's lemma there is a maximal family S of mutually pairwise disjoint nonatomic positive normalized measures containing $\{\mu_F : F\in\mathscr{F}\}$ and each $\mu\in S$ is such that $L_1(\mu,\mathbb{R}) = L_1([0,1]^n,\mathbb{R})$ for some cardinal $n \geqslant \aleph_0$. Since $\dim C(T,R) = m$, it follows that $n<m$, and since $\operatorname{card} C(T,\mathbb{R})^* = 2^m$, the result follows. ∎

Theorem 7. *Let m be a cardinal with $m \geqslant 2^{\aleph_0}$ and suppose T is a compact Hausdorff space which can be mapped continuously onto G^m. If Ω is an extremally disconnected compact Hausdorff space of weight $\leqslant m$, then Ω is homeomorphic to a subspace of T. Moreover, if $n^{\aleph_0} \leqslant m$ for all $n\in C_m$ and* $\dim C(T,\mathbb{R}) = m$, *then* $C(T,\mathbb{R})^*$ *is linearly isometric to* $C(G^m,\mathbb{R})^*$.

Proof. Since the weight of $\Omega \leqslant m$, Ω can be mapped homeomorphically into G^m and since T can be mapped onto G^m by a continuous map h, there is a minimal subspace of T which can be mapped onto $\Omega\subset G^m$. Thus h is a homeomorphism when restricted to this subspace.

Let $\mathscr{F}_n$ be a family of closed sets in G^m each homeomorphic to G^m and $\operatorname{card}\mathscr{F}_n = 2^m$ and $\mathscr{F} = \bigcup_{n\in C_m} \mathscr{F}_n$ consist of mutually pairwise disjoint sets. Then $\mathscr{F}_n' = \{h^{-1}(F) : F\in\mathscr{F}_n\}$ is a family of mutually pairwise disjoint sets in T with $\operatorname{card}\mathscr{F}_n' = 2^m$ for all $n\in C_m$.

Now for $n\in C_m$, $L_\infty([0,1]^n,\mathbb{R}) = C(\Omega_n,\mathbb{R})$ for some extremally disconnected compact Hausdorff space Ω_n and $\dim C(\Omega_n,R) = n^{\aleph_0} \leqslant m$.

By theorem 10 of section 11 there is a regular Borel measure μ_n on Ω_n such that $L_1([0,1]^n,\mathbb{R}) = L_1(\mu_n,\mathbb{R})$. Since Ω_n can be embedded in G^m, by the above argument, Ω_n can be homeomorphically embedded in F for each $F\in\mathscr{F}_n'$. Thus condition (2) of theorem 6 is satisfied and the result follows. ∎

As an application of the above theorem we note that the Stone-Čech compactification βT of a discrete space of cardinal $m \geqslant \aleph_0$ is of weight 2^m and can be mapped continuously onto G^{2^m} so that $C(\beta T,\mathbb{R})^* = l_\infty(T,\mathbb{R})^* = C(G^{2^m},\mathbb{R})^*$. Also, if Ω_m is the extremally disconnected compact Hausdorff space such that $C(\Omega_m,\mathbb{R}) = L_\infty([0,1]^m,\mathbb{R})$ then the weight of Ω_m is $m^{\aleph_0}$ and Ω_m can be mapped continuously onto $G^{m^{\aleph_0}}$ so that $C(\Omega_m,\mathbb{R})^* = (L_\infty[0,1]^m,\mathbb{R})^*$. The case βT is easy to see. The proof of the case Ω_m can be found in [245].

We now wish to investigate some further Banach space properties of abstract L_1 spaces.

Let X, Y, Z be Banach spaces with $Y \subset Z$. If $A: X \to Z/Y$ is a bounded linear operator, we say that A has a *lifting* to Z if there is a bounded linear operator $B: X \to Z$ such that $\pi B = A$, where $\pi: Z \to Z/Y$ is the natural map.

The following theorem is an exercise in adjoints.

Theorem 8. *Let X be a Banach space. Then X^* is a $\mathscr{P}_\lambda(\mathbb{R})$ space for some $\lambda \geqslant 1$ if and only if for each pair Y, Z of Banach spaces with $Y \subset Z$ and each bounded linear operator $A: X \to Z/Y$ there is a lifting $B: X \to Z^{**}$ (consider $Z/Y \subset Z^{**}/Y^{\perp\perp}$ in the natural fashion) such that $\|B\| \leqslant \lambda \|A\|$.* ∎

Corollary. *Let X be a Banach space. Then X is linearly isometric to an abstract L_1 space if and only if there is a lifting (as above) such that $\|B\| = \|A\|$.* ∎

Let us note that we can replace $\|B\| = \|A\|$ with the condition: for each $\varepsilon > 0$ there is a lifting B_ε such that $\|B_\varepsilon\| \leqslant (1+\varepsilon)\|A\|$. To do this all we have to show is that if X^* is a $\mathscr{P}_{1+\varepsilon}(\mathbb{R})$ space for all $\varepsilon > 0$, then X^* is a $\mathscr{P}_1(\mathbb{R})$ space. For, let Y, Z be Banach spaces with $Y \subset Z$ and $A: Y \to X^*$. Then for each $\varepsilon > 0$ there is an extension A_ε of A with $A_\varepsilon: Z \to X^*$ and $\|A_\varepsilon\| \leqslant (1+\varepsilon)\|A\|$. Since the closed unit ball of X^* is compact in the weak* topology, there is a subnet $\{A_{\varepsilon'}\}$ of $\{A_\varepsilon\}$ such that $A_{\varepsilon'}(z) \to B(z)$ for some bounded linear operator $B: Z \to X^*$. Moreover, $\|B\| \leqslant \|A_{\varepsilon'}\|$ for all ε' and $B|Y = A$. Hence $\|B\| = \|A\|$ and it follows that X^* is a $\mathscr{P}_1(\mathbb{R})$ space. (We note that Lindenstrauss has proved this without the assumption that the space is a conjugate space. The reader can see [182] for the details.)

We now characterize when we can get a lifting of X into Z always with some norm preserving conditions.

Theorem 9. *Let X be a Banach space. Then X is linearly isometric to $l_1(\Gamma, \mathbb{R})$ for some index set Γ if and only if for each pair Y, Z of Banach spaces with $Y \subset Z$, each $\varepsilon > 0$, and each bounded linear operator $A: X \to Z/Y$ there is a lifting $B: X \to Z$ such that $\|B\| \leqslant (1+\varepsilon)\|A\|$.*

Proof. Suppose $X = l_1(\Gamma, \mathbb{R})$. Let $e_\gamma \in X$ denote the function which is one at γ and 0 elsewhere. Now suppose Y and Z are Banach spaces with $Y \subset Z$ and let $A: X \to Z/Y$ be a bounded linear operator, and let $\varepsilon > 0$ be given. For each $\gamma \in \Gamma$ choose $z_\gamma \in Z$ with $\|z_\gamma\| \leqslant (1+\varepsilon)\|\pi(z_\gamma)\|$ and $\pi(z_\gamma) = A(e_\gamma)$. Then $B: X \to Z$ defined by $B(\sum a_\gamma e_\gamma) = \sum a_\gamma z_\gamma$ is the required mapping.

By the above corollary if X has the lifting property, then $X = L_1(\mu, \mathbb{R})$ for some measure μ. Now, let Γ be some index set so that there is a bounded linear mapping of $l_1(\Gamma, \mathbb{R})$ onto X. Thus there is an isomorphism A of X onto $l_1(\Gamma, \mathbb{R})/Y$ for some closed subspace Y. Hence there is a lifting $B: X \to l_1(\Gamma, \mathbb{R})$ of A. It is easy to see that B has a bounded inverse. Thus by theorem 4, $X = l_1(\Delta, \mathbb{R})$ for some index set Δ. ∎

We now wish to show that each complemented subspace of a space of the type $l_1(\Gamma, \mathbb{R})$ is isomorphic to a space of the type $l_1(\Delta, \mathbb{R})$. We note that from the above proof an immediate consequence of this is that a Banach space X is isomorphic to $l_1(\Delta, \mathbb{R})$ for some index set Δ if and only if for each pair Y, Z of Banach spaces with $Y \subset Z$ and each bounded linear operator $A: X \to Z/Y$ there is a lifting $B: X \to Z$ of A.

The following lemma is easy to verify.

Lemma 6. *Let T be an index set of cardinal m and $\{T_i\}_{i \in I}$ be a nonempty pairwise disjoint collection of finite subsets of T. Suppose for each $i \in I$, $0 \neq f_i \in l_1(T, \mathbb{R})$ with $f_i(t) = 0$ for all $t \notin T_i$. Then the closed linear span Y of the f_i's is linearly isometric to $l_1(T, \mathbb{R})$ where $n = \operatorname{card} I$. Moreover, there is a contractive projection of $l_1(T, \mathbb{R})$ onto Y.* ∎

The following theorem due to Rosenthal [245] is a generalization of a result due to Köethe [156] which, in turn, is a generalization of theorem 8 of section 12.

Theorem 10. *Let X, Y be Banach spaces with $Y \subset X$ and let Γ be an index set and suppose m is an infinite cardinal number. If $L: X \to l_1(\Gamma, \mathbb{R})$ is a bounded linear operator and $\delta > 0$ is such that $\operatorname{card}\{\gamma \in \Gamma$: there is a $y \in Y$ with $\|y\| \leqslant 1$ and $Ly(\gamma)| > \delta\} = m$, then Y contains a subspace Y_0 isomorphic to $l_1(\Delta, \mathbb{R})$ $(\operatorname{card} \Delta = m)$ and complemented in X. Moreover, $L|Y_0$ is an isomorphism.*

Proof. We first prove there is a family Δ of pairwise disjoint finite subsets of Γ with $\operatorname{card} \Delta = m$ and for each $F \in \Delta$ there is $k_F \in K = \{Ly: y \in Y, y \leqslant 1\}$ so that $\|k_F\| \geqslant \delta/4$ and

$$\sum_{\gamma \notin F} |k_F(\gamma)| < \frac{\delta}{32}.$$

Since $\operatorname{card}\{\gamma \in \Gamma$: there is a $k \in K$ with $|k(\gamma)| \geqslant \delta\} = m$, it follows that if $K' \subset K$ is a maximal family with the property that $\|k_1 - k_2\| \geqslant \delta/2$ whenever $k_1, k_2 \in K'$ and $k_1 \neq k_2$, then $\operatorname{card} K' \geqslant m$.

Consider all pairs $(\mathscr{F}, f)$ where $\mathscr{F}$ is a nonempty family of finite pairwise disjoint subsets of Γ and $f: \mathscr{F} \to K$ is a function such that $\|f(F)\| \geqslant \delta/4$ and $\sum_{\gamma \notin F} |f(F)(\gamma)| < \delta/32$. This family of all such pairs is partially ordered by $(\mathscr{F}_1, f_1) \leqslant (\mathscr{F}_2, f_2)$ if and only if $\mathscr{F}_1 \subset \mathscr{F}_2$ and $f_2|\mathscr{F}_1 = f_1$. Clearly each chain has an upper bound and by Zorn's lemma there is a maximal element (Δ, f). Suppose $\operatorname{card} \Delta < m$. Then for $\Gamma_1 = \bigcup_{F \in \Delta} F$, $\operatorname{card} \Gamma_1 < m$. Let $K'' = \{k|\Gamma_1 : k \in K'\}$. If $\operatorname{card} K'' < m$, then since $\operatorname{card} K' = m$, there are distinct elements k_1, k_2 in K' such that $k_1|\Gamma_1 = k_2|\Gamma_1$. If $\operatorname{card} K'' = m$, let $S_k = \{g \in l_1(\Gamma_1, \mathbb{R}) : \|g - k|\Gamma_1\| < \delta/64\}$ for $k \in K'$. Since $\dim l_1(\Gamma_1, \mathbb{R}) = \operatorname{card} \Gamma_1 < m$, two of these spheres must intersect. Hence in both cases there are distinct k_1, k_2 in K' with $\|k_1|\Gamma_1 - k_2|\Gamma_1\| < \delta/32$. Let $F_0 \subset \Gamma \setminus \Gamma_1$ be a finite set such that $\sum_{\gamma \notin F} |(k_1 - k_2)(\gamma)| < \delta/16$. Then for $k = \frac{1}{2}(k_1 - k_2)$ and $\mathscr{F} = \Delta \cup \{F_0\}$, let g be defined by $g(F) = f(F)$ for $F \in \Delta$ and $g(F_0) = k$. Then $(\mathscr{F}, g)$ is a pair larger than (Δ, f) which is a contradiction to the maximality of (Δ, f). Thus $\operatorname{card} \Delta \geqslant m$.

By restricting to a subset of Δ (if necessary) we may assume $\operatorname{card} \Delta = m$.

Now, for each $F \in \Delta$ let $e_F \in l_1(\Gamma, \mathbb{R})$ be defined by $e_F(\gamma) = k_F(\gamma)$ if $\gamma \in F$. Then by lemma 6 the closed linear span W of $\{e_F : F \in \Delta\}$ is linearly isometric to $l_1(\Delta, \mathbb{R})$ and there is a contractive projection P of $l_1(\Gamma, \mathbb{R})$ onto W. But, this implies that the closed linear span Z of $\{k_F : F \in \Delta\}$ is isomorphic to W and in fact, $P|Z$ is an isomorphism of Z onto W. Thus, $Q = (P|Z)^{-1} P$ is a projection of $l_1(\Gamma, \mathbb{R})$ onto Z.

Now, for each $F \in \Delta$ let $y_F \in Y$ with $\|y_F\| \leqslant 1$ and $L y_F = k_F$. Then it is easy to see that the closed linear span Y_0 of $\{y_F : F \in \Delta\}$ is isomorphic to $l_1(\Delta, \mathbb{R})$. Moreover, $L|Y_0$ is an isomorphism of Y_0 onto Z. A projection of X onto Y_0 is given by $(L|Y_0)^{-1} PL$. ∎

Corollary 1. *Let Γ be an infinite index set and X be a closed subspace of $l_1(\Gamma, \mathbb{R})$ with $\dim X = \operatorname{card} \Gamma$. Then there is a subspace Y of X isomorphic to $l_1(\Gamma, \mathbb{R})$ and complemented in $l_1(\Gamma, \mathbb{R})$.* ∎

Corollary 2. *Let X be a complemented subspace of $l_1(\Gamma, \mathbb{R})$. Then X is isomorphic to $l_1(\Delta, \mathbb{R})$ for some index set Δ.* ∎

This is an immediate consequence of the above theorem and lemma 1 of section 12.

Corollary 3. *Let X be a Banach space. Then X is isomorphic to $l_1(\Gamma, \mathbb{R})$ for some index set Γ if and only if for each pair Y, Z of Banach spaces with $Y \subset Z$ and each bounded linear operator $A : X \to Z/Y$, there is a lifting $B : X \to Z$ of A.* ∎

Exercises. 1. Prove that an abstract L_1 space X is linearly isometric to $l_1(\Gamma, \mathbb{R})$ for some index set Γ if and only if it does not contain an isomorphic copy of an infinite dimensional Hilbert space.

2. Let X be an abstract L_1 space. Show that if X is linearly isomorphic to $l_1(\Gamma, \mathbb{R})$ for some index set, it is linearly isometric to $l_1(\Gamma, \mathbb{R})$.

3. Let T be a compact metrizable space with a nonempty perfect set. Show that $C(T, \mathbb{R})^*$ is linearly isomorphic to $\left(\oplus \sum_{2^{\aleph_0}} L_1([0,1], \mathbb{R})\right)_1$.

4. Let $m \geqslant \aleph_0$. Prove that $C(G^m, \mathbb{R})^*$ is isomorphic to $\left(\oplus \sum_{2^m} L_1([0,1]^m, \mathbb{R})\right)_1$.

5. Let Γ be a nonempty index set and for each $\gamma \in \Gamma$, let $m_\gamma \geqslant \aleph_0$ be given. Show that there is a compact Hausdorff space T so that $C(T, \mathbb{R})^*$ is isomorphic to $\left(\oplus \sum_{\gamma \in \Gamma} \sum_{2^m} L_1([0,1]^{m_\gamma}, \mathbb{R})\right)_1$.

Chapter 7

L_1-Predual Spaces

In this chapter we shall study both real and complex Banach spaces X such that X^* is linearly isometric to an abstract L_1 space. We call such spaces L_1-*predual spaces*. There are several classes of classical Banach spaces which have this property (e.g. abstract M spaces and spaces of the type $C(T,\mathbb{C})$). Lindenstrauss was the first one to undertake a systematic study of them (although Grothendieck did study certain types of them, the so called G spaces). In [182] he developed the beginnings of a general structure theory for the real case and has also made many other contributions to the area (see [175], [176], [186], [187], [190], and [193]). Recently interest in the complex case has brought about some significant results (see [289], [291], [292], [293], and [294]). We shall also present some of these, especially in section 23.

Sections 19 and 20 are devoted to the study of partially ordered real L_1-predual spaces. We prove, for example, in section 19 that if K is a compact convex set, then $A(K,\mathbb{R})$ is an L_1-predual space if and only if $A(K,\mathbb{R})^*$ is an abstract L_1 space in its natural ordering. Section 20 is concerned mainly with the concept of a compact Choquet simplex and its role in the representation theory of continuous linear functionals on L_1-predual spaces. We do give some results here for the complex case. Notably, we prove the theorem of Hirsberg which says that if A is a linear subspace of $C(T,\mathbb{C})$ which contains the constants and separates the points of T, then for each continuous linear functional x^* on A there is a boundary measure μ on T which represents x^* and has the same norm as x^* (i.e. $\|x^*\|=\|\mu\|$).

In section 21 we study the general structure of real L_1-predual spaces. For example, we prove that X is an L_1-predual space if and only if each finite collection of closed balls in X which intersects in pairs has nonvoid intersection.

In section 22 we investigate embeddings into real L_1-predual spaces and characterize these spaces for which $X^*=l_1(\Gamma,\mathbb{R})$ for some index set Γ.

Section 23 is devoted primarily to the study of complex L_1-predual spaces, although some of the characterizations presented are valid for both the real and complex case. For example, X is an L_1-predual space if and only if it is an $\mathscr{L}_{\infty,\lambda}$ space for all $\lambda > 1$.

§ 19. Partially Ordered L_1-Predual Spaces

When we have a partially ordered Banach space, the natural ordering on the dual having certain properties forces properties on the space (and conversely). Some of these were noted in section 2. These properties are studied here when the dual space is an abstract L_1 space in the dual ordering which occurs, for example, when the space is an abstract M space. In this and the next section we study another class of such spaces, namely, the spaces of affine continuous real valued functions on a compact Choquet simplex. This section is devoted to developing some basic abstract properties of such spaces.

Definition 1. A partially ordered Banach space (X,C) is said to be a *simplex space* if C is 1 normal, (X,C) is regular and has the decomposition property, and the open unit ball $b_0(X)$ of X is directed upwards.

The terminology comes from the theory of compact Choquet simplexes. A *simplex* K is a compact convex set in some locally convex Hausdorff space so that $A(K,\mathbb{R})^*$ is an abstract L_1 space. From the results of section 2 we note that a simplex space (X,C) has the property that (X^*,C^*) is an abstract L_1 space. In this section we show that the converse is true, namely that a partially ordered Banach space such that (X^*,C^*) is an abstract L_1 space is a simplex space. The reader is referred to [1], [64], and [237] for other treatments of simplexes.

From section 2 it follows that if (X^*,C^*) is an abstract L_1 space, then $b_0(X)$ is directed udwards, (X,C) is regular and C is 1 normal (see lemma 1 and theorems 2 and 4 of section 2). Hence we have only to show that if (X,C) is a partially ordered Banach space such that (X^*,C^*) is an abstract L_1 space, then (X,C) has the decomposition property. We accomplish this in the lemmas and theorems below.

Lemma 1. *Let X be a linear space and C a cone in X such that (X,C) has the decomposition property. If $-f,g$ are subadditive and positive homogeneous functionals on C such that $f \geqslant g$, then there is an additive positive homogeneous functional h on C such that $f \geqslant h \geqslant g$.*

Proof. Let h be defined on C by $h(x) = \inf\left\{\sum_{i=1}^{n} f(x_i) : x = \sum_{i=1}^{n} x_i,\, x_i \in C\right\}$ for all $x \in C$. Then clearly h is positive homogeneous, subadditive and $f \geqslant h \geqslant g$. If $x = y + z$ with x,y,z in C and $\varepsilon > 0$, we choose $x_i \in C$ $(i = 1, \ldots, n)$

such that $x=\sum_{i=1}^{n} x_i$ and $\sum_{i=1}^{n} f(x_i) \leqslant h(x)+\varepsilon$. Then there are $z_{ij} \in C$ such that $y=\sum_{i=1}^{n} z_{i1}$, $z=\sum_{i=1}^{n} z_{i2}$ and $z_{i1}+z_{i2}=x_i$ for $i=1,\ldots,n$. Hence $h(x) \geqslant \sum_{i=1}^{n} f(x_i)-\varepsilon \geqslant \sum_{i=1}^{n} f(z_{i1})+\sum_{i=1}^{n} f(z_{i2})-\varepsilon \geqslant h(y)+h(z)-\varepsilon$ and it follows that h is additive. ∎

Lemma 2. *Let X be a linear space and C a cone in X. If p,f are functions from X to $\mathbb{R}$ with p being positive homogeneous on C and f being affine on X and $f(x) \leqslant p(x)$ for all $x \in C$, then $g=f-f(0) \leqslant p$.*

Proof. Suppose there is an $x \in C$ with $g(x)>p(x)$. Let $\varepsilon=-f(0)$ and $\delta=f(x)-p(x)$. Then there is an $r \geqslant 1$ such that $r(\delta+\varepsilon)>\varepsilon$ and since f is affine, $f(x)=\frac{1}{r} f(rx)+\left(1-\frac{1}{r}\right) f(0)$. Thus $f(rx)-p(rx)=r(f(x)-p(x)) +(r-1)=r(\delta+\varepsilon)-\varepsilon>0$ which is a contradiction. ∎

The following theorem is due to Asimov and Ellis [20].

Theorem 1. *Let (X^*, C^*) be an abstract L_1 space. If $-f,g$ are weak* continuous positive homogeneous subadditive functionals on C^* such that $f \geqslant g$, then there is an $x \in X$ such that $f(x^*) \geqslant x^*(x) \geqslant g(x^*)$ for all $x^* \in C^*$.*

Proof. Let $K=\{x^* \in X^*: 0 \leqslant x^*, \|x^*\| \leqslant 1\}$. Recall that K is convex and weak* compact and the map $x \to \hat{x}$ where $\hat{x}(x^*)=x^*(x)$ for $x \in X$, $x^* \in K$ is an order preserving linear isometry of X onto $A_0(K,\mathbb{R}) =\{h \in A(K,\mathbb{R}): h(0)=0\}$ (see theorem 5 of section 2). Hence we assume that $X=A_0(K,\mathbb{R})$.

Let G be the weak* closed convex hull of the graph of f in $K \times \mathbb{R} \subset X^* \times \mathbb{R}$. If $\hat{f}$ is defined by $\hat{f}(x^*)=\sup\{h(x^*): h \in A(K,\mathbb{R}), h \leqslant f\}$ for all $x^* \in K$, then $\hat{f}(x^*) \leqslant \inf\{a \in \mathbb{R}: (x^*, a) \in G\}$ for all $x^* \in K$. If $x^* \in K$, $b<\inf\{a \in \mathbb{R}: (x^*, a) \in G\}$, then by the second basic separation theorem there is an $h \in A(K,\mathbb{R})$ such that $h \leqslant f$ and $h(x^*)>b$. Thus $\hat{f}(x^*)=\inf\{a \in \mathbb{R}: (x^*, a) \in G\}$ for all $x^* \in K$. Let $\varepsilon>0$ be given and for each $x^* \in K$ let U_{x^*} be a weak* compact convex neighborhood of x^* such that $|f(x^*)-f(y^*)|<\varepsilon$ for each $y^* \in U_{x^*}$ and let $K \subset U_{x_1^*} \cup \cdots \cup U_{x_n^*}$. Then for each $x^* \in K$, $(x^*, \hat{f}(x^*)) \subset \operatorname{co}\left\{\bigcup_{i=1}^{n} U_{x_i^*} \times [f(x_i^*)-\varepsilon, f(x_i^*)+\varepsilon]\right\}$ and thus $x^*=\sum_{i=1}^{n} a_i y_i^*$ and $\hat{f}(x^*)=\sum_{i=1}^{n} a_i r_i$ with $y_i \in U_{x_i^*}$, $r_i \in [f(x_i^*)-\varepsilon, f(x_i^*)+\varepsilon]$, $a_i \geqslant 0$, and $\sum_{i=1}^{n} a_i=1$. If $\bar{f}$ is defined by $\bar{f}(x^*) =\inf\left\{\sum_{i=1}^{n} f(x_i^*): x_i^* \in C, x^*=\sum_{i=1}^{n} x_i^*\right\}$ for each $x^* \in C$, then for $x^* \in K$,

$\bar{f}(x^*) \leqslant \sum_{i=1}^{n} a_i f(y_i^*) \leqslant \sum_{i=1}^{n} a_i f(x^*) + \varepsilon \leqslant \sum_{i=1}^{n} a_i r_i + 2\varepsilon$. Thus $\bar{f}(x^*) \leqslant \hat{f}(x^*)$ for all $x^* \in K$. If $a>0$ and $\hat{f}_a$ is defined by $\hat{f}_a(x^*) = \sup\{h(x^*): h \in A(aK, \mathbb{R}), h \leqslant f\}$ for $x^* \in C^*$, then the argument above shows that $\bar{f}(x^*) \leqslant \hat{f}_a(x^*)$ for all $x^* \in aK$. In particular, $\hat{f}_a(0) = 0 = \bar{f}(0)$. Suppose $0 \neq x^* = \sum_{i=1}^{n} x_i^*$ where $x_i^* \in C$ and $a = \sum \|x_i^*\| = \|x^*\|$. Let $y_i^* = 0$ if $x_i^* = 0$ and $y_i^* = a \frac{x_i^*}{\|x_i^*\|}$ if $x_i^* \neq 0$. Then $y_i^* \in aK$ and since f_a is convex, $\hat{f}_a(x) \leqslant \sum_{i=1}^{n} \frac{\|x_i^*\|}{a} \hat{f}_a(y_i^*) \leqslant \sum_{i=1}^{n} \frac{\|x_i^*\|}{a} \hat{f}(y_i^*) = \sum_{i=1}^{n} f(x_i^*)$. Hence $\hat{f}_a \leqslant \bar{f}$ on C^*.

By lemma 1, $\bar{f}$ is additive on C^* and the above argument shows that $\bar{f}$ is weak* lower semi continuous on aK for all $a>0$. The set $\{x^* \in C^*: \bar{f}(x^*) \leqslant r\}$ is convex and its intersection with each multiple of the unit sphere of X^* is weak* closed. Hence $\bar{f}$ is weak* lower semi-continuous additive and positive homogeneous on C^* and $g \leqslant \bar{g} \leqslant \bar{f} \leqslant f$. Let $\varepsilon>0$ and $r>a$, then by separating the sets $\{(x^*, t) \in C^* \times \mathbb{R}: t > \bar{f}(x^*)\}$ and $\{(y^*, s-\varepsilon/r) \in K \times \mathbb{R}: s \leqslant \bar{g}(y^*)\}$ and applying lemma 2, we obtain a w_ε in $A_0(K, \mathbb{R})$ such that $w_\varepsilon \leqslant f$ and $w_\varepsilon(x^*) > g(x^*) - \varepsilon/r$ for all $x^* \in K$. Hence if $z_\varepsilon = (g - w_\varepsilon) \vee 0$, then z_ε is positive homogeneous, subadditive and weak* continuous on C^* and $\|z_\varepsilon\| < \varepsilon/r$. The above argument shows that $\bar{z}_\varepsilon$ is weak* upper semicontinuous on C^* and that $\|\bar{z}_\varepsilon\| < a\|z_\varepsilon\| < \varepsilon$. Since $K \times \{\varepsilon/r\}$ is disjoint from the weak* closed cone $\{(x^*, t): x^* \in C^*, t \leqslant \bar{z}_\varepsilon(x^*)\}$, the second basic separation theorem gives a $p_\varepsilon \in A_0(K, \mathbb{R})$ such that $p_\varepsilon \geqslant \bar{z}_\varepsilon \geqslant (g - w_\varepsilon) \vee 0$, and $\|p_\varepsilon\| \leqslant \varepsilon$. Using the above procedure we choose $f_1, g_1 \in A_0(K, \mathbb{R})$ such that $f_1 \leqslant f$, $g_1 \geqslant 0$, $g \leqslant f_1 + g_1$ and $\|g_1\| < 1/2$. In particular, $f \wedge (f_1 + g_1) \geqslant g \vee f_1$. By induction there are sequences $\{f_n\}$, $\{g_n\}$ in $A_0(K, \mathbb{R})$ such that (i) $g_n \geqslant 0$ and $\|g_n\| < 2^{-n}$, (ii) $g \vee f_n \leqslant f_{n+1} + g_{n+1}$, and (iii) $f_{n+1} \leqslant f \wedge (f_n + g_n)$. Properties (ii) and (iii) yield $-g_{n+1} \leqslant f_{n+1} - f_n \leqslant g_n$ so that $\|f_{n+1} - f_n\| < 2^{-n}$. Thus $\{f_n\}$ converges to some $h \in A_0(K, \mathbb{R})$ such that $h \leqslant f$ (by (iii)) and $h \geqslant g$ (by (ii)). ∎

We come to the main theorem of this section. In the following K is a compact convex set in some locally convex Hausdorff topological linear space E.

Theorem 2. *The following statements are equivalent.*

(1) *$A(K, \mathbb{R})$ is a simplex space, i.e., K is a simplex,*

(2) *$A(K, \mathbb{R})^*$ is a vector lattice (in its natural ordering),*

(3) *$A(K, \mathbb{R})^*$ is an abstract L_1 space (in its natural ordering),*

(4) *$A(K, \mathbb{R})^*$ is linearly isometric to an abstract L_1 space, that is, $A(K, \mathbb{R})$ is an L_1-predual space,*

(5) *for each pair* $-f,g$ *of convex upper semicontinuous functions on* K *with* $f \geqslant g$ *there is an* $h \in A(K,\mathbb{R})$ *with* $f \geqslant h \geqslant g$,

(6) $A(K,\mathbb{R})$ *has the decomposition property.*

Proof. Clearly (1) implies (2), (1) implies (3), (3) implies (2), (3) implies (4), and (6) implies (2). The proof that (5) implies (6) is an easy argument which is left to the reader. The fact that (4) implies (3) comes from theorem 2 of section 18 and the fact that $K\hat{} = \{x^* \in A(K,\mathbb{R})^* : x^*(1) = 1 = \|x^*\|\}$ is a maximal face in the surface of the closed unit ball of $A(K,\mathbb{R})^*$ and it generates the cone in the dual partial ordering on $A(K,\mathbb{R})^*$. (Moreover, K and $K\hat{}$ are affinely homeomorphic so we shall consider $K = K\hat{}$.) The proof of (2) implies (3) is left to the reader.

If $A(K,\mathbb{R})^*$ is an abstract L_1 space, then the positive cone in $A(K,\mathbb{R})$ is 1 normal, $A(K,\mathbb{R})$ is regular, and the open unit sphere of $A(K,\mathbb{R})$ is upwards directed (see the remarks following definition 1). Thus it remains only to show that (3) implies (5).

Suppose $-f,g$ are as in (5) and are bounded by $M>0$ and that $f(x^*) > g(x^*)$ for all $x^* \in K$. Then $G = \{(x^*,t): -M \geqslant t \geqslant g(x^*)\}$ is compact in $K \times \mathbb{R}$ and contained in the relativity open convex set $H = \{(x^*,t): t < f(x^*)\}$ in $K \times \mathbb{R}$. Thus the closed convex hull of G is contained in H and for each $x^* \in K$ there is an $f_{x^*} \in A(K,\mathbb{R})$ and a neighborhood U_{x^*} of x^* such that $g(y^*) < f_{x^*}(y^*) < f(y^*)$ for all $y^* \in U_{x^*}$. Since K is compact, $K \subset \bigcup_{i=1}^{n} U_{x_i^*}$ and if $f' = f_{x_1^*} \wedge \cdots \wedge f_{x_n^*}$, then f' is weak* continuous and concave on K with $g < f' < f$. Similarly, there is a weak* continuous convex function g' on K such that $g < g' < f' < f$. The functions have natural extensions g', f' to weak* continuous subadditive positive homogeneous functions on C^* (the positive cone in $A(K,\mathbb{R})^*$) such that $g' \leqslant f'$. Hence by theorem 1, there is an $h' \in A(K,\mathbb{R})$ such that $g < g' \leqslant h' \leqslant f' < f$.

In the general case there is an $h_1 \in A(K,\mathbb{R})$ with $f+1 > g-1$. Consider $(f \wedge h) + 1/2$, $(g \vee h) - 1/2$. We can obtain an $h_2 \in A(K,\mathbb{R})$ with $f + 1/2 > h_2 > g - 1/2$ and $\|h_2 - h_1\| < 1/2$. Proceeding by induction we obtain a sequence $\{h_n\}$ in $A(K,\mathbb{R})$ with $\|h_{n+1} - h_n\| < \frac{1}{2}n$ and $f + \frac{1}{2^{n-1}} > h_n > g - 1/2^{n-1}$ for all n. Thus $\{h_n\}$ converges to some $h \in A(K,\mathbb{R})$ $f \geqslant h \geqslant g$. ∎

The equivalence of (1), (2), and (3) above is due to Semadeni [257], the equivalence of (3) and (6) is due to Lindenstrauss [182] and the equivalence of (1) and (5) is due to Edwards [94].

Recall that we have already shown that if (X,C) is a partially ordered Banach space and (X^*,C^*) is an abstract L_1 space, then (X,C) is regular,

C is 1 normal, and $b_0(X)$ is upwards directed. To show that (X, C) has the decomposition property we show that $A(K,\mathbb{R})$ does where $K = \{x^* \in C^* : \|x^*\| \leqslant 1\}$. Since $X = A_0(K,\mathbb{R}) = \{h \in A(K,\mathbb{R}) : h(0) = 0\}$, this will suffice. By theorem 2 it suffices to show that K is the base of a lattice cone. To this end let $Y = X^* \times \mathbb{R}$ and $D = \{(x^*, a) : x^* \geqslant 0, a \geqslant 0\}$. Then D is a lattice cone since C^* is. In fact, $(x^*, a) \wedge (y^*, b) = (x^* \wedge y^*, \min(a, b))$. Let $B = \{(x^*, a) \in D : \|x^*\| + a = 1\}$. Then B is a base for D and B is affinely isomorphic to K under the map $x^* \to (x^*, 1 - \|x^*\|)$. Hence K is affinely isomorphic to a base for a lattice cone and it follows that $A(K,\mathbb{R})^*$ is a lattice in its natural order.

Thus we have the following theorem due to Davies [76].

Theorem 3. *Let (X, C) be a partially ordered Banach space. Then (X, C) is a simplex space if and only if (X^*, C^*) is an abstract L_1 space.* ∎

We now wish to establish a Banach Stone type theorem for simplex spaces which is a generalization of a theorem due to Lazar [171].

Theorem 4. *Let (X_1, C_1) and (X_2, C_2) be two simplex spaces. If X_1 is linearly isometric to X_2, then (X_1, C_1) is linearly order isometric to (X_2, C_2).*

Proof. Let $K_i = \{x^* \in X_i^* : 0 \leqslant x^*, \|x^*\| \leqslant 1\}$ for $i = 1, 2$. Since $X_i = A_0(K_i, R)$ for $i = 1, 2$, it suffices to show that K_1 is affinely homeomorphic to K_2 (recall that K_i has the induced weak* topology for $i = 1, 2$). Let $L_i = \{x^* \in K_i : \|x^*\| = 1\}$ for $i = 1, 2$. Then L_i is a face in unit sphere of X_i^* and the extreme points of the unit sphere of X_i^* are contained in $-L_i \cup L_i$. Suppose that α is a linear isometry of X_1 onto X_2. Then clearly α^* is a linear isometry of X_2^* onto X_1^* and $F = \alpha^*(K_2) \cap K_1$, $G = \alpha^*(-K_2) \cap K_1$ are convex and weak* compact. Moreover, $K_1 = \mathrm{co}(F \cup G)$. Now, let $M = \alpha^*(L_2) \cap L_1$ and $M' = \alpha^*(-L_2) \cap L_1$. Then since α is a linear isometry, $M = F \cap L_1$ and $M' = G \cap L_1$. Clearly $\mathrm{co}(M \cup M') \subset L_1$. If $x^* \in L_1$, then $x^* = t x_1^* + (1-t) x_2^*$ with $x_1^* \in F$, $x_2^* \in G$ and $0 \leqslant t \leqslant 1$. Since $\|x^*\| = 1$, $\|x_1^*\| = \|x_2^*\| = 1$ and since L_1 is a face of the unit sphere of X_1^*, x_1^*, x_2^* are in L_1 and $L_1 = \mathrm{co}(M \cup M')$. Since α^* is an isometry it is easy to see that M and M' are faces in L_1 and, clearly, $M \cap M' = \emptyset$. Thus, since $L_1 = \mathrm{co}(M \cup M')$, it follows that M, M' are complementary faces in L_1 (see theorem 3 of section 18). From this it follows that each element of K_1 can be uniquely expressed as $x^* = t\alpha^*(y_1^*) + (1-t)\alpha^*(y_2^*)$ where $0 \leqslant t \leqslant 1$, y_1^*, y_2^* in K_2. Thus we let $\beta(x^*) = t y_1^* + (1-t) y_2^*$ for each $x^* \in K_1$. Then using the compactness of K_1 and K_2 it is routine but tedious to show that β is an affine homeomorphism of K_1 onto K_2. ∎

Corollary 1 (Lazar). *If K_1, K_2 are compact convex sets in some locally convex Hausdorff topological vector spaces E_1, E_2 respectively and*

$A(K_i,\mathbb{R})$ are simplex spaces which are linearly isometric, then K_1 is affinely homeomorphic to K_2. That is, $A(K_1,\mathbb{R})$ is linearly order isometric to $A(K_2,\mathbb{R})$. ∎

Corollary 2. *If X_1, X_2 are abstract M spaces which are linearly isometric then they are linearly order isometric.* ∎

We now wish to give a general method for constructing simplex spaces. The theorem which we prove here has a converse which will be established in section 21.

Definition 2. Let T be a compact Hausdorff space. A function $\rho: T \to C(T,\mathbb{R})^*$ is said to be a *barycentric mapping* if

(1) for each $f \in C(T,\mathbb{R})$, the function f_ρ defined by $f_\rho(t) = \int f\, d\rho(t)$ for all $t \in T$ is integrable with respect to each $\mu \in C(T,\mathbb{R})^*$,

(2) $\|\rho(t)\| \leqslant 1$ for all $t \in T$,

(3) if $\mu, \nu \in C(T,\mathbb{R})^*$ and $\int f\, d\mu = \int f\, d\nu$ for all $f \in C(T,\mathbb{R})$ with $f = f_\rho$, then $\int f_\rho\, d\mu = \int f_\rho\, d\nu$ for all $f \in C(T,\mathbb{R})$.

One of the main reasons for the above definition is given in the following theorem.

Theorem 6. *Let T be a compact Hausdorff space and $\rho: T \to C(T,\mathbb{R})^*$ a barycentric mapping. Then $A_\rho = \{f \in C(T,\mathbb{R}) : f = f_\rho\}$ is an L_1-predual space.*

Proof. Let $\mu \in C(T,\mathbb{R})^*$. Then the mapping defined by $f \to \int f_\rho\, d\mu$ is a continuous linear functional on $C(T,\mathbb{R})$ and hence there is a unique $P\mu \in C(T,\mathbb{R})^*$ such that $\int f\, dP\mu = \int f_\rho\, d\mu$ for all $f \in C(T,\mathbb{R})$. Clearly the correspondence $\mu \to P\mu$ is linear and $\|P\| \leqslant 1$. Since for $f_\rho = f \in C(T,\mathbb{R})$ we have that $\int f\, dP\mu = \int f_\rho\, d\mu = \int f\, d\mu$, it follows from (3) of the definition that $\int f\, dP(P\mu) = \int f_\rho\, dP\mu = \int f\, dP\mu$ for all $f \in C(T,\mathbb{R})$. That is, $P \circ P = P$ and P is a contractive projection on $C(T,\mathbb{R})^*$. Moreover, the range N of P is $\{\mu \in C(T,\mathbb{R})^* : \int (f - f_\rho)\, d\mu = 0$ for all $f \in C(T,\mathbb{R})\}$ and the kernel of P is weak* closed.

Now, $A_\rho^\perp = \{\mu \in C(T,\mathbb{R})^* : \mu | A_\rho = 0\}$ is the kernel of P. For, if $\mu \in A_\rho^\perp$, then $0 = \int f\, d\mu$ for all $f = f_\rho \in C(T,\mathbb{R})$ and by (3) of the definition, $\int f_\rho\, d\mu = 0$ for all $f \in C(T,\mathbb{R})$. Hence $P\mu = 0$. Clearly $P\mu = 0$ implies $\mu \in A_\rho^\perp$. Moreover the restriction mapping $\mu \to \mu | A_\rho$ is a linear isometry of N onto A_ρ^*. For, let $x^* \in A_\rho^*$ and $\nu \in C(T,\mathbb{R})^*$ with $\|\nu\| = \|x^*\|$ and $\nu | A_\rho = x^*$. Then for $\mu = P\nu$, $\mu | A_\rho = \nu | A_\rho = x^*$ since $A_\rho^\perp$ is the kernel of P and for $f \in C(T,\mathbb{R})$, $|\int f\, d\mu| = |\int f\, dP\nu| \leqslant \|P\|\, |\int f\, d\nu| \leqslant \|f\|\, \|\nu\|$ and, thus, $\|\mu\| = \|\nu\|$. Hence the restriction mapping is a linear isometry of N onto A_ρ^*. By theorem 3 of section 17 N is an abstract L_1 space and we are done. ∎

We leave the proof of the following theorem to the reader.

Theorem 7. *Let T be a compact Hausdorff space and $\rho: T \to C(T, \mathbb{R})$ be a barycentric mapping with $\rho(t) \geqslant 0$ and $\|\rho(t)\| = 1$ for all $t \in T$. Then $A\rho$ is a simplex space with strong order unit.* ▮

We shall show later (in section 21) that each simplex space is generated by some barycentric mapping.

Exercises. 1. Let T be a compact Hausdorff space and $t_1, \ldots, t_n$ be distinct limit points in T. Show that there is a subspace $A \subset C(T, \mathbb{R})$ containing 1 and separating the points of T such that A is a simplex space in its natural ordering and $\varepsilon_t | A$ is an extreme point of the unit sphere of A^* if and only if $t \neq t_1, \ldots, t_n$.

2. Prove that an abstract M space is a simplex space.

3. Prove that a simplex space which is a lattice in its natural ordering in an abstract M space.

4. Let X be a simplex space with strong order unit. Prove that X is linearly isometric to $C(T, \mathbb{R})$ for some compact Hausdorff space T if and only if X is a lattice in its natural ordering. In particular, show that in such a case T is homeomorphic to the set of positive extreme points of the unit sphere of X^*.

5. Give an example of a simplex space without a strong order unit which is not an abstract M space. Give an example of a simplex space with a strong order unit which is not a space of the type $C(T, \mathbb{R})$.

6. Let T be a compact Hausdorff space and $\rho: T \to C(T, \mathbb{R})^*$ be a weak* continuous barycentric mapping. Show that A_ρ is equivalent to $C_\sigma(S, \mathbb{R})$ for some compact Hausdorff space S and some involuntary homeomorphism σ on S.

§ 20. Compact Choquet Simplexes

In this section we develop some intrinsic characterizations of compact Choquet simplexes. Recall that in the last section we defined a compact Choquet simplex as a compact convex set K in some locally convex Hausdorff topological vector space E such that $A(K, \mathbb{R})^*$ is an abstract L_1 space. In this section we study properties of compact convex sets $K \subset E$ to determine when they are simplexes. *Throughout this section we assume that K is a compact convex set in some locally convex Hausdorff topological vector space E.*

Definition 1. A probability measure μ (a positive normalized regular Borel measure) on K is said to *represent* $x \in K$ if $x^*(x) = \int x^* \, d\mu$ for all $x^* \in E^*$ and x is said to be the *resultant* of μ, $x = r(\mu)$.

We note that since the collection $\{x^*|K+a: x^*\in E^*, a\in \mathbb{R}\}$ is uniformly dense in $A(K,\mathbb{R})$ (see the appendix), it follows that μ represents x if and only if $h(x)=\int h\,d\mu$ for all $h\in A(K,\mathbb{R})$.

Note that each $x\in K$ is represented by the point mass ε_x. On the other hand, each probability measure μ on K represents some point of K. The easiest way to see this is to recall that K can be completely identified with $K\hat{\ }=\{x^*\in A(K,\mathbb{R})^*: x^*(1)=1=\|x^*\|\}$ in the weak* topology and μ is in $C(K,\mathbb{R})^*$. Since $\mu|A(K,\mathbb{R})$ is positive and norm 1, it is in $K\hat{\ }$.

Theorem 1. *Let $x\in K$. Then x is in the set $\operatorname{ext}(K)$ of extreme points of K if and only if the only probability measure μ on K which represents x is ε_x.*

Proof. Suppose x is not an extreme point of K. Then there are distinct points y and z in K such that $x=\frac{1}{2}y+\frac{1}{2}z$. Clearly x is represented by $\frac{1}{2}\varepsilon_y+\frac{1}{2}\varepsilon_z$ which is distinct from ε_x.

Suppose $x\in\operatorname{ext}(K)$ and μ represents x. Let $F\subset K$ be a compact set with $x\notin F$. If $\mu(F)>0$, then there is a $y\in F$ and a closed convex neighborhood V of y such that $x\notin V$ and $\mu(V\cap K)>0$. Since $x\notin V\cap K$, $\mu(V\cap K)=a>1$. For, if $\mu(V\cap K)=1$, then $v=\mu|V\cap K$ is a probability measure on the compact convex set $V\cap K$ and hence would represent a point in $V\cap K$. This is impossible since there are enough functionals in E^* to separate the points of K. Let μ_1 and μ_2 be defined by $\mu_1(B)=\dfrac{1}{a}\mu(B\cap V\cap K)$ and $\mu_2(B)=\dfrac{1}{1-a}\mu(B\cap(K\setminus V))$. Then μ_1,μ_2 are probability measures on K which represent points x_1, x_2 of K respectively. Moreover, $x_1\in V\cap K$ and $x_1\neq x$. Clearly $x=ax_1+(1-a)x_2$ since $\mu=a\mu_1+(1-a)\mu_2$ which is a contradiction. Hence $\mu(F)=0$ and by the regularity of μ, $\mu=\varepsilon_x$. ∎

We can refine representation slightly as follows. Let $\operatorname{ext}K\subset T\subset K$ with T closed. Then the restriction mapping $H\to h|T$ is a linear isometry of $A(K,\mathbb{R})$ into $C(T,\mathbb{R})$, and each $x\in K$ is represented by some probability measure μ on T and each probability measure μ on T represents some unique point of K. For, if $x\in K$, let $x^*\in A(K,\mathbb{R})$ be defined by $x^*(h)=h(x)$ for all $h\in A(K,\mathbb{R})$. By the Hahn-Banach theorem there is a $\mu\in C(T,\mathbb{R})^*$ such that $\|\mu\|=\|x^*\|$ and $\mu(h|T)=x^*(h)$ for all $h\in A(K,\mathbb{R})$. Clearly $\mu\geqslant 0$ and $\|\mu\|=1$ and μ represents x. On the other hand, if μ is a probability measure on T, then $x^*(h)=\int h|T\,d\mu$ for $h\in A(K,\mathbb{R})$ defines a unique element of $K\hat{\ }$ which then corresponds to some (unique) element of K.

Let $\boldsymbol{K}$ denote the set of all continuous concave functions on K. Clearly $\boldsymbol{K}+\boldsymbol{K}\subset\boldsymbol{K}$ and $a\boldsymbol{K}\subset\boldsymbol{K}$ for all $a\geqslant 0$. Also, if $u,v\in\boldsymbol{K}$, then so is $u\wedge v$ and $A(K,\mathbb{R})=-\boldsymbol{K}\cap\boldsymbol{K}$.

We define the *upper* and *lower envelopes* of an arbitrary function $f:K\to R$ as follows:

$$\bar{f}(x)=\inf\{u(x):\quad u\in\boldsymbol{K}, u\geqslant f\},$$

$$\underline{f}(x)=\sup\{v(x):-v\in\boldsymbol{K}, v\leqslant f\}\quad \text{for all } x\in K.$$

Clearly $\underline{f}=-(\overline{-f})$ and $\bar{f}$ is concave and upper semicontinuous while $\underline{f}$ is convex and lower semicontinuous. The following properties of these envelopes are easily established.

(1) If $f\leqslant g$, then $\bar{f}\leqslant\bar{g}$,

(2) $\overline{f+g}\leqslant\bar{f}+\bar{g}$,

(3) for $a\geqslant 0$, $\overline{af}=a\bar{f}$,

(4) $\underline{f}\leqslant f\leqslant\bar{f}$ for all functions $f:K\to\mathbb{R}$,

(5) $\bar{f}(x)=\inf\{h(x):h\in A(K,\mathbb{R}), h\geqslant f\}$ for all $x\in K$,

(6) $\bar{f}$ is the smallest upper semicontinuous concave function dominating f on K.

Lemma 1. *Let $x\in K$, $f\in C(K,\mathbb{R})$, and μ a positive measure on K representing x. Then*

(a) $\underline{f}(x)\leqslant\int f\,d\mu\leqslant\bar{f}(x)$,

(b) if $\underline{f}(x)\leqslant a\leqslant\bar{f}(x)$, *then there is a μ in $C(K,\mathbb{R})^*$ representing x and $\int f\,d\mu=a$.*

Proof. (a) Let $h\in A(K,\mathbb{R})$ with $h\geqslant f$. Then $\int f\,d\mu\leqslant\int h\,d\mu=h(x)$ and it follows that $\int f\,d\mu\leqslant\bar{f}(x)$. Similarly $\underline{f}(x)\leqslant\int f\,d\mu$.

(b) Let $a\in\mathbb{R}$ be such that $\underline{f}(x)\leqslant a\leqslant\bar{f}(x)$ and x^* be defined on the span of f by $x^*(bf)=ba$ for all $b\in\mathbb{R}$. Thus x^* has an extension μ on $C(K,\mathbb{R})$ such that $\mu(g)\leqslant\bar{g}(x)$ for all $g\in C(K,\mathbb{R})$. Since if $g\leqslant 0$, $\bar{g}(x)\leqslant 0$, it follows that μ is positive and since $\mu(1)=1$, μ is normalized.

Now for $h\in A(K,\mathbb{R})$, $\bar{h}=h$ and $\mu(h)\leqslant h(x)$, $\mu(-h)\leqslant -h(x)$ and, thus, $\mu(h)=h(x)$. ∎

Corollary. *The following statements are equivalent.*

(1) $x\in\operatorname{ext}K$,

(2) $\bar{f}(x)=f(x)$ *for all* $f\in C(K,\mathbb{R})$,

(3) $\bar{f}(x)=f(x)$ *for all* $f\in-\boldsymbol{K}$. ∎

For each $f\in C(K,\mathbb{R})$ the *boundary of f*, B_f, is given by $B_f=\{x\in K:f(x)=\bar{f}(x)\}$. Clearly from the above corollary $\operatorname{ext}K=\bigcap\{B_f:f\in-\boldsymbol{K}\}$ and since $\bar{f}$ is upper semicontinuous B_f is a G_δ-set.

The boundary is useful in characterizing nice representing measures as developed below.

Definition 2. Let K be a compact convex set. For μ, ν positive and in $C(K,\mathbb{R})$ we say that $\mu \prec \nu$ if and only if $\int f\,d\mu \leqslant \int f\,d\nu$ for all $f \in -\boldsymbol{K}$.

The above partially ordering is called *Choquet's ordering*. We note that it is a partially ordering since $\boldsymbol{K}-\boldsymbol{K}$ is dense in $C(K,\mathbb{R})$ and that if $\mu \prec \nu$, then $\int h\,d\mu = \int h\,d\nu$ for all $h \in A(K,\mathbb{R})$ and $\|\mu\| = \|\nu\|$. In particular, if $1 = \|\mu\| = \|\nu\|$, then they represent the same element of K. Also, if μ represents $x \in K$, then $\varepsilon_x \prec \mu$. For, if $f \in K$, then $\bar{f} = f$ and $f(x) = \inf\{h(x): h \in A(K,\mathbb{R}), h \geqslant f\} = \inf\{\int h\,d\mu: h \in A(K,\mathbb{R}), h \geqslant f\} \geqslant \int f\,d\mu$.

By a *maximal measure* μ on K we shall always mean maximal with respect to Choquet's ordering.

Lemma 2. *If* $0 \leqslant \mu \in C(K,\mathbb{R})^*$, *then* μ *is dominated by some maximal measure* ν.

Proof. This follows immediately from the fact that $\{\nu \in C(K,\mathbb{R})^* : \mu \prec \nu\}$ is bounded by $\|\nu\|$ and from Zorn's lemma. ∎

Corollary. *Any point* $x \in K$ *is represented by some maximal measure.* ∎

We now wish to develop some characteristic properties of maximal measures.

Lemma 3. *Let* μ *be a maximal measure on* K. *Then* $\mu(\{x\}) = 0$ *for all* $x \in K \setminus (\operatorname{ext} K)$.

Proof. Let $x \in K \setminus (\operatorname{ext} K)$ and ν be a positive measure on K representing x and $\nu(\{x\}) = 0$. Suppose $\mu(\{x\}) = b > 0$ and let $\mu_1 = \mu - b\,\varepsilon_x + b\,\nu$. Then $\mu \prec \mu_1$ and $\mu_1 \neq \mu$ which is a contradiction. ∎

Clearly for any $x \in K$, ε_x is maximal if and only if $x \in \operatorname{ext} K$.

We now associate with each positive $\mu \in C(K,\mathbb{R})^*$ a gauge on $C(K,\mathbb{R})$ as follows: $\bar{\mu}(f) = \int \bar{f}\,d\mu$ for all $f \in C(K,\mathbb{R})$ (recall that $\bar{f}$ is a bounded upper semicontinuous function on K).

Lemma 4. Let $\mu \in C(K,\mathbb{R})^*$ and ν be a linear form on $C(K,\mathbb{R})$. Then $\nu \leqslant \bar{\mu}$ if and only if ν is positive and $\mu \prec \nu$.

Proof. Suppose $\nu \leqslant \bar{\mu}$. Let $f \in C(K,\mathbb{R})$ with $f < 0$. Then $\nu(f) \leqslant \bar{\mu}(f) = \int f\,d\mu \leqslant 0$. Now, suppose $f \in \boldsymbol{K}$. Then $\nu(f) \leqslant \bar{\mu}(f) = \int \bar{f}\,d\mu$ and it follows that $\mu \prec \nu$.

On the other hand, if ν is positive and $\mu \prec \nu$, then for any $f \in C(K,\mathbb{R})$ and $h \in \boldsymbol{K}$ with $h \geqslant f$, $\nu(f) \leqslant \nu(h) \leqslant \mu(h)$. Thus $\nu(f) \leqslant \inf\{\mu(h): h \in \boldsymbol{K}, h \geqslant f\} = \int f\,d\mu$. ∎

The following theorem is a composite of results due to Choquet, Meyer, Bishop, and de Leeuw (see [44] and [66]).

Theorem 2. *Let K be a compact convex set and $\mu \in C(K,\mathbb{R})^*$ be positive. Then the following statements are equivalent.*

(1) *μ is maximal,*

(2) *$\mu = \bar{\mu}$ on $C(K,\mathbb{R})$,*

(3) *$\bar{\mu}$ is linear on $C(K,\mathbb{R})$,*

(4) *$\mu(K \setminus B_f) = 0$ for each $f \in -\boldsymbol{K}$.*

(5) *for each $f \in C(K,\mathbb{R})$ and each $\varepsilon > 0$ there are $g \in -\boldsymbol{K}$ and $h \in \boldsymbol{K}$ with $g \leqslant f \leqslant h$ and $\mu(h-g) < \varepsilon$.*

Proof. (1) implies (2). Let $f \in C(K,\mathbb{R})$. By the Hahn-Banach theorem there is a linear form ν on $C(K,\mathbb{R})$ such that $\nu \leqslant \bar{\mu}$ and $\nu(f) = \bar{\mu}(f)$. Hence ν is positive and $\mu \prec \nu$. Since μ is maximal, $\mu = \nu$. Thus for any $f \in C(K,\mathbb{R})$, $\mu(f) = \bar{\mu}(f)$. (2) implies (3) is trivial, (3) implies (1). Suppose $\nu \in C(K,\mathbb{R})^*$ is positive and $\mu \prec \nu$. Then $\nu \leqslant \bar{\mu}$ and since $\bar{\mu}$ is linear, $\bar{\mu} = \nu$. Since $\mu = \bar{\mu}$ on $\boldsymbol{K}$, $\mu = \bar{\mu}$ on $C(K,\mathbb{R})$ and $\mu = \nu$, i.e., μ is maximal. If $\mu(f) = \bar{\mu}(f) = \mu(\bar{f})$, then $\mu(\bar{f} - f) = 0$. (2) implies (4). Let $f \in -\boldsymbol{K}$. Hence $\mu(K \setminus B_f) = 0$. (4) implies (2). If $\mu(K \setminus B_f) = 0$ for $f \in -\boldsymbol{K}$, then $\mu(f) = \mu(\bar{f}) = \bar{\mu}(f)$ for $f \in -\boldsymbol{K}$. Hence $\mu = \bar{\mu}$ on $-\boldsymbol{K} \cup \boldsymbol{K}$. Since $\bar{\mu}$ is positive and $\mu \leqslant \bar{\mu}$ it follows that $\mu = \bar{\mu}$ on $C(K,\mathbb{R})$. (5) is equivalent to (2). For, let $\underline{\mu}$ be defined by $\underline{\mu}(f) \equiv \mu(\underline{f}) = -\mu(\overline{-f}) = -\bar{\mu}(-f)$ for $f \in C(K,\mathbb{R})$. Then clearly $\bar{\mu} = \mu$ if and only if $\underline{\mu} = \mu$ if and only if $\underline{\mu} = \bar{\mu}$. Now, $\underline{\mu}(f) = \sup\{\int g\,d\mu : g \in -\boldsymbol{K}, g \leqslant f\}$ and $\bar{\mu}(f) = \inf\{\int h\,d\mu : h \in \boldsymbol{K}, h \geqslant f\}$. Thus, $\underline{\mu}(f) = \bar{\mu}(f)$ if and only if for each $\varepsilon > 0$ there are $g \in -\boldsymbol{K}$, $h \in \boldsymbol{K}$ with $g \leqslant f \leqslant h$ and $\mu(h-g) < \varepsilon$. ∎

We now wish to characterize simplexes in terms of the ideas developed in this section. A preliminary lemma is needed.

Let K be a compact convex set. For $x \in K$, we let $M_x = \{\mu \in C(K,\mathbb{R})^* : 0 \leqslant \mu$ and μ represents $x\}$. Let $M_d = \left\{\sum_{i=1}^{n} a_i \varepsilon_{x_i} : x_i \in K, a_i \geqslant 0, \sum_{i=1}^{n} a_i = 1, n = 1, 2, \ldots\right\}$.

Lemma 5. *For each $f \in C(K,\mathbb{R})$ and $x \in K$, $\bar{f}(x) = \sup\{\mu(f) : \mu \in M_d \cap M_x\}$.*

Proof. Let $\varepsilon > 0$ be given and $\mu \in M_x$. There are compact convex sets $V_1, \ldots, V_n$ in K such that $K = \bigcup_{i=1}^{n} V_i$ and $|f(x) - f(y)| < \varepsilon$ for all $x, y \in V_i$ for $i = 1, \ldots, n$. Let $W_1 = V_1$ and $W_i = V_i \cap [K \setminus (V_1 \cup \cdots \cup V_{i-1})]$ for $i = 2, \ldots, n$. Then $K = \bigcup_{i=1}^{n} W_i$ and $\mu(W_1) + \cdots + \mu(W_n) = 1$. Let $S = \{i : 1 \leqslant i \leqslant n, \mu(W_i) > 0\}$. Now let ν_i be defined by $\nu_i(f) = \dfrac{1}{\mu(W_i)} \displaystyle\int f\,d\mu$

for all $i\in S$. Then $W_i \subset V_i$ and v_i represents some point $x_i \in V_i$ for $i=1,\ldots,n$. Let $v=\sum_{i\in S}\mu(W_i)\varepsilon_{x_i}$. Then for $h\in A(K,\mathbb{R})$.

$$v(h)=\sum_{i\in S}\mu(W_i)h(x_i)=\sum_{i\in S}\mu(W_i)v_i(h)=\sum_{i\in S}W_i\int h\,d\mu=\int h\,d\mu \quad\text{and } v\in M_x.$$

Now,

$$|\mu(f)-v(f)|=\left|\sum_{i\in S}\int_{W_i} f\,d\mu-\sum_{i\in S}\mu(W_i)f(x_i)\right|=\left|\sum_{i\in S}\int_{W_i}(f-f(x_i))\,d\mu\right|$$
$$\leqslant \varepsilon\sum_{i\in S}\mu(W_i)\leqslant\varepsilon. \quad ▮$$

For the following theorem the reader ia again referred to [44] and [66] and the comments on theorem 2 of section 19.

Theorem 3. *Let K be a compact convex set. Then the following statements are equivalent.*

(1) *K is a simplex,*

(2) *$\overline{f}$ is affine for all $f\in -\boldsymbol{K}$,*

(3) *if $\mu\in M_x$ is maximal, then $\mu(f)=\overline{f}(x)$ for all $f\in -\boldsymbol{K}$,*

(4) *$f\to\overline{f}$ is additive on $-\boldsymbol{K}$,*

(5) *for each $x\in K$ there is a unique maximal measure $\mu\in M_x$,*

(6) *$\overline{f}$ is affine for each upper semicontinuous convex function on K,*

(7) *if f,g are bounded lower semicontinuous concave functions on K with $f\leqslant g$, then there is an $h\in A(K,\mathbb{R})$ with $f\leqslant h\leqslant g$,*

(8) *if u_1,u_2,v_1,v_2 are in $A(K,\mathbb{R})$ with $u_1\vee u_2\leqslant v_1\wedge v_2$, then there is an $h\in A(K,\mathbb{R})$ with $u_1\vee u_2\leqslant h\leqslant v_1\wedge v_2$.*

(9) *$A(K,\mathbb{R})$ has the decomposition property,*

(10) *the space $A_u(K,\mathbb{R})$ of all bounded upper semicontinuous affine functions on K is a lattice (in its natural ordering).*

Proof. As before we identify K as $\{x^*\in A(K,\mathbb{R})^*: x^*(1)=1=\|x^*\|\}$ and let $C^*=\{x^*\in A(K,\mathbb{R})^*: x^*\geqslant 0\}$.

(1) implies (2). Let $f\in -\boldsymbol{K}$. Then for $x^*\in K$, $\overline{f}(x^*)=\sup\{\mu(f): \mu\in M_d\cap M_{x^*}\}=\sup\left\{\sum_{i=1}^n a_i f(x_i^*): a_i\geqslant 0, \sum_{i=1}^n a_i=1, x_i^*\in K, \text{ and } x^*=\sum_{i=1}^n a_i x_i^*\right\}$ $=\sup\left\{\sum_{i=1}^n y_i^*(f): y^*\in C^*, x^*=\sum_{i=1}^n y_i^*\right\}$. Now the unique positive homogeneous extension of f to C^* is given by $\overline{f}(x^*)=\sup\left\{\sum_{i=1}^n y_i^*(f): x^*=\sum_{i=1}^n y_i^*, y_i^*\in C^*\right\}$.

Since C^* is a lattice, if x^*,y^*,u_i^* are in C^* with $\sum_{i=1}^n u_i^*=x^*+y^*$, then there are x_i^*,y_i^* in C^* such that $u_i^*=x_i^*+y_i^*$ for $i=1,\ldots,n$. Thus

$\overline{f}(x^*+y^*) = \sup\left\{\sum_{i=1}^{n} u_i^*(f): u_i^* \in C^*, \sum_{i=1}^{n} u_i^* = x^*+y^*\right\} = \sup\left\{\sum_{i=1}^{n} x_i^*(f) + y_i^*(f): x_i^*, y_i^* \in C^*, \sum_{i=1}^{n} x_i^* = x^*, \sum_{i=1}^{n} y_i^* = y^*\right\} \leqslant \overline{f}(x^*)+\overline{f}(y^*)$. Thus $\overline{f}$ is convex on C^*.

Since $\overline{f}$ is concave, it is affine on K.

(2) implies (3). Let $x^* \in K$ and $\mu \in M_{x^*}$. For $f \in -\boldsymbol{K}$, $\mu(\overline{f}) = \overline{f}(x^*)$ since $\overline{f} = \inf\{h: h \in A(K,\mathbb{R}), h \geqslant f\}$. Since μ is maximal, $\mu(f) = \mu(\overline{f}) = \overline{f}(x^*)$.

(3) implies (4). Let $f, g \in -\boldsymbol{K}$ and $x^* \in K$. Choose a maximal measure $\mu \in M_{x^*}$. Then $(\overline{f+g})(x^*) = \mu(f+g) = \mu(f) + \mu(g) = \overline{f}(x^*) + \overline{g}(x^*)$.

(4) implies (5). Let $x^* \in K$. By (4) there is a unique positive normalized measure μ such that $\mu(f) = \overline{f}(x^*)$ for all $f \in -\boldsymbol{K}$. Clearly μ represents x^*.

Now suppose $v \in M_{x^*}$. Then for $f \in -\boldsymbol{K}$, $v(f) \leqslant v(\overline{f}) \leqslant \overline{f}(x^*) = \mu(f)$ and μ is maximal. Clearly we have shown μ is the unique maximal measure representing x^*.

(5) implies (1). The collection N of all positive maximal measures on K is clearly a closed lattice ordered cone and $N-N$ is thus a closed sublattice of $C(K,\mathbb{R})$. By (5) the transformation $x^* \in C^*$ goes into the unique maximal measure μ such that $x^* = \mu$ on $A(K,\mathbb{R})$ is one to one, affine, and onto. Thus, $A(K,\mathbb{R})^*$ is a lattice in its natural ordering and is an abstract L_1 space (in fact, it is order isometric to $N-N$).

Clearly (6) implies (2). Suppose K is a simplex and f is upper semicontinuous and convex on K. Then $f = \inf\{g \in -\boldsymbol{K}: f < g\}$ is directed downwards (see the appendix, theorem 9).

Let $x \in K$ and $\mu \in M_x$. Then for $g \in -\boldsymbol{K}$, $g > f$, $\mu(g) = \overline{g}(x) = \mu(\overline{g})$ and $\overline{f}(x) \geqslant \mu(\overline{f}) \geqslant \mu(f) = \inf\{\mu(g): g \in -\boldsymbol{K}, g > f\} = \inf\{g(x): g \in -\boldsymbol{K}, g > f\} \geqslant \overline{f}(x)$. Since each $\overline{g}$ is affine, so is $\overline{f}$.

(7), (8), and (9) are equivalent to (1) by theorem 2 of section 19.

(6) implies (10). Let $f, g \in A_u(K,\mathbb{R})$. Then $f \vee g = h$ is convex and upper semicontinuous. Thus $\overline{h} \in A_u(K,\mathbb{R})$ and $\overline{h}$ is the supremum of $\{f, g\}$ in $A_u(K,\mathbb{R})$.

(10) implies (5). Let $X = \{f_1 \vee \cdots \vee f_n: f_i \in A(K,\mathbb{R}), 1 \leqslant i \leqslant n, n = 1, 2, \ldots\}$. We first show that for $f \in X$, $\overline{f}$ is affine. Then the proof follows as in the proofs of (2) implies (3), (3) implies (4), and (4) implies (5) using X instead of $-\boldsymbol{K}$ and the fact that $X - X$ is dense in $C(K,\mathbb{R})$. Let $f = f_1 \vee \cdots \vee f_n$ with $f_i \in A(K,\mathbb{R})$ for $1 \leqslant i \leqslant n$. By (10) there is a $g \in A_u(K,\mathbb{R})$ such that g is the supremum of $f_1, \ldots, f_n$ in $A_u(K,\mathbb{R})$. Hence $g = \overline{g} \geqslant \overline{f}$. If $h \in A(K,\mathbb{R})$ and $h \geqslant f$, then $h \geqslant g$. Thus $\overline{f} \geqslant g$ and it follows that $\overline{f} = g$ is affine. ∎

A convex function f on K is said to be *strongly convex* if for each pair $x \neq y$ for distinct points in K and each $0 < a < 1$, $f(ax + (1-a)y) < af(x) + (1-a)f(y)$.

The following lemmas and theorems are left to the reader to prove.

Lemma 6. *If f is continuous and strongly convex on K, then $B_f = \operatorname{ext} K$.* ▮

Theorem 4. *A compact convex set K is metrizable if and only if there is a continuous strongly convex function f on K.*

Proof. For the necessity consider a sequence of norm one elements in $A(K, \mathbb{R})$ which is dense in the norm one functions in $A(K, \mathbb{R})$.

For the sufficiency consider $F(x,y) = \frac{1}{2}(f(x) + f(y)) - f\left(\frac{x+y}{2}\right)$ for x, y in K. ▮

Theorem 5 (Choquet). *If K is a metrizable compact convex set, then each point $x \in K$ is represented by a probability measure μ on K such that $\mu(K \setminus \operatorname{ext} K) = 0$.* ▮

The following theorem is useful at times to determine when maximal measures are almost supported $\operatorname{ext} K$.

Theorem 6. *Let μ be a maximal measure on the compact convex set K and $\{T_n\}$ a sequence of closed subsets of K such that $\operatorname{ext} K \subset \bigcup_{n=1}^{\infty} T_n$. Then $\mu\left[K \setminus \bigcup_{n=1}^{\infty} T_n\right] = 0$.*

Proof. It suffices to show that $\mu(T) = 0$ for each compact set $T \subset K \setminus \bigcup_{n=1}^{\infty} T_n$. For each n let $f_n \in C(K, \mathbb{R})$ with $0 \leqslant f_n \leqslant 1$, $f_n = 1$ on T and $f_n = 0$ on T_n. Then the pointwise infimum f of the f_n's is 0 on $\operatorname{ext} K$. Let $\varepsilon > 0$ be given. Then by an easy modification of theorem 2(5) there is a continuous convex function $g \leqslant t$ such that $\mu(f) < \mu(g) + \varepsilon$. For $h = g \vee 0$, $0 \leqslant h \leqslant f$ and $\mu(T) \leqslant \mu(f) \leqslant \mu(g) + \varepsilon$. Since $h = 0$ on $\operatorname{ext} K$ it follows that $h = 0$ and $\mu(T) < \varepsilon$. ▮

We now wish to characterize $A(K, \mathbb{R})$ as a subspace of $C(T, \mathbb{R})$ where T is the closure of $\operatorname{ext} K$ and $A(K, \mathbb{R})$ is naturally embedded into $C(T, \mathbb{R})$ under the restriction mapping $f \to f|T$ for all $f \in A(K, \mathbb{R})$. The formulation here follows Effros [97].

Theorem 7. *Let K be a compact convex set and B a compact subset of K. If $C = \{(\mu, \nu): \mu, \nu$ are probability measures on K, ν is maximal, $\mu \prec \nu$, and $S(\mu) \subset B\}$ and $D = \{(\varepsilon_x, \mu): x \in B$, μ is a maximal probability measure on K representing $x\}$, then C is contained in the weak* closed convex hull of D.*

Proof. Suppose C is not contained in the weak* closed convex hull of D. Then by the basic separation theorem there is weak* continuous linear functional L on $C(K,\mathbb{R})^* \times C(K,\mathbb{R})^*$ a pair $(\mu,\nu)\in C$, and a constant a such that $L(\varepsilon_x,\lambda)\leqslant a<L(\mu,\nu)$ for all $(\varepsilon_x,\lambda)\in D$. Thus there are f,g in $C(K,\mathbb{R})$ such that $L(\sigma,\tau)=\sigma(f)-\tau(g)$ for all σ,τ in $C(K,\mathbb{R})^*$ and, hence, $f(x)-\lambda(g)\leqslant a<\mu(f)-\nu(g)$ for all $(\varepsilon_x,\lambda)\in D$.

Now, for all $(\varepsilon_x,\lambda)\in D$, $f(x)-\lambda(\overline{g})\leqslant a<\mu(f)-\nu(\overline{g})$ since λ and ν are maximal measures. Let $u\in \boldsymbol{K}$ with $u\geqslant g$ and $f(x)-\lambda(u)\leqslant a<\mu(f)-\nu(u)$ for all $(\varepsilon_x,\lambda)\in D$. Moreover, for $x\in B$, $\underline{u}(x)=\inf\{\lambda(u):(\varepsilon_x,\lambda)\in T\}$ and, thus, $f(x)-a\leqslant\underline{u}(x)$ for all $x\in B$. Since $\underline{u}$ is convex, $\mu<\nu$, and ν is maximal, it follows that $\mu(f)-a\leqslant\mu(\underline{u})\leqslant\nu(\underline{u})$ which is a contradiction. ∎

For the lemma and theorem below let K be a compact convex set and T the closure of $\operatorname{ext} K$. Note that by theorem 7 if μ is a maximal on K, then $\mu(K\setminus T)=0$. Let $A(T,\mathbb{R})=\{f\in C(T,\mathbb{R}): f(t)=\mu(f)$ for each maximal probability measure μ on T representing $x\}$.

Lemma 6. *If* $f\in A(T,\mathbb{R})$ *and* μ,ν *are probability measures on* T *with* $\mu\prec\nu$, *then* $\mu(f)=\nu(f)$.

Proof. Clearly we may assume that ν is maximal. Let $P(T,\mathbb{R})$ denote the norm one positive elements of $C(T,\mathbb{R})^*$ and let $F: P(T,\mathbb{R})\times P(T,\mathbb{R})\to R$ be defined by $F(\sigma,\tau)=\sigma(f)-\nu(f)$ for all σ,τ in $P(T,\mathbb{R})$. Then F is affine and weak* continuous. Since $f\in A(T,\mathbb{R})$, f is 0 on D (as in theorem 7 where $B=T$) and, thus, on the convex hull of D. Hence F is 0 on C and the result follows. ∎

Theorem 8. *A function* $f\in C(T,\mathbb{R})$ *is in* $A(K,\mathbb{R})$ *if and only if*

(1) $f\in A(T,\mathbb{R})$

and

(2) *if* ν_1 *and* ν_2 *are maximal probability measures on* K *which represent the same point in* K, *then* $\nu_1(f)=\nu_2(f)$.

Proof. The necessity is clear. Suppose $f\in C(T,\mathbb{R})$ satisfies (1) and (2). It suffices to show that if $0\neq\sigma\in C(T,\mathbb{R})^*$ annihilates $A(K,\mathbb{R})$, then $\sigma(f)=0$. Now $\sigma(1)=0$ implies $\sigma^+(1)=\sigma^-(1)$ and $\dfrac{\sigma}{\|\sigma^+\|}=\mu_1-\mu_2$ where μ_1,μ_2 are in $P(T,\mathbb{R})$. Since $\mu_1(h)=\mu_2(h)$ for all $h\in A(K,\mathbb{R})$, it follows that μ_1 and μ_2 represent the same in K. Let ν_1,ν_2 be maximal measures dominating μ_1 and μ_2 respectively. Then $\mu_1(f)-\mu_2(f)=\nu_1(f)-\nu_2(f)=0$ and it follows that $\sigma(f)=0$. ∎

We will need the following theorem briefly in section 21. The proof of it can be found in [237, p. 100].

Theorem 9. *Let K be a compact convex set and $f: K \to R$ an affine Borel measurable function such that for each closed set $B \subset K$, $f|B$ has at least one point of continuity. Then $\mu(f) = f(x)$ for each probability measure representing x for each $x \in K$.* ∎

We mention here an unsolved problem which is related to theorem 5. We state three different aspects of it.

Problem 1. Given a separable complete metric space S is it homeomorphic to the set of extreme points of some (necessarily metrizable) compact convex set K *in some locally convex Hausdorff topological vector space E?*

Problem 2. Same as problem 1, except K *is now a simplex.*

Problem 3. Let T be a metrizable compact space and suppose A is a closed linear subspace of $C(T, \mathbb{R})$ containing the constant functions and separating the points of T. Does there exist a closed subspace X of $C(T, \mathbb{R})$ containing A such that X is a simplex space and whenever $x^* \in X^*$ is an extreme point of the closed unit ball of X^*, $x^*|A$ is also an extreme point of the closed unit ball of A^*?

Problem 1 seems to have first been proposed by Choquet. It is also mentioned in [191] along with a comment that problem 2 has been solved by Choquet (which is an error). Problem 3 is stated as a theorem (without proof) in [64].

We now turn our attention to the use of Choquet theory in the representation of continuous linear functionals. The real case is straightforward and we leave the details to the reader after outlining the procedure. The complex case will occupy the rest of this section and we shall explore it further in section 23.

Let T be a compact Hausdorff space and A be a linear subspace of $C(T, \mathbb{R})$ which contains the constants and separates the points of T. As before, we put $K = \{x^* \in A^* : x^*(1) = 1 = \|x^*\|\}$ (K is sometimes called the *state space* of A) so that K is a weak* compact convex set and the mapping $f \to \hat{f}$ where $\hat{f}(x^*) = x^*(f)$ for all $f \in A$ and $x^* \in K$ is a linear isometry from A onto a dense set in $A(K, \mathbb{R})$ (see theorem 6 of section 2). Moreover, since A separates the points of T, the mapping $\varphi: T \to K$ defined by $\varphi(t)(f) = f(t)$ for all $t \in T$ and $f \in A$ is a homeomorphism into K and $\varphi(T) \supset \operatorname{ext} K$. In particular, any $\mu \in M(T, \mathbb{R})$ corresponds to the element $\mu \circ \varphi^{-1}$ of $M(K, \mathbb{R})$ defined by the formula $(\mu \circ \varphi^{-1})(B) = \mu(\varphi^{-1}(B))$ for all Borel sets B in K. Moreover, any $\mu \in M(\varphi(T), \mathbb{R})$ corresponds to the element $\mu \circ \varphi \in M(T, \mathbb{R})$ defined by the formula, $(\mu \circ \varphi)(B) = \mu(\varphi(B))$ for all Borel sets B in T. Now recall that any maximal measure μ on K has the property that $\mu(K \setminus \varphi(T)) = 0$ (theorem 6) so that μ can correspond to the element $\mu \circ \varphi$ of $M(T, \mathbb{R})$. Hence an element μ of $M(T,\mathbb{R})$ is said to be a *boundary measure* on T (with respect to $\bar{A}$) if $|\mu|\circ\varphi^{-1}$ is maximal as a measure on K.

Thus we can state a representation theorem as follows (we leave the proof to the reader).

Theorem 10. *For each $x^*\in A^*$ there is a boundary measure μ on T such that $x^*(f)=\int f\,d\mu$ for all $f\in A$ and $\|x^*\|=\|\mu\|$. Moreover, for each $x^*\in A^*$ there is a unique such boundary measure if and only if K is a simplex.* ∎

We wish to establish a similiar representation theorem in the complex case and we investigate the uniqueness statements in section 23. Some additional terminology is required.

Let A be a linear subspace of $C(T,\mathbb{C})$ which contains the constants and separates the points of T. As above we put $K=\{x^*\in A^*:x^*(1)=1=\|x^*\|\}$. It is again easy to see that K is a weak* compact convex set and, hence, we may apply the theory of maximal measures to it (using real functions of course). We again define $\varphi:T\to K$ by $\varphi(t)(f)=f(t)$ for all $t\in T$ and $f\in A$ and φ is a homeomorphic embedding with $\varphi(T)\supset \operatorname{ext} K$.

Let U^* be the closed unit ball of A^*. Then U^* is a weak* compact convex set (and K is a closed face of U^*) and the mapping $f\to\hat{f}$ where $\hat{f}(x^*)=x^*(f)$ for all $x^*\in U^*$ and $f\in A$ is a linear isometry of A onto a dense set of $A_0(U^*,\mathbb{C})=\{h\in A(U^*,\mathbb{C}):h(0)=0\}$ and, moreover, one can easily see that $f\to\hat{f}|K$ is a linear isometry of A onto a dense set of $A(K,\mathbb{C})$ (although we do not use the denseness property), where we mean the affine continuous complex valued functions on U^* and K respectively.

Now let $\Gamma=\{\alpha\in\mathbb{C}:|\alpha|=1\}$ define $\Phi:\Gamma\times T\to U^*$ by $(\Phi(\alpha,t))(f)=\alpha f(t)$ for all $(\alpha,t)\in\Gamma\times T$ and $f\in A$. It is also easily seen that Φ is a homeomorphic embedding and that $\Phi(\Gamma\times T)\supset \operatorname{ext} U^*$ (see problem 20 at the end of chapter 2).

Now let $L:C(T,\mathbb{C})\to C(\Gamma\times T,\mathbb{C})$ be defined by $(Lf)(\alpha,t)=\alpha f(t)$ for all $(\alpha,t)\in\Gamma\times T$ and $f\in C(T,\mathbb{C})$. Then L is a linear isometric embedding and $L^*:M(\Gamma\times T,\mathbb{C})\to M(T,\mathbb{C})$ is given by $\int_T f\,d(L^*\mu)=\int_{\Gamma\times T}\alpha f(t)\,d\mu(\alpha,t)$ for all $f\in C(T,\mathbb{C})$ and $\mu\in M(\Gamma\times T,\mathbb{C})$. Finally, for $\mu\in M(\Gamma\times T,\mathbb{C})$ we define $P\mu\in M^+(T,\mathbb{C})$ by $\int_T f\,dP\mu=\int_{\Gamma\times T} f(t)\,d|\mu|(\alpha,t)$ for all $f\in C(T,\mathbb{C})$. Note that for $\mu\in M(\Gamma\times T,\mathbb{C})$, $|(L^*\mu)f|=|\int_{\Gamma\times T}\alpha f(t)\,d\mu(\alpha,t)|\leqslant\int_{\Gamma\times T}|f(t)|\,d|\mu|(\alpha,t)=\int_T|f|\,dP\mu$ and since $|L^*\mu|\,|f|=\sup\{|(L^*\mu)(hf)|: h\in C(T,\mathbb{C}), \|h\|\leqslant 1\}$, we obtain that $|L^*\mu|\,|f|\leqslant(P\mu)|f|$ for all $f\in C(T,\mathbb{C})$. That is, $P\mu-|L^*\mu|$ is a positive measure.

We are now in a position to define what we shall call *Hustad's map* $H:M(\Phi(\Gamma\times T),\mathbb{C})\to M(T,\mathbb{C})$ as follows. For $\mu\in M(\Phi(\Gamma\times T),\mathbb{C})$, $H\mu=L^*(\mu\circ\Phi)$. Clearly H is a linear operator and $\|H\mu\|\leqslant\|\mu\|$ for all μ.

In [294] Hustad defined such a mapping in order to prove a representation theorem for $x^* \in A^*$. We shall use the version due to Hirsberg [292] so that we obtain a boundary measure. Analogous to the real case, we shall say that $\mu \in M(T, \mathbb{C})$ is a *boundary measure* (with respect to $\bar{A}$) if $|\mu| \circ \varphi^{-1}$ is a maximal measure on K (in the sense of Choquet as in the first part of this section).

Theorem 11 (Hirsberg). *Let $x^* \in A^*$. Then there is a boundary measure v on T such that $x^*(f) = \int f dv$ for all $f \in C(T, \mathbb{C})$ and $\|x^*\| = \|v\|$.*

Proof. Clearly without loss of generality we may assume that $\|x^*\| = 1$. The idea is as follows. There is a maximal probability measure μ on U^* which represents x^*, that is, for all $f \in A$, $\int \hat{f} d\mu = x^*(f)$. Now μ is supported on $\Phi(\Gamma \times T)$ so that $\mu \in M(\Phi(\Gamma \times T), \mathbb{C})$. Hence we put $v = H(\mu)$. Now, for $f \in A$, $\int_T f dv = \int_T f dL^*(\mu \circ \Phi) = \int_{\Gamma \times T} \alpha f(t) d(\mu \circ \Phi)(\alpha, t)$ $= \int_{\Phi(\Gamma \times T)} \hat{f} d\mu = x^*(f)$ and, clearly, $\|v\| = 1 = \|x^*\|$. So, we need only show that $|v \circ \varphi^{-1}|$ is a maximal measure on K (Hustad showed that $|v \circ \varphi^{-1}|$ vanishes on each compact $\mathscr{F}_\sigma$ set disjoint from $\operatorname{ext} K$ while Hirsberg showed that $|v \circ \varphi^{-1}|$ is actually a maximal measure).

By theorem 2 we need only show that $\int_K (\bar{f} - f) d|v \circ \varphi^{-1}| = 0$ for all $f \in C(K, \mathbb{R})$ and since $\overline{f + \lambda} = \bar{f} + \lambda$ for any $\lambda > 0$, we can assume that f is strictly positive. Let $g: U^* \to R$ be defined by

$$g(\alpha x^*) = \begin{cases} f(x^*) & \text{for } (\alpha, x^*) \in \Gamma \times K, \\ 0 & \text{otherwise}. \end{cases}$$

Then since f is strictly positive, g is upper semicontinuous. Moreover, $\bar{g}(\alpha x^*) = \bar{g}(x^*)$ for all $(\alpha, x^*) \in \Gamma \times K$. For, let $(\alpha_0, x_0^*) \in \Gamma \times K$ and $\varepsilon > 0$ be given. There is an $h \in A(U^*, \mathbb{R})$ such that $\bar{g}(\alpha_0 x_0^*) + \varepsilon \geqslant h(\alpha_0 x_0^*)$ and $h \geqslant g$. We define $h_0 \in A(U^*, \mathbb{R})$ by $h_0(y^*) = h(\alpha_0 y^*)$ for all $y^* \in U^*$. Then $h_0(x_0^*) \leqslant \bar{g}(\alpha_0 y^*) + \varepsilon$, $h_0 \geqslant g$ and, hence, $\bar{g}(x_0^*) \leqslant h_0(x_0^*) \leqslant \bar{g}(\alpha x_0^*) + \varepsilon$, that is, $\bar{g}(x_0^*) \leqslant \bar{g}(\alpha_0 x_0^*)$. Since $(1/\alpha_0, \alpha_0 x_0^*) \in \Gamma \times K$, we obtain $\bar{g}(\alpha_0 x_0^*) \leqslant \bar{g}(x_0^*)$ from the above.

Now by lemma 5 and the fact that K is a face of U^*, we get that for all $x^* \in K$, $\bar{g}(x^*) = \sup\{\mu(g): \mu \in M_d \cap M_{x^*}\} = \sup\{\mu(f): \mu \in M_d \cap M_{x^*}\} = \bar{f}(x^*)$. Hence we have that

$$\begin{aligned} 0 &\leqslant \int_K (\bar{f} - f) d|v \circ \varphi^{-1}| = \int_T (\bar{f} - f) \circ \varphi d|L^*(\mu \circ \Phi)| \\ &\leqslant \int_T (\bar{f} - f) \circ \varphi dP(\mu \circ \Phi) = \int_{\Gamma \times T} [\bar{g}(\varphi(t)) - g(\varphi(t))] d\mu \circ \Phi(\alpha, t) \\ &= \int_{\Phi(\Gamma \times T)} (\bar{g} - g) d\mu = \int_{U^*} (\bar{g} - g) d\mu = 0. \quad \blacksquare \end{aligned}$$

We devote the rest of this section to further investigation of L^* (and hence of H). The notation and terminology is preserved as above and these results are due to Fuhr and Phelps [291]. Recall that if $v \in M(T, \mathbb{C})$, then there is a Borel function h of modules one such that

$\int f\,dv = \int fh\,d|v|$ for all $f \in C(T,\mathbb{C})$ (see theorem 6 of section 8). In such a case we write $v = h|v|$.

Lemma 7. *Suppose $v \in M(T,\mathbb{C})$ and write $v = h|v|$ where h is a Borel function of modulus one. Then there is a Borel set $B \subset T$ such that $|v|(T \setminus B) = 0$ and for any $g \in C(U^*,\mathbb{C})$, $t \to g(h(t)\phi(t))$ is Borel measurable on B.*

Proof. From Lusin's theorem it is possible to choose a Borel set $B \subset T$ such that $|v|(T \setminus B) = 0$ and a sequence $f_n \in C(T,\mathbb{C})$ with $f_n(t) \to h(t)$ for all $t \in B$. Thus for $g \in C(T,\mathbb{C})$, $\lim g(f_n(t)\varphi(t)) = g(h(t)\varphi(t))$ for all $t \in B$. That is, $t \to g(h(t)\varphi(t))$ defines a function which is a pointwise limit of continuous functions and hence is Borel measurable. ∎

We now define an operator $L_0: M(T,\mathbb{C}) \to M(\Gamma \times T,\mathbb{C})$ which will be the inverse of L^* on a certain set of measures. For $\mu \in M(T,\mathbb{C})$, put $\mu = h|\mu|$ where h is a (unique except for a set of $|\mu|$ measure zero) Borel measurable function with $|h| = 1$ and for $f \in C(\Gamma \times T,\mathbb{C})$, $(L_0\mu)(f) = \int_T f(h(t),t)\,d|\mu|(t)$. (The above lemma guarantees that $L_0\mu$ is definable in this manner.)

Lemma 8. *Let $v \in M(T,\mathbb{C})$ with $\|v\| = 1$ and let $x^* = v|A$. Then $\mu = L_0(v)$ is positive and norm one. Moreover, $\int_{U^*} \hat{f}\,d\mu \circ \Phi^{-1} = x^*(f)$ for all $f \in A$, that is, x^* is the resultant of $\mu \circ \Phi^{-1}$.*

Proof. Clearly μ is a probability measure on $\Gamma \times T$. Let $f \in A$. Then

$$\begin{aligned}\int_{U^*} \hat{f}\,d\mu \circ \Phi^{-1} &= \int_{\Gamma \times T} \hat{f}\,\Phi(h(t),\varphi(t))\; d\mu(h(t),\varphi(t))\\ &= \int_T h(t) f(t)\,d|v|(t) = \int f\,dv = x^*(f)\,.\end{aligned}$$ ∎

For the next three lemmas we assume that $\mu \in M(\Phi(\Gamma \times T),\mathbb{C})$ is a positive measure of norm one and that the functional x^* defined by $x^*(f) = \int \hat{f}\,d\mu$ also has norm one (that is, x^* is the resultant of μ in U^*). Furthermore, we put $v = H\mu = L^*(\mu \circ \Phi)$.

Lemma 9. *For each bounded complex valued Borel measurable function g on T,*

$$\int_T g\,dv = \int_{\Gamma \times T} \alpha g(t)\,d\mu \circ \Phi(\alpha,t)\,.$$

Proof. We first observe that the right hand integral indeed exists since $(\alpha,t) \to \alpha g(t)$ is the composition of a Borel function $((\alpha,t) \to (\alpha,g(t)))$ followed by a continuous function $((\alpha,\beta) \to \alpha\beta)$. Let λ be the probability measure on T defined by $\int f\,d\lambda = \int_{\Gamma \times T} f(t)\,d\mu \circ \Phi(\alpha,t)$ for all $f \in C(T,\mathbb{C})$. We shall show that $\lambda = |v|$. For, if $f \in C(T,\mathbb{C})$ and $f \geqslant 0$, then $|v|(f) = \sup\{v(g) : g \in C(T,\mathbb{C}),\ |g| \leqslant f\}$. Moreover, if $g \in C(T,\mathbb{C})$ and $|g| \leqslant f$, then $|v(g)| = |\int_{\Gamma \times T} \alpha g(t)\,d\mu \circ \Phi(\alpha,t)| \leqslant \int_{\Gamma \times T} |g(t)|\,d\mu \circ \Phi(\alpha,t) = \lambda(|g|) \geqslant \lambda(f)$. Thus $|v| \leqslant \lambda$. On the other hand, since $\|x^*\| = 1$, for any $\varepsilon > 0$ there is an $f \in A$ with $\|f\| = 1$ and $x^*(f) > 1 - \varepsilon$. Since $\hat{f}(\alpha\varphi(t)) = \alpha f(t)$ for

$(\alpha,t)\in\Gamma\times T$, we have that $1-\varepsilon<x^*(f)=\int\hat{f}\,d\mu=\int_{\Gamma\times T}\alpha f(t)\,d\mu\circ\Phi(\alpha,t)$ $=v(f)$. Hence, $\|v\|=1$ and it follows that $|v|=\lambda$.

Now from Lusin's theorem we can find a sequence $\{g_n\}$ in $C(T,\mathbb{C})$ such that $\|g_n\|\leqslant\|g\|$ (supremum norms) and $\{g_n\}$ converges pointwise $|v|$ almost everywhere to g. Moreover, $\alpha g_n(t)\to\alpha g(t)$ μ almost everywhere. For, let $\pi:\Gamma\times T\to T$ be the natural projection mapping. Then the corresponding map from $M(\Gamma\times T,\mathbb{C})$ to $M(T,\mathbb{C})$ is given by $\sigma\to\sigma\circ\pi^{-1}$ is readily seen that $\mu\circ\Phi\circ\pi^{-1}=\lambda=|v|$. Thus for any Borel set $B\subset T$, $|v|(B)=\mu(\Phi(\pi^{-1}(B)))=\mu(\Phi(\Gamma\times B))$. In particular, if B is a Borel set in T with $|v|(B)=1$ and $\lim g_n(t)=g(t)$ for all $t\in B$, then $\lim\alpha g_n(t)$ $=\alpha g(t)$ for all $(\alpha,t)\in\Gamma\times B$ and $\mu(\Phi(\Gamma\times B))=1$ so that $\lim\alpha g_n(t)=\alpha g(t)$ μ almost everywhere. By the Lebesgue bounded convergence theorem we have that $\int g\,dv=\lim\int g_n\,dv=\lim\int_{\Gamma\times T}\alpha g_n(t)\,d\mu\circ\Phi(\alpha,t)$ $=\int_{\Gamma\times T}\alpha g(t)\,d\mu\circ\Phi(\alpha,t)$. ∎

Lemma 10. *Let $v=h|v|$ where h is a Borel function with $|h|=1$ and put $B=\{h(t)\varphi(t):t\in T\}$. Then B is Borel measurable and $\mu(B)=1$.*

Proof. Let $k:\Gamma\times T\to\Gamma$ be defined by $k(\alpha,t)=\alpha\overline{h(t)}$. Then k is Borel measurable and $k^{-1}(1)=\{(\alpha,t):\alpha=h(t)\}=\{(h(t),t):t\in T\}$ is Borel measurable. Hence $\Phi(k^{-1}(1))$ is Borel measurable. Since $|v|$ is a probability measure, we have that $1=\int_T d|v|=\int_T\bar{h}\,dv=\int_{\Gamma\times T}\alpha\bar{h}(t)\,d\mu\circ\Phi(\alpha,t)$ $=\int_{\Gamma\times T}k\,d\mu\circ\Phi$. Thus $1=\mu(\Phi(k^{-1}(1)))=\mu(B)$. ∎

Lemma 11. *For any bounded Borel function g on T, $\int_T g\,d|v|$ $=\int_{\Gamma\times T}g(t)\,d\mu\circ\Phi(\alpha,t)$.*

Proof. Clearly $v=h|v|$ implies that $|v|=\bar{h}v$ and hence

$$\begin{aligned}\int_T g\,d|v|&=\int_T g\bar{h}\,dv=\int_{\Gamma\times T}\alpha g(t)\overline{h(t)}\,d\mu\circ\Phi(\alpha,t)\\&=\int_{\Gamma\times T}h(t)g(t)\overline{h(t)}\,d\mu\circ\Phi(\alpha,t)=\int_{\Gamma\times T}g(t)\,d\mu\circ\Phi(\alpha,t).\end{aligned}$$ ∎

We are now ready to prove that L_0 is the inverse of H on an appropriate set of measures. We shall use $r(\mu)$ to denote the resultant in U^* of $\mu\in M(\Phi(\Gamma\times T),\mathbb{C})$.

Theorem 12. *If $\mu\in M^+(\Phi(\Gamma\times T),\mathbb{C})$ and $\|\mu\|=1=\|r(\mu)\|$, then $L_0(H\mu)$ $=\mu\circ\Phi$. Furthermore, if $v\in M(T,\mathbb{C})$ and $\|v\|\leqslant1$, then $L^*(L_0(v))=v$.*

Proof. Let $v=H\mu$ and write $v=h|v|$ where h is a Borel measurable function of modulus one and put $B=\{h(t)\varphi(t):t\in T\}$. Then for any $g\in C(U^*,\mathbb{C})$, $(L_0v)(g\circ\Phi)=\int_T g(h(t)\varphi(t))\,d|v|(t)=\int_T g(h(t)\varphi(t))\overline{h(t)}\,dv(t)$ $\int_{\Gamma\times T}\alpha g(h(t)\varphi(t))\overline{h(t)}\,d\mu\circ\Phi(\alpha,t)=\int_{\Gamma\times T}h(t)g(h(t)\varphi(t))\overline{h(t)}\,d\mu\circ\Phi(\alpha,t)$ $=\int_{\Gamma\times T}g(h(t)\varphi(t))\,d\mu\circ\Phi(\alpha,t)=\int_B g(h(t)\varphi(t))\,d\mu(h(t)\varphi(t))=\int_{U^*}g\,d\mu$ $=\int_{\Gamma\times T}g\circ\Phi\,d\mu\circ\Phi$.

For the second part, let $\mu=L_0(v)$ and $v=h|v|$ as above. Then μ is positive and of norm one. By the definition of L_0 we have that for all $g \in C(\Gamma \times T, \mathbb{C})$, $\int_{\Gamma \times T} g\, d\mu = \int_T g(h(t),t)\, d|v|(t)$. Thus if $f \in C(T,\mathbb{C})$, we can define g by $g(\alpha,t) = \alpha f(t)$ in $C(\Gamma \times T, \mathbb{C})$ so that $(L\mu)(f) = \int_{\Gamma \times T} \alpha f(t)\, d\mu(\alpha,t)$ $= \int_T h(t) f(t)\, d|v|(t) = \int_T f\, dv$. ∎

Exercises. 1. Let $\{K_i\}_{i \in I}$ be a family of simplexes. Call a simplex K the *product* of the family $\{K_i\}$ if and only if there is a family $\{f_i\}_{i \in I}$ of affine continuous mappings of K onto K_i such that

(1) for every simplex L and every pair g and h of affine continuous mappings of L to K, if $f_i g = f_i h$ for all $i \in I$, then $g = h$,

(2) for every simplex L and every family $\{g_i\}_{i \in I}$ of affine continuous mappings, $g_i: L \to K_i$, there is an affine continuous mapping $g: L \to K$ such that $f_i g = g_i$ for all $i \in I$.

Show that every family of simplexes has a unique product.

2. Let K be a compact convex set. Show that $A(K,\mathbb{R})^{**}$ is linearly isometric to the space $A_b(K,\mathbb{R})$ of all bounded affine functions on K in the supremum norm.

3. Let K be a compact Choquet simplex and regard K as $\{x^* \in A(K,\mathbb{R})^* : x^*(1) = 1 = \|x^*\|\}$. Let F be a face in K. Show that the following are equivalent.

(i) F is complemented in K,

(ii) F is closed in the norm topology.

(iii) there is an $h \in A_b(K,\mathbb{R})$ such that $F = \{x^* \in K : h(x^*) = \|h\|\}$,

(iv) there is a norm closed hyperplane M in $A(K,\mathbb{R})^*$ such that $F = K \cap M$.

4. Let K be a compact Choquet simplex. Show that there is a one to one correspondence between the complemented faces of K and the closed lattice ideals in $A(K,\mathbb{R})^*$.

5. Let K be a compact Choquet simplex and let $\mathscr{B}$ be the set of complemented faces together with the empty set and K. Show that $\mathscr{B}$ is a complete Boolean algebra under some natural operations.

6. Let $K, \mathscr{B}$ be as in 5. Show that if T is the Stone space of $\mathscr{B}$ then $A(K,\mathbb{R})^{**} = C(T,\mathbb{R})$.

7. Let K be a compact convex set, $h_1, \ldots, h_n$ in $A(K,\mathbb{R})$ and $f = h_1 \vee \cdots \vee h_n$. Then for each $x \in K$, $x = \sum_{i=1}^{n} a_i x_i$ where $a_i \geq 0$, $\sum_{i=1}^{n} a_i = 1$, $f(x_i) = h_i(x_i)$, and $f(x) = \sum_{i=1}^{n} a_i h_i(x_i)$.

8. Let K be a compact convex set. A *selection* $\rho: K \to C(K,\mathbb{R})^*$ is a mapping such that $\rho(x)$ is a probability measure on K representing x for each $x \in K$. Prove that K is a simplex if and only if it admits an affine selection. Show that in such a case, ρ is unique. Show that if ρ is also weak* continuous, then the set of extreme points of K is closed.

9. Let K be a compact convex set and ρ a selection such that f_ρ is universally integrable ($f_\rho(x) = \int f(d\rho(x))$ for all $x \in K$) and if μ, v

are probability measures on K representing the same point of K, then $\int f_\rho d\mu = \int f_\rho d\nu$ for all $f \in C(K, \mathbb{R})$. Show that K is a simplex.

10. Let $A \subset C(T, \mathbb{C})$ be a linear subspace which contains constants and separates the points of T and put $\operatorname{Re} A = \{\operatorname{Re} f : f \in A\}$. Let $K = \{x^* \in A^* : x^*(1) = 1 = \|x^*\|\}$ and $L = \{y^* \in (\operatorname{Re} A)^* : y^*(1) = 1 = \|y^*\|\}$. Show that K and L are affinely homeomorphic with respect to the weak* topologies on A^* and $(\operatorname{Re} A)^*$ respectively. Let $t \in T$. Show that $\varepsilon_t | \operatorname{Re} A$ is an extreme point of the closed unit ball of $(\operatorname{Re} A)^*$.

Show that if A_1, A_2 are as above and $\overline{\operatorname{Re} A_1} = \overline{\operatorname{Re} A_2}$, then the state space of A_1 is affinely homeomorphic to the state space of A_2.

11. (Fuhr and Phelps). Let A be a closed linear subspace of $C(T, \mathbb{C})$ which contains the constants and separates the points of T and put $K = \{x^* \in A^* : x^*(1) = 1 = \|x^*\|\}$ (the state space of A). Show that if for each $x^* \in A^*$ there is a unique boundary measure μ on T such that $\|x^*\| = \|\mu\|$ and $x^*(f) = \int f d\mu$ for all $f \in A$, then K is a simplex.

12. Let A be a closed linear subspace of $C(T, \mathbb{C})$ which contains the constants and separates the points of T and put $K = \{x^* \in A^* : x^*(1) = 1 = \|x^*\|\}$ (the state space of A). Show that if for each $x^* \in A^*$ there is a unique boundary measure μ on T such that $\|x^*\| = \|\mu\|$ and $x^*(f) = \int f d\mu$ for all $f \in A$, then K is a simplex. Give an example to show that K can be a simplex without such unique representations. Prove that if A is self-adjoint and K is a simplex, then such representations are unique.

§ 21. Characterizations of Real L_1-Predual Spaces

We have already seen several classes of Banach spaces X such that X^* is linearly isometric to $L_1(\mu, \mathbb{R})$ for some measure μ. These are spaces of the type $C(T, \mathbb{R})$, $C_0(LT, \mathbb{R})$, $c(\Gamma, \mathbb{R})$, $c_0(\Gamma, \mathbb{R})$, $A(K, \mathbb{R})$, $A_0(K, \mathbb{R})$ and abstract M spaces. All of these Banach spaces are partially ordered and the L_1 structure of the dual is in terms of the dual ordering. In this section and the next we investigate the general properties of real L_1-predual spaces. In particular, we show that every L_1-predual space has associated with it a barycentric mapping and investigate the simplicial type properties of the closed unit ball of X^*.

For technical reasons we need to introduce the notion of the *metric approximation property* (M.A.P) due to Grothendieck. A Banach space X is said to have the M.A.P. if for each compact set $K \subset X$ and each $\varepsilon > 0$ there is a bounded linear operator $L: X \to X$ with finite dimensional range and such that $\|T\| = 1$ and $\|Tx - x\| < \varepsilon$ for all $x \in K$. (The example due to Enflo [290] does not have the M.A.P.) Some of the equivalences below are valid without the assumption of the M.A.P.

The reader can see [190] for the details. Basically we use the M.A.P. in order to obtain the $\lambda+\varepsilon$ bounds below for all $\varepsilon>0$. Since we prove later that an L_1-predual space has the M.A.P., we can apply it to these spaces.

Theorem 1 (Lindenstrauss). Let X be a Banach space with the metric approcimation property. Then the following statements are equivalent.

(1) *X^{**} is a $\mathscr{P}_\lambda(\mathbb{R})$.*

(2) *For every pair Y, Z of Banach spaces with $Z \supset Y$ and every bounded linear operator $A: Y \to X$ there is a bounded linear operator $B: Z \to X^{**}$ such that $B|Y = JA$ and $\|B\| \leqslant \lambda \|A\|$, where J is the natural map of X into X^{**}.*

(3) *For every pair Y, Z of Banach spaces with $Z \supset Y$ and every compact linear operator $A: Y \to X$ and each $\varepsilon>0$ there is a compact linear operator $B: Z \to X$ such that $B|Y = A$ and $\|B\| \leqslant (\lambda+\varepsilon)\|A\|$.*

(4) *The same as* (3) *except that* $\dim Z/Y < \infty$.

(5) *For every pair Y, Z of Banach spaces with $Z \supset X$ and every bounded linear operator $A: X \to Y^*$ there is a bounded linear operator $B: Z \to Y^*$ such that $B|X = A$ and $\|B\| \leqslant \lambda \|A\|$.*

(6) *For every pair Y, Z of Banach spaces with $Z \supset X$ and every compact linear operator $A: X \to Y$ there is a compact linear operator $B: Z \to Y$ with $B|X = A$ and $\|B\| \leqslant \lambda \|A\|$.*

(7) *Same as* (6) *with compact replaced by weakly compact.*

(8) *For every Banach space $Z \supset X$ with $\dim Z/X < \infty$, and every $\varepsilon>0$, and every compact linear operator $A: X \to X$ there is a compact linear operator $B: Z \to X$ with $B|X = A$ and $\|B\| \leqslant (\lambda+\varepsilon)\|A\|$.*

(9) *For every Banach space $Z \supset X$ there is a bounded linear operator $A: Z \to X^{**}$ such that $A|X = J$ and $\|A\| \leqslant \lambda$.*

(10) *For every pair Y, Z of Banach spaces with $Z \supset Y$, every $\varepsilon>0$, and every compact linear operator $A: Y \to X$ there is a compact linear operator $B: Z \to X$ such that $\|B|Y - A\| < \varepsilon$ and $\|B\| \leqslant (\lambda+\varepsilon)\|A\|$.*

Proof. We note that (1) implies (2), (3) implies (4), (7) implies (8), (6) implies (8), (2) implies (9), and (5) implies (9) are all immediate. (4) implies (9). Let $Z \supset X$ and $\mathscr{B} = \{(Y, \varepsilon): Y$ is a finite dimension subspace of Z and $0<\varepsilon\leqslant 1\}$. Clearly $\mathscr{B}$ is partially ordered by $(Y_1, \varepsilon_1) \leqslant (Y_2, \varepsilon_2)$ if and only if $Y_1 \subset Y_2$ and $\varepsilon_1 \geqslant \varepsilon_2$. By (4), for each pair (Y, ε) in $\mathscr{B}$ there is a compact linear operator $A_{Y,\varepsilon}: Y \to X$ such that $Z_{Y,\varepsilon}|Y \cap X$ is the identity and $\|A_{Y,\varepsilon}\| \leqslant \lambda+\varepsilon$. Let $T = \prod_{z \in Z} \|z\|(\lambda+1)U$, where U is the closed unit ball in X^{**} in the weak* topology. Then T is compact in the product topology. For each $A_{Y,\varepsilon}$ let $t_{Y,\varepsilon}(z)=\begin{cases}A_{Y,\varepsilon}(z) & \text{if } z\in Y.\\ 0 & \text{if } z\notin Y.\end{cases}$ Then $t_{Y,\varepsilon}=\{t_{Y,\varepsilon}(z)\}_{z\in Z}$ is an element of T. Let t be a limit point of the net $\{t_{Y,\varepsilon}\}$. Then $t(x)=x$ for all $x\in X$, $t(az_1+bz_2)=at(z_1)+bt(z_2)$ for $z_1,z_2\in Z$ and $a,b\in\mathbb{R}$ and $\|t(x)\|\leqslant\lambda\|z\|$ for $z\in Z$. The operator B defined by $B(z)=t(z)$ has the required properties.

(9) implies (1). Let Z be a $\mathscr{P}_1(\mathbb{R})$ space containing X^{**}. By (9) there is a bounded linear operator $A:Z\to Z$ with $A(Z)\subset X^{**}$, $\|A\|\leqslant\lambda$, and $A|JX$ is the identity. Thus $A^{**}Z^{**}\subset X^{**\perp\perp}$ (where $X^{**\perp\perp}\subset Z^{****}$ is the double annihalator of X^{**} in Z^{****}). Let $P=QJ^*$ where $J:X\to X^{**}$ and $Q:X^*\to X^{***}$ are the natural maps. Then P is a contractive projection of X^{***} onto QX^* and $Px^{***}=Q(x^{***}|JX)$. Hence $P^{-1}(0)=(JX)^\perp$ and P^* is a projection of X^{****} onto $(JX)^{\perp\perp}$. Using the natural linear isometry between X^{****} and $(X^{**})^{\perp\perp}$ we obtain a contractive projection π of $(X^{**})^{\perp\perp}$ onto $X^{\perp\perp}\subset Z^{**}$. Thus πA^{**} is a projection Z^{**} onto $X^{\perp\perp}$ and $\|\pi A^{**}\|\leqslant\lambda$ since Z^{**} is a $\mathscr{P}_1(\mathbb{R})$ space, X^{**} is a $\mathscr{P}_\lambda(\mathbb{R})$ space. (9) implies (5). Let Y,Z be Banach spaces with $Z\supset X$ and $A:X\to Y^*$ a bounded linear operator. By (9) there is a bounded linear operator $A_0:Z\to X^{**}$ such that $A_0|X=J$. Let $Q:Y\to Y^{**}$ be the natural map and $B=Q^*A^{**}A_0$. Then $B|X=A$ and $\|B\|\leqslant\lambda\|A\|$.

The proofs of (9) implies (6) and (9) implies (7) are similar to the above. Hence it remains only to show that (8) implies (10) and (10) implies (3). The only place that the hypothesis that X has the metric approximation property is used is in proving that (8) implies (10) which we now give. Let Y,Z be Banach spaces with $Z\supset Y$, $\varepsilon>0$, and $A:Y\to X$ a compact linear operator. Since A is compact and X has the metric approximation property there is a bounded linear operator $A_0:X\to X$ such that $\|A_0\|=1$, $X_0=A_0(X)$ is finite dimensional, and $\|A_0A-A\|<\varepsilon$. Let W be a Banach space such that $Z\supset W\supset Y$ and $\dim W/Y<\infty$ and Z_0 be a $\mathscr{P}_1(\mathbb{R})$ space containing X. Then there is a bounded linear operator $A_1:W\to Z_0$ such that $A_1|Y=A$ and $\|A_1\|=\|A\|$. By letting $V=X$ (plus a finite dimensional space if necessary) we obtain $V\supset X$, $A_1:W\to V$, $\|A_1\|=\|A\|$, and $\dim V/X\leqslant\dim W/Y<\infty$. By (8) there is an extension B_0 of A_0 from X to V such that $\|B_0\|\leqslant\lambda+\varepsilon$. Thus $A_W=A_0B_0A_1$ is a bounded linear operator from W to X_0 with finite dimensional range and such that $\|A_W|Y-A\|\leqslant 2\varepsilon$ and $\|A_W\|\leqslant(\lambda+\varepsilon)\|A\|$. Since $\dim X_0<\infty$, an argument similar to the one given in proving that (9) implies (4) gives an operator $B:Z\to X_0$ satisfying $\|B|Y-A\|\leqslant 2\varepsilon$ and $\|B\|\leqslant(\lambda+\varepsilon)\|A\|$.

(10) implies (3). Let Y,Z be Banach spaces with $Y\subset Z$ and $A:Y\to X$ a compact linear operator. For each $\varepsilon>0$, by (10) there is a sequence $\{A_n\}$ of compact linear operators $A_n:Z\to X$ such that $\|A_1\|\leqslant(\lambda+\varepsilon)\|A\|$,

$\|A_1|Y-A\|<\varepsilon/2$, and for $n\geqslant 2$, $\|A_n\|\leqslant(\lambda+1)\|A-(A_1+\cdots+A_{n-1})|Y\|$, and $\|A_n|Y-(A-A_1-\cdots-A_{n-1})|Y\|\leqslant\varepsilon/2^n$. In particular, $A_0=\sum_{n=1}^{\infty}A_n$ is compact, $A_0|Y=A$, $\|A_0\|\leqslant\|A_1\|+\sum_{n=2}^{\infty}\frac{(\lambda+1)\varepsilon}{2^{n-1}}\leqslant(\lambda+\varepsilon)\|A\|+(\lambda+1)\varepsilon$. ▮

We say that a Banach space X is an N_λ space if there is a collection $\{X_\gamma\}$ of finite dimensional subspaces of X which are upwards directed, their union is dense in X, and each X_γ is a $\mathscr{P}_\lambda(\mathbb{R})$ space.

Lemma 1. *Let X be a Banach space which is an N_λ space and suppose $\lambda'>\lambda$. Then there is a family $\{B_\tau\}$ of finite dimensional subspaces of X directed upwards by inclusion such that $X=\bigcup B_\tau$ and each B_τ is a $\mathscr{P}_{\lambda'}(\mathbb{R})$ space.* ▮

This follows immediately from theorem 6 of section 17.

Theorem 2. *Let X be an N_λ space. Then X satisfies property* (10) *of theorem* 1.

Proof. Let Y,Z be Banach spaces with $Z\supset Y$, $A:Y\to X$ a nonzero compact linear operator, and $\varepsilon>0$ be given. Let $K=A(U)$ and $\delta>0$, where $U=b(Y)$. Since $\bar{K}$ is compact, there are $x_1,\ldots,x_n$ in X such that for each $x\in K$ there is an x_i with $\|x-x_i\|<\delta$. By the above lemma there is a finite dimensional space $B\subset X$ such that $\{x_1,\ldots,x_n\}\subset B$ and B is a $\mathscr{P}_{\lambda+\delta}(\mathbb{R})$ space. Let P be a projection of X onto B with $\|P\|\leqslant\lambda+\delta$. Then for each $y\in U$ there is an $x\in B$ with $\|Ay-x\|\leqslant\delta$. Thus $\|PAy-Ay\|\leqslant\|PAy-Px\|+\|x-Ay\|\leqslant\delta(\lambda+1+\delta)$. Since B is a $\mathscr{P}_{\lambda+\delta}$ space there is an extension A' of PA from Z to B satisfying $\|A'\|\leqslant(\lambda+\delta)\|PA\|\leqslant(\lambda+\delta)(\|A\|+(\lambda+1+\delta)\delta)$. Hence for small enough δ, $\|A'\|\leqslant(\lambda+\varepsilon)\|A\|$ and $\|A'|Y-A\|\leqslant\varepsilon$. ▮

From theorem 11 of section 11 it follows that X is an L_1-predual space if and only if X^{**} is a $\mathscr{P}_1(\mathbb{R})$ space. We shall use this fact and the equivalence in theorem 1 above to give other characterizations of L_1-predual spaces. Some of these characterizations are given in terms of intersection properties which we discuss now. The development follows [182].

A Banach space X is said to have the n.2.I.P. of for each set of closed balls $S_1,\ldots,S_n$ in X (with varying centers and radii) such that $S_i\cap S_j\neq\emptyset$ for all i and j, it follows that $\bigcap_{i=1}^{n}S_i\neq\emptyset$. If we require all the radii to be the same, we refer to it as the R.n.2.I.P. If X has the n.2.I.P. for all n, we say that X has the F.2.I.P. *(finite binary intersection property)*.

It is necessary to develop some lemmas and theorems in order to be able to relate the intersection properties to theorem 1. The value of the first theorem below is clear.

Theorem 3. *For any Banach space X,* 4.2.I.P. *implies* F.2.I.P.

Proof. We show that if $n \geqslant 4$, then n.2.I.P. implies $(n+1)$.2.I.P. Let $S_1, \ldots, S_{n+1}$ be closed balls in X which intersect in pairs. Let $x \in X$ and $\theta = \max d(x, S_i)$. If $\varepsilon > 0$ is given, then $S_1, \ldots, S_{n+1}$, $S(x, \theta+\varepsilon) = S$ intersect in pairs. Let A denote the set of all subsets of $\{1, \ldots, n+1\}$ consisting of $n-1$ numbers. Then the number of elements of A is $n(n+1)/2$ and for each $\alpha \in A$, there is an $x_\alpha \in X$ such that $x_\alpha \in \bigcap_{i \in \alpha} S_i \cap S(x, \theta+\varepsilon)$. Let $y = \dfrac{2}{n(n+1)} \sum_{\alpha \in A} x_\alpha$. Then $y \in S$. Furthermore, for $1 \leqslant i \leqslant n+1$, $d(y, S_i) \leqslant \dfrac{2}{(n+1)n} \sum_{\alpha \in A} d(x_\alpha, S_i) = \dfrac{2}{(n+1)n} \sum_{i \notin \alpha} d(x_\alpha, S_i)$ $\leqslant \dfrac{2}{(n+1)n} \sum_{i \notin \alpha} (d(x, S_i) + \|x_\alpha - x\|) \leqslant \left(\dfrac{2}{(n+1)n}\right) n(2\theta+\varepsilon) = (\theta+\varepsilon/2) \dfrac{4}{n+1}$.
Since $n \geqslant 4$, $\dfrac{4}{n+1} > 1$. If $\dfrac{4}{n+1} < c < 1$ and ε is small, $c\theta \geqslant \max d(y, S_i)$. Hence there is a sequence $z_0, z_1, \ldots$, with $z_0 = x$ and $\|z_{m+1} - z_m\| \leqslant 2c^m \theta$ and $c^m \theta \geqslant \max d(z_m, S_i)$. Thus $\{z_m\}$ is a Cauchy sequence and $z_m \to z$ for some $z \in X$. Since $d(z, S_i) = 0$ for all $1 \leqslant i \leqslant n+1$, $z \in \bigcap_{i=1}^{n+1} S_i$. ∎

The following lemma is easy to prove.

Lemma 2. *Let X be a Banach space with the* 4.2.I.P. *and $Y = X \oplus \cdots \oplus X$ with $\|(x_1, \ldots, x_p)\| = \max \|x_i\|$. Then Y has the* 4.2.I.P. ∎

The following lemma says that if the intersection property is valid under small increases in radii, then it is valid.

Lemma 3. *Let X be a Banach space and $n \geqslant 3$. If for each $\varepsilon > 0$ and each collection $\{S(x_1, r_1), \ldots, S(x_n, r_n)\}$ of pairwise intersecting closed balls there is an $x \in X$ such that $\|x - x_i\| \leqslant r_i + \varepsilon$ for $i = 1$, $i = 1, \ldots, n$, then X has the* n.2.I.P.

Proof. Let $\varepsilon > 0$ and $S(x_1, r_1), \ldots, S(x_n, r_n)$ be a collection of closed balls which intersect in pairs. Let $x \in X$ with $\|x - x_i\| \leqslant r_i + \varepsilon$ for $i = 1, \ldots, n$. Then for each i there is a $y_i \in X$ such that $\|y_i - x\| \leqslant \varepsilon + \varepsilon/6$ and $\|y_i - x_j\| \leqslant r_j + \varepsilon/6$ for $j \neq i$.

Let $y = \dfrac{1}{n} \sum_{i=1}^{n} y_i$. Then $\|y - x\| \leqslant \varepsilon + \varepsilon/6$ and $\|y - x_j\| \leqslant 1/n$ $\left(\sum_{i \neq j} \|y_i - x_j\| + \|y_i - x_i\|\right) \leqslant \dfrac{1}{n} [(n-1) r_j + \varepsilon/6 + \|y_i - x\| + \|x - x_j\|] \leqslant r_j$ $+ \varepsilon/6 + \dfrac{2\varepsilon}{n} \leqslant r_j + \dfrac{5\varepsilon}{6}$ since $n \geqslant 3$.

Hence by letting $z_0 = x$ and constructing z_m inductively we obtain $\|z_{m+1} - z_m\| \leqslant 2(\frac{5}{6})^m \varepsilon$ and $\|z_{m+1} - x_j\| \leqslant r_j + (\frac{5}{6})^m \varepsilon$ for $j = 1, \ldots, n$. Hence $\lim z_m$ exists and is in $\bigcap_{i=1}^{n} S(x_i, r_i)$. ∎

Finally we have that the restricted intersection property is the same as the intersection property.

Lemma 4. *Let X be a Banach space. If X has the* R.4.2.I.P., *then it has the* 4.2.I.P.

Proof. Suppose $S(x_1, r_1), \ldots, S(x_4, r_4)$ are closed balls which intersect in pairs. Suppose there is an $\varepsilon > 0$ such that $\bigcap_{i=1}^{4} S(x_i, r_i + \varepsilon) = \emptyset$. Let $r > \max(r_1, r_2, r_3, r_4)$. We shall construct four balls $S(y_i, r) \supset S(x_i, r_i)$ and $\bigcap_{i=1}^{4} S(y_i, r) = \emptyset$. Let y_1 be chosen so that $S(y_1, r) \supset S(x_1, r_1)$ and $\emptyset = S(y_1, r) \cap \bigcap_{i=2}^{4} S(x_i, r_i + \varepsilon)$. Let $A = S(y_1, r) \cap S(x_3, r_3 + \varepsilon) \cap S(x_4, r_4 + \varepsilon)$. Then A is a closed convex set and $A \cap S(x_2, r_2 + \varepsilon) = \emptyset$. Hence there is an $x^* \in X^*$ such that $\|x^*\| \leqslant \dfrac{1}{r_2 + \varepsilon}$ and for all $x \in A$, $x^*(x - x_2) \geqslant 1$. To see how to obtain y_2 let $\delta > 0$ (to be choosen precisely later) and $z \in X$ with $\|z\| = 1$ and $x^*(x) \leqslant -\|x^*\| + \delta$. Let $y_2 = x_2 - (r - r_2) z$. For $x \in S(x_2, r_2)$, $\|x - y_2\| \leqslant \|x - x_2\| + r - r_2 \leqslant r$. Hence $S(y_2, r) \supset S(x_2, r_2)$. Now for $x \in S(y_2, r)$, $r\|x^*\| \geqslant x^*(x - y_2) = x^*(x - y_2) - (r - r_2) x^*(z)$ or $x^*(x - x_2) \leqslant r\|x^*\| + (r - r_2)(-\|x^*\| + \delta) \leqslant \dfrac{r_2}{r_2 + \varepsilon} + \delta(r - r_2)$. Thus if δ is small enough, $x^*(x - x_2) < 1$ for all $x \in S(y_2, r)$, and $A \cap S(y_2, r) = \emptyset$. The choices of y_3 and y_4 are similar. ∎

The intersection property is preserved under contractive projections.

Lemma 5. *Let X, Y be Banach spaces with $X \subset Y$ and suppose there is a contractive projection of Y onto X. Then if $\{S_Y(x_i, r_i)\}_{i \in I}$ is any family of closed balls in Y with $x_i \in X$ and $\bigcap_{i \in I} S_Y(x_i, r_i) \neq \emptyset$, then $\bigcap_{i \in I} S_X(x_i, r_i) \neq \emptyset$. Moreover, if Y has the* n.2.I.P. (*or* R.n.2.I.P.), *then so does X.* ∎

Recall that from theorem 8 of section 11 the following statement is true. If X, Y, Z are Banach spaces and $Z \supset X$ with $\dim Z/X = 1$ and $z \in Z \setminus X$, then a bounded linear operator $A: X \to Y$ has a norm preserving extension to Z if and only if $\bigcap_{x \in X} S(Tx, \|T\| \, \|z - x\|) \neq \emptyset$.

The finite binary intersection property implies an infinite binary intersection property if the set of centers is conditionally compact.

Theorem 4. *Let X be a Banach space with the* F.2.I.P. *Then for any collection $\{S(x_i,r_i)\}_{i\in I}$ of closed balls in X which pairwise intersect and $\{x_i\}_{i\in I}$ is compact, $\bigcap_{i\in I} S(x_i,r_i)\neq\emptyset$.*

Proof. There is a sequence $\{x_n\}$ in $\{x_i\}_{i\in I}$ such that $\{x_n\}$ is dense in $\{x_i\}_{i\in I}$. Let $a_1=\sup_{i\in I}(\|x_1-x_i\|)$ and for $j\geqslant 2$,

$$a_j=\max_{1\leqslant k<j}\left\{\sup_{i\in I}\|x_j-x_i\|-r_i,\ \|x_j-x_k\|-a_k\right\}.$$

Since $r_{i_1}+r_{i_2}\geqslant\|x_{i_1}-x_{i_2}\|$ for all $i_1,i_2\in I$, $a_j\leqslant r_j$ for $j=1,2,\dots$. If for some j, $a_j\leqslant 0$, then $x_j\in\bigcap_{i\in I}S(x_i,r_i)$. Without loss of generality we can assume that $a_j=r_j$ for $j=1,2,\dots$ and hence if $\varepsilon>0$, for each j there is an $i_{\varepsilon,j}$ such that $r_j+r_{i_{\varepsilon,j}}\leqslant\|x_j-x_{i_{\varepsilon,j}}\|+\varepsilon$. Moreover, $\{x_i\}_{i\in I}\subset\bigcap_{j=1}^{n}S(x_j,\varepsilon)$ for some n. Since X has the F.2.I.P., there is a $y\in X$ with $\|y-x_j\|\leqslant r_j$ for $1\leqslant j\leqslant n$.

Let $i\in I$ and suppose that $\|x_j-x_i\|\leqslant\varepsilon$. Then $r_{i_{\varepsilon,j}}+r_i\geqslant\|x_{i_{\varepsilon,j}}-x_i\|\geqslant\|x_{i_{\varepsilon,j}}-x_j\|-\varepsilon\geqslant r_i+r_{i_{\varepsilon,j}}-2\varepsilon$ and, thus, $r_i\geqslant r_j-2\varepsilon$. Therefore for all $i\in I$, $\|y-x_i\|\leqslant\|y-x_j\|+\|x_j-x_i\|\leqslant r_i+3\varepsilon$.

Let $\varepsilon_k\to 0$ and for each k let n_k be a positive integer such that $\{x_i\}_{i\in I}\subset\bigcap_{j=1}^{n_k}S(x_j,\varepsilon_k)$. Let $z_1\in\bigcap_{j=1}^{n_1}S(x_j,r_j)$. Then $S(z_1,3\varepsilon_1)\cap S(x_i,r_i)\neq\emptyset$ for all $i\in I$. Choose $z_k\in\bigcap_{j=1}^{n_k}S(x_j,r_j)\cap\bigcap_{j=1}^{k-1}S(z_j,3\varepsilon_j)$. Then $\|z_k-z_h\|\leqslant 3\varepsilon_k$ for $h>k$ and thus $\{z_k\}$ is a Cauchy sequence. Let $z=\lim z_k$. Thus $z\in\bigcap_{i\in I}S(x_i,r_i)$ since $\|z_k-x_i\|\leqslant 3\varepsilon_k+r_i$ for all $i\in I$. ∎

We can always find intersections by enlarging X.

Lemma 6. *Let X be a Banach space and $\{S(x_i,r_i)\}_{i\in I}$ a family of pairwise intersecting closed balls in X. Then there is a Banach space $Z\supset X$ with $\dim Z/X=1$ and $\bigcap_{i\in I}S_Z(x_i,r_i)\neq\emptyset$.*

Proof. We can consider that $X\subset l_\infty(\Gamma,\mathbb{R})$ for some index set Γ and use the fact that $l_\infty(\Gamma,\mathbb{R})$ has the binary intersection property (since $l_\infty(\Gamma,\mathbb{R})=C(\beta\Gamma,\mathbb{R})$ is order complete, see theorem 7 of section 11). ∎

The following theorem ties the intersection properties to condition 8 of theorem 1 for $\lambda=1$.

Theorem 5. *Let X be a Banach space. Then X has the* F.2.I.P. *if and only if for each pair of Banach spaces Y,Z with $Z\supset Y$ and $\dim Z/Y=1$, each $\varepsilon>0$, and each compact linear operator $A\colon Y\to X$, there is a bounded linear operator $B\colon Z\to X$ such that $B|Y=A$ and $\|B\|\leqslant(1+\varepsilon)\|A\|$.*

Proof. Suppose X has the F.2.I.P. and Y, Z, A, ε are given with $\|A\|=1$. Let $z \in Z \setminus Y$, $\|z\|=1$. Let $M>0$ and consider $\{S(Ay, \|y-z\|): \|y\| \leqslant M\}$. Any two of these balls intersect and since A is compact, by theorem 4 there is an $x_M \in X$ with $\|Ay-x_M\| \leqslant \|y-z\|$ for all $y \in Y$ with $\|y\| \leqslant M$. Hence, $\|x_M\| \leqslant \|z\|=1$. Let $y \in X$ with $\|y\| \geqslant M$. Then $\|Ay-x_M\| \leqslant \|Ay\| + \|x_M\| \leqslant \|y\|+1$ and $\|y-z\| \geqslant \|y\| - \|z\| = \|y\|-1$. Thus $\dfrac{\|Ay-x_M\|}{\|y-z\|} \leqslant \dfrac{\|y\|+1}{\|y\|-1} \leqslant \dfrac{M+1}{M-1}$. Thus for M large enough, $x_M \in \bigcap_{y \in Y} S(Ay, (1+\varepsilon)\|z-y\|)$, and A has a norm preserving extension to Z.

Suppose $S(x_1, r_1), \ldots, S(x_4, r_4)$ are four pairwise intersecting closed balls, $\varepsilon>0$, and $x_1=0$. Let Y be a three dimensional subspace containing x_1, x_2, x_3, x_4 and $Z \supset Y$ with $\dim Z/Y=1$ and a point $z \in Z$ with $\|z-x_i\| \leqslant r_i$ for $i=1,2,3,4$. Let A be a bounded linear operator from Z to X so that $A|Y$ is the identity and $\|A\| \leqslant 1+\varepsilon$. Then $x=Az$ satisfies $\|x-x_i\| = \|A(z-x_i)\| \leqslant (1+\varepsilon)\|z-x_i\| \leqslant (1+\varepsilon) r_i$ for $i=1,2,3,4$. ∎

Finally, we tie intersection properties in X to those in X^{**}.

Lemma 7. *Let X be a Banach space and $J: X \to X^{**}$ the natural map. Then for any collection $S_X(x_1, r_1), \ldots, S_X(x_n, r_n)$ of closed balls in X, $\bigcap_{i=1}^n S_{X^{**}}(Jx_i, r_i) \neq \emptyset$ if and only if $\bigcap_{i=1}^n S_X(x_i, r_i+\varepsilon) \neq \emptyset$ for all $\varepsilon>0$.*

Proof. If $\bigcap_{i=1}^n S_X(x_i, r_i+\varepsilon) \neq \emptyset$ for all $\varepsilon>0$, then by the weak* compactness of $S_{X^{**}}(Jx_i, r_i+\varepsilon)$, $\bigcap_{i=1}^n S_{X^{**}}(Jx_i, r_i) \neq \emptyset$.

Suppose that for some $\varepsilon>0$, $\bigcap_{i=1}^n S_X(x_i, r_i+\varepsilon)=\emptyset$. Then there is a quotient space Y of X with $\dim Y \leqslant n-1$ such that $\bigcap_{i=1}^n S_X(\pi x_i, r_i)=\emptyset$, where $\pi: X \to Y$ is the quotient map. For by Klee's theorem (see the appendix), there is a closed subspace Z of X such that $\dim X/Z \leqslant n-1$. and no translate of Z meets all the spheres $S_X(x_i, r_i+\varepsilon/2)$. Thus $Y=X/Z$. Now, suppose there is an $x^{**} \in \bigcap_{i=1}^n S_{X^{**}}(x_i, r_i)$. Then $\|\pi^{**}(x^{**})-\pi x_i\| \leqslant r_i$ for all $i=1, \ldots, n$. That is, $\pi^{**} x^{**} \in \bigcap_{i=1}^n S_Y(x_i, r_i)$ which is a contradiction. ∎

Corollary. *If X is a Banach space and X^{**} has the* n.2.I.P., *then X has the* n.2.I.P. ∎

Let us summarize the characterizations we have so far of L_1-predual spaces. These are all taken from [182].

Theorem 6 (Lindenstrauss). *Let X be a Banach space. Then the following are equivalent.*

(1) *X is an L_1-predual space,*

(2) *X^{**} is a $\mathscr{P}_1(\mathbb{R})$ space,*

(3) *X has the* F.2.I.P.

(4) *X has the* 4.2.I.P.

(5) *X has the* R.4.2.I.P.

(6) *X has the* M.A.P. *and any one of the equivalent conditions of theorem* 1 *for* $\lambda=1$.

Proof. The equivalence of (1) and (2) has already been established in theorem 11 of section 11. Since a $\mathscr{P}_1(\mathbb{R})$ space has the F.2.I.P., by the above corollary, (2) implies (3). Moreover, (3), (4), and (5) have been shown to be equivalent in this section.

Now, suppose X has the F.2.I.P. Then X has the metric approximation property. For, let $Y\subset X$ be a finite dimensional subspace and $\varepsilon>0$. Let Y_1 be a finite dimensional space whose unit sphere is a polyhedron approximating the unit sphere of Y suitably close and embed Y_1 in $l_\infty(n,\mathbb{R})$ for some n. Then there is a finite dimensional space $Z\supset Y$ with Z a $\mathscr{P}_{1+\varepsilon}(\mathbb{R})$ space. By theorem 5 the identity operator of Y into X has an extension A from Z into X with $\|A\|\leqslant 1+\varepsilon$. Since Z is a $\mathscr{P}_{1+\varepsilon}(\mathbb{R})$ space there is a bounded linear operator $A_1: X\to Z$ such that $A_1|Y$ is the identity and $\|A_1\|\leqslant 1+\varepsilon$. Thus $B=\pi A_1$ maps X into itself, $B|Y$ is the identity and $\|B\|\leqslant(1+\varepsilon)^2$. Hence X has the M.A.P.

By theorem 5, (3) implies property (4) of theorem 1 for $\lambda=1$.

Clearly (6) implies (2). ∎

We now give a characterization of real L_1-predual spaces in terms of maximal measures and the notions of compact Choquet simplex theory as developed in section 20.

The setting is given as follows. Let X be a Banach space, U^* the closed unit ball of X^* (in the weak* topology), $E=\operatorname{ext} U^*$, and T the weak* closure of E. There is a natural affine homeomorphism σ of U^* onto U^* given by $\sigma(x^*)=-x^*$ for all $x^*\in U^*$. Clearly $\sigma^2=$ identity and $\sigma|E$, $\sigma|T$ are onto E and T respectively. Moreover, if $\mu\in C(U^*,\mathbb{R})^*$ (or $C(T,\mathbb{R})^*$), then it is easy to see that $\mu(B)=\mu(\sigma(B))$ for all Borel sets $B\subset U^*$ (or $B\subset T$) if and only if $\int f\,d\mu=\int f\circ\sigma\,d\mu$ for all $f\in C(U^*,\mathbb{R})$ (or $f\in C(T,\mathbb{R})$).

We define $\{f\in C(U^*,\mathbb{R}): f=-f\circ\sigma\}=C_\sigma(U^*,\mathbb{R})$ and similarly, $A_\sigma(U^*,\mathbb{R})=\{f\in A(U^*,\mathbb{R}): f=-f\circ\sigma\}$, $C_\sigma(T,\mathbb{R})=\{f\in C(T,\mathbb{R}): f=-f\circ\sigma\}$. By $A(T,\mathbb{R})$ we mean $\{f\in C(T,\mathbb{R})$: if $x^*\in T$ and μ is a maximal measure on U^* representing x^*, then $f(x^*)=\int f\,d\mu\}$ (see section 20). Thus, $A_\sigma(T,\mathbb{R})=\{f\in A(T,\mathbb{R}): f=-f\circ\sigma\}$.

The following lemma is easily verified.

Lemma 8. *Let X be a Banach space. Then X is linearly isometric to $A_\sigma(U^*, \mathbb{R})$ under the mapping $x \to \hat{x}$ where $\hat{x}(x^*) = x^*(x)$ for $x \in X$ and $x^* \in U^*$.* ▮

Recalling the definition of the upper envelope as given in section 20, for $f: U^* \to \mathbb{R}$ we define f' as follows, $f' = \bar{f} - \bar{f} \circ \sigma$. The functions f' will play the role that $\bar{f}$ does in the characterization of simplexes.

If $\mu \in C(U^*, \mathbb{R})^*$, we define $\operatorname{odd}\mu$ by $\operatorname{odd}\mu = \frac{1}{2}[\mu - \mu \circ \sigma]$.

The following theorem, due to Lazar [173], should be compared with theorem 3 of section 20. We preserve the notation above.

Theorem 7 (Lazar). *Let X be a Banach space. Then the following are equivalent.*

(1) *X is an L_1-predual space.*

(2) *For any continuous convex function f on U^*, f' is affine.*

(3) *If μ is a maximal measure on U^* representing $x^* \in U^*$ and f is a continuous convex function on U^*, then $f'(x^*) = \int (f - f \circ \sigma) d\mu$.*

(4) *If μ_1 and μ_2 are maximal measures on U^* representing $x^* \in U^*$, then $\operatorname{odd}\mu_1 = \operatorname{odd}\mu_2$.*

(5) *For each continuous convex function f on U^*, $\bar{f}(0) = \frac{1}{2}\sup\{f(x^*) + f(-x^*) : x^* \in U^*\}$.*

Proof. (1) implies (2). Let f be a continuous convex function on U^*, x^*, y^* be in U^*, $0 \leqslant a \leqslant 1$, and $z^* = a x^* + (1-a) y^*$. We are going to show that $\bar{f}(z^*) + a\bar{f}(-x^*) + (1-a)\bar{f}(-y^*) \leqslant \bar{f}(-z^*) + a\bar{f}(x^*) + (1-a)\bar{f}(y^*)$. Since this is equivalent to $f'(z^*) \leqslant a f'(x^*) + (1-a) f'(y^*)$ and $f' = -f' \circ \sigma$, it follows that f' is affine.

From lemma 4 of section 20 it is enough to show that $\sum_{i=1}^{n} b_i f(z_i^*) + a \sum_{i=1}^{n} c_i f(-x_i^*) + (1-a) \sum_{i=1}^{n} d_i f(-y_i^*) \leqslant \bar{f}(-z^*) + (1-a)\bar{f}(y^*)$, where $z^* = \sum_{i=1}^{n} b_i z_i^*$, $x^* = \sum_{i=1}^{n} c_i x_i^*$, $y^* = \sum_{i=1}^{n} d_i y_i^*$, and $0 \leqslant b_i, c_i, d_i$, $\sum_{i=1}^{n} b_i = \sum_{i=1}^{n} c_i = \sum_{i=1}^{n} d_i = 1$, and x_i^*, y_i^*, z_i^* are in U^* for $i = 1, \ldots, n$.

We shall now assume $X^* = L_1(\nu, \mathbb{R})$ for some measure ν and use the L_1 structure of X^*. There are positive elements $z_i', z_i'', x_i', x_i'', y_i', y_i''$ in U^* and scalars b_i', c_i', d_i' in $[0,1]$ for $i = 1, \ldots, n$ so that $z_i^* = b_i' z_i' + (1 - b_i')(-z_i'')$, $x_i^* = c_i' x_i' + (1 - c_i')(-x_i'')$, and $y_i^* = d_i' y_i' + (1 - d_i')(-y_i'')$ for $i = 1, \ldots, n$. By the convexity of f we have that

$\sum_{i=1}^{n} b_i f(z_i^*) \leqslant \sum_{i=1}^{n} b_i b_i' f(z_i') + \sum_{i=1}^{n} b_i (1 - b_i') f(-z_i'')$, $\sum_{i=1}^{n} c_i f(x_i^*) \leqslant \sum_{i=1}^{n} c_i c_i' f(x_i') + \sum_{i=1}^{n} c_i (1 - c_i') f(-x_i'')$, and $\sum_{i=1}^{n} d_i f(y_i^*) \leqslant \sum_{i=1}^{n} d_i d_i' f(y_i') + \sum_{i=1}^{n} d_i (1 - d_i') f(-y_i'')$.

Also, $\sum_{i=1}^{n} b_i b'_i z'_i + a \sum_{i=1}^{n} c_i(1-c'_i)x''_i + (1-a) \sum_{i=1}^{n} d_i(1-d'_i)y''_i = \sum_{i=1}^{n} b_i(1-b'_i)z''_i + a \sum_{i=1}^{n} c_i c'_i x'_i + (1-a) \sum_{i=1}^{n} d_i d'_i y'_i$.

Using the decomposition theorem for vector lattices again, there are elements u_{ij} in X^* of norm one and positive scalars a_{ij} for $i,j=1,\dots,3n$ such that

$$
\begin{aligned}
b_i b'_i z'_i &= \sum_{j=1}^{n} a_{ij} u_{ij}, \\
a c_i(1-c'_i)x''_i &= \sum_{j=1}^{3n} a_{n+i,j} u_{n+i,j}, \qquad (1)\\
(1-a) d_i(1-d'_i)y''_i &= \sum_{j=1}^{3n} a_{2n+i,j} u_{2n+i,j}
\end{aligned}
$$

for $1 \leqslant i \leqslant n$ and

$$
\begin{aligned}
b_j(1-b'_j)z''_j &= \sum_{i=1}^{3n} a_{ij} u_{ij}, \\
a c_j c'_j x'_j &= \sum_{i=1}^{3n} a_{i,n+j} u_{i,n+j}, \qquad (2)\\
(1-a) d_j d'_j y'_j &= \sum_{i=1}^{3n} a_{i,2n+j} u_{i,2n+j}
\end{aligned}
$$

for $1 \leqslant j \leqslant n$. The additivity of the norm in the positive cone for X^* yields

$$
\begin{aligned}
& b_i b'_i = \sum_{j=1}^{3n} a_{ij}, \quad a c_i(1-c'_i) = \sum_{j=1}^{3n} a_{i+n,j}, \\
& (1-a) d_i(1-d'_i) = \sum_{j=1}^{3n} a_{2n+i,j} \quad \text{when } 1 \leqslant i \leqslant n \quad \text{and} \qquad (3)\\
& b_j(1-b'_j) = \sum_{i=1}^{3n} a_{ij}, \quad a c_j c'_j = \sum_{i=1}^{3n} a_{i,n+j}, \\
& (1-a) d_j d'_j = \sum_{i=1}^{3n} a_{i,2n+j} \quad \text{if } 1 \leqslant j \leqslant n.
\end{aligned}
$$

Using the fact that f is a convex function and the above equalities we obtain

$$\sum_{i=1}^{n} b_i f(z_i^*) + \sum_{i=1}^{n} c_i f(-x_i^*) + \sum_{i=1}^{n} d_i f(-y_i^*)$$
$$\leqslant \sum_{i=1}^{n} \sum_{j=1}^{3n} a_{ij} f(u_{ij}) + \sum_{j=1}^{n} \sum_{i=1}^{3n} a_{ij} f(-u_{ij}) + \sum_{i=1}^{n} \sum_{j=1}^{3n} a_{n+i,j} f(u_{n+i,j})$$
$$+ \sum_{j=1}^{n} \sum_{i=1}^{3n} a_{i,n+j} f(-u_{i,n+j}) + \sum_{i=1}^{n} \sum_{j=1}^{3n} a_{2n+i,j} f(u_{2n+i,j})$$
$$+ \sum_{j=1}^{n} \sum_{i=1}^{3n} a_{i,2n+j} f(-u_{i,2n+j}).$$

Regrouping terms in the right side of the above inequality we have

$$\sum_{i=1}^{n} b_i f(z_i^*) + \sum_{i=1}^{n} c_i f(-x_i^*) + \sum_{i=1}^{n} d_i f(-y_i^*)$$
$$\leqslant \sum_{j=1}^{n} \sum_{i=1}^{3n} a_{ij} f(-u_{ij}) + \sum_{j=1}^{n} \sum_{i=1}^{n} a_{ij} f(u_{ij}) + \sum_{j=1}^{n} \sum_{i=1}^{3n} a_{i,n+j} f(u_{i,n+j})$$
$$+ \sum_{i=1}^{n} \sum_{j=1}^{3n} a_{n+i,j} f(-u_{n+i,j}) + \sum_{j=1}^{n} \sum_{i=1}^{3n} a_{i,2n+j} f(u_{i,2n+j})$$
$$+ \sum_{i=1}^{n} \sum_{j=1}^{3n} a_{2n+i,j} f(-u_{2n+i,j}).$$

Now

$$\sum_{i=1}^{n} \sum_{j=1}^{3n} a_{ij}(-u_{ij}) + \sum_{j=1}^{n} \sum_{i=1}^{3n} a_{ij} u_{ij} = -z^*,$$
$$\sum_{j=1}^{n} \sum_{i=1}^{3n} a_{i,n+j} u_{i,n+j} + \sum_{i=1}^{n} \sum_{j=1}^{3n} a_{n+i,j}(-u_{n+i,j}) = a x^*,$$
$$\sum_{j=1}^{n} \sum_{i=1}^{3n} a_{i,2n+j} u_{i,2n+j} + \sum_{i=1}^{n} \sum_{j=1}^{3n} a_{2n+i,j}(-u_{2n+i,j}) = (1-a) y^*.$$

Thus by the equalities (1), (2), (3), and (4) and lemma 5 of section 20

$$\sum_{i=1}^{n} \sum_{j=1}^{3n} a_{ij} f(-u_{ij}) + \sum_{j=1}^{n} \sum_{i=1}^{3n} a_{ij} f(u_{ij}) \leqslant \bar{f}(-z^*),$$
$$\sum_{j=1}^{n} \sum_{i=1}^{3n} a_{i,n+j} f(u_{i,n+j}) + \sum_{i=1}^{n} \sum_{j=1}^{3n} a_{n+i,j} f(-u_{n+i,j}) \leqslant a \bar{f}(x^*),$$
$$\sum_{j=1}^{n} \sum_{i=1}^{3n} a_{i,2n+j} f(u_{i,2n+j}) + \sum_{i=1}^{n} \sum_{j=1}^{3n} a_{2n+i,n} f(-u_{2n+i,j}) \leqslant (1-a) \bar{f}(y^*).$$

The last three inequalities yield the desired result.

(2) implies (3). Let μ, x^*, f be as in (3). Since $\bar{f}$ and $\bar{f}\circ\sigma$ are upper semicontinuous and $f'=\bar{f}-\bar{f}\circ\sigma$ is affine, by theorems 2 and 9 of section 20 $f'(x^*)=\int f' d\mu=\int(\bar{f}-\bar{f}\circ\sigma)d\mu=\int(f-f\circ\sigma)d\mu$.

(3) implies (4). Suppose μ_1, μ_2 are two maximal measures on U^* representing $x^*\in U^*$. Let f be a continuous convex function on U^*. Then $f'(x^*)=\int(f-f\circ\sigma)d\mu_1=\int(f-f\circ\sigma)d\mu_2$. Thus $\int f\, d(\operatorname{odd}\mu_1)=\int f\, d(\operatorname{odd}\mu_2)$ and it follows that $\operatorname{odd}\mu_1=\operatorname{odd}\mu_2$.

(4) implies (5). Let μ be a maximal measure on U^* representing 0. Since for any $x^*\in E$, $\mu_1=\frac{1}{2}(\varepsilon_{x^*}+\varepsilon_{-x^*})$ is a maximal measure representing 0, $\operatorname{odd}\mu=\operatorname{odd}\mu_1$. Thus for any Borel set $B\subset U^*$, $\mu(B)-\mu(\sigma B)=\frac{1}{2}(\varepsilon_{x^*}(B)-\varepsilon_{-x^*}(B))-\frac{1}{2}(\varepsilon_{x^*}(\sigma B)-\varepsilon_{-x^*}(\sigma B))=0$. Hence for any continuous convex function f on U^*, $f(0)=\int f\, d\mu=\int f\circ\sigma\, d\mu$ and $\int f\, d\mu=\frac{1}{2}\int(f+f\circ\sigma)d\mu \leqslant\frac{1}{2}\|f+f\circ\sigma\|$. From this and lemma 4 of section 20, $f(0)\leqslant\frac{1}{2}\|f+f\circ\sigma\|$. The converse inequality is clear.

(5) implies (1). Let $S(x_i, r_i)$, $i=1,2,3,4$, be closed pairwise intersecting balls in $X=A_\sigma(U^*, \mathbb{R})$. We shall show that for any $\varepsilon>0$, $\bigcap_{i=1}^{4} S(x_i, r_i+\varepsilon)\neq\emptyset$. Thus by theorem 6 and lemma 3, X is an L_1-predual space.

Let g be defined by $g(x^*)=\max\{x^*(x_i)-r_i : 1\leqslant i\leqslant 4\}$. Then g is continuous and convex on U^* and since $|x^*(x_i-x^*(x_j))|\leqslant r_i+r_j$ for $x^*\in U^*$, $1\leqslant i,j\leqslant 4$ we have that $g(x^*)+g(-x^*)\leqslant 0$ for any $x^*\in U^*$. Thus by (5) $\bar{g}(0)\leqslant 0$. From the definition of the upper envelope there is a continuous affine function h on U^* such that $h\geqslant g$ and $0\leqslant h(0)<\varepsilon$. Now, $f=h-h(0)\in A_\sigma(U^*, \mathbb{R})$ and $r_i+\varepsilon\geqslant f(-x^*)-x^*(x_i)=f(x^*)-x^*(x_i) \geqslant -r_i-\varepsilon$ for any $x^*\in U^*$ and $1\leqslant i\leqslant 4$. Thus $f\in\bigcap_{i=1}^{4} S(x_i, r_i+\varepsilon)$ and we are done. ∎

We can get a representation for X as a space of functions. For this corollary to the above theorem we preserve the notation.

Corollary. *Let X be an L_1-predual space. Then $X=A_\sigma(T, \mathbb{R})$.*

Proof. Let $f\in A_\sigma(T, \mathbb{R})$ and suppose μ_1, μ_2 are two maximal measures on U^* representing $x^*\in U^*$. Then $\operatorname{odd}\mu_1=\operatorname{odd}\mu_2$ and since $f=f\circ\sigma$, $\int f\, d\mu_1=\int f\, d(\operatorname{odd}\mu_1)=\int f\, d(\operatorname{odd}\mu_2)=\int f\, d\mu_2$. Thus by theorem 8 of section 20 $f\in A_\sigma(U^*, \mathbb{R})$. Clearly $X=A_\sigma(U^*, \mathbb{R})\subset A_\sigma(T, \mathbb{R})$. ∎

We now wish to give the converse to theorem 6 of section 19.

Theorem 8 (Bednar-Lacey). *Let X be an L_1-predual space. Then there is a compact Hausdorff T and a barycentric mapping $\rho: T\to C(T, \mathbb{R})^*$ so that X is linearly isometric to $A_\rho=\{f\in C(T, \mathbb{R}): f(t)=\int f\, d\rho(t)$ for all $t\in T\}$.*

Proof. As before, T is the weak* closure of the set of extreme points of the unit ball U^* of X^*. We define ρ as follows. For any $x^* \in T$ let μ be a maximal measure representing x^* and put $\rho(x^*) = \text{odd}\,\mu$. By theorem 8 the mapping ρ is well defined. We will show that ρ is a barycentric mapping and $X = A_\rho$. Clearly $\|\rho(x^*)\| \leqslant 1$ for all $x^* \in T$. If f is a continuous convex function on U^*, then by theorem 7, for $x^* \in T$, $\frac{1}{2} f'(x^*) = \frac{1}{2}\int (f - f\circ\sigma)\,d\mu = \frac{1}{2}\int f\,d(\text{odd}\,\mu) = \int f\,d\rho(x^*)$ for any maximal measure μ on U^* representing x^*. Thus f_ρ is universally integrable for f the restriction to T of a convex continuous function on U^*. Since these functions are uniformly dense in $C(T, \mathbb{R})$, f is universally integrable for all $f \in C(T, \mathbb{R})$.

Let the operator $P: C(T, \mathbb{R})^* \to C(T, \mathbb{R})^*$ be defined as follows. If $\nu \in C(T, \mathbb{R})^*$ and $\nu \geqslant 0$ with $\|\nu\| = 1$, let μ be any maximal measure dominating ν (in the ordering of Choquet) and put $P\nu = \text{odd}\,\mu$. By theorem 7 $P\nu$ is well defined. Clearly P is affine on the probability measures in $C(T, \mathbb{R})^*$ and hence has a unique linear extension to $C(T, \mathbb{R})^*$. Clearly $\|P\| \leqslant 1$ and since $\text{odd}(\text{odd}\,\mu) = \text{odd}\,\mu$ for any $\mu \in C(T, \mathbb{R})^*$, P is a projection.

The kernel of P is $\{\mu \in C(T, \mathbb{R})^* : \int f_\rho\,d\mu = 0$ for all $f \in C(T, \mathbb{R})\}$. For, if $P\mu = 0$ and f is continuous and convex on U^* then $\int f_\rho\,d\mu = \frac{1}{2}\int(\bar{f} - \bar{f}\circ\sigma)\,d\mu = \frac{1}{2}\int(\bar{f} - \bar{f}\circ\sigma)\,d\mu^+ - \frac{1}{2}\int(\bar{f} - \bar{f}\circ\sigma)\,d\mu^-$. Now, let μ_1, μ_2 be two positive maximal measures dominating μ^+ and μ^- respectively in Choquet's ordering. Then $\int f_\rho\,d\mu = \frac{1}{2}\int(\bar{f} - \bar{f}\circ\sigma)\,d\mu_1 - \frac{1}{2}\int(\bar{f} - \bar{f}\circ\sigma)\,d\mu_2 = \frac{1}{2}\int(f - f\circ\sigma)\,d\mu_1 - \frac{1}{2}\int(f - f\circ\sigma)\,d\mu_2 = \int f\,dP\mu^+ - \int f\,dP\mu - = \int f\,dP\mu = 0$. Thus $P\mu = 0$ implies $\int f_\rho\,d\mu = 0$ for all $f \in C(T, \mathbb{R})$. On the other hand, suppose that $\int f_\rho\,d\mu = 0$ for all $f \in C(T, \mathbb{R})$. Let x^*, y^* be the elements of X^* represented by μ^+, μ^- respectively and μ_1, μ_2 be two maximal measures dominating μ^-, μ^- respectively in the ordering of Choquet. If $x^* \neq y^*$, then for some $x \in X$, $x^*(x) \neq y^*(x)$. Thus $\int \hat{x}\,d(\mu_1 - \mu_2) = x^*(x) - y^*(x) = \int \hat{x}\,d(\mu^+ - \mu^-) \neq 0$ (where $\hat{x}(z^*) = z^*(x)$ for $z^* \in T$). This is a contradiction since $\hat{x}_\rho = \hat{x}$, hence $x^* = y^*$. Thus $\text{odd}\,\mu_1 = \text{odd}\,\mu_2$ and $P\mu = 0$.

We now show that the range of P is $\{\mu \in C(T, \mathbb{R}) : \int f\,d\mu = \int f_\rho\,d\mu$ for all $f \in C(T, \mathbb{R})$. Suppose $\nu \in C(T, \mathbb{R})^*$ is positive and of norm 1. Let μ be a maximal measure on U^* dominating ν and f a continuous convex function on U^*. Then $f_\rho = \frac{1}{2}(\bar{f} - \bar{f}\circ\sigma)$, $\int f_\rho\,dP\nu = \frac{1}{4}[\int \bar{f}\,d\mu - \int \bar{f}\,d\mu\circ\sigma - \int \bar{f}\circ\sigma\,d\mu + \int f\circ\sigma\,d\mu\circ\sigma] = \frac{1}{2}[\int f\,d\mu - \int f\,d\mu\circ\sigma] = \int f\,dP\nu$. From this it follows that $\int f_\rho\,dP\nu = \int f\,dP\nu$ for all $f \in C(T, \mathbb{R})$.

The reverse inclusion is easily verified.

Clearly the mapping $x \to \hat{x}$ is a linear isometry of X into A_ρ. On the other hand, if $\mu \in C(T, \mathbb{R})^*$ with $\|\mu\| \leqslant 1$ $\mu \in \hat{X}^\perp$, then μ^+, μ^- represent the same point in U^* and, thus, $P\mu = 0$. Hence $\hat{X}^\perp = A^\perp = $ kernel of P.

From this it follows easily that ρ is a barycentric map and $\hat{X}=A_\rho$ and we are done. ∎

Corollary. *Let X be an L_1-predual space and suppose that T is the weak* closure of the closed unit ball U^* of X^*.*

(1) *If $T=\operatorname{ext} U^*$, then $X=C_\Sigma(T,\mathbb{R})$, where $\Sigma(x^*)=-x^*$ for all $x^*\in T$,*

(2) *If $T=\operatorname{ext} U^*\cup\{0\}$, then $X=C_\sigma(T,\mathbb{R})$, where $\sigma(x^*)=-x^*$ for all $x^*\in T$.* ∎

This follows immediately from the above theorem ((1) was proved by Lindenstrauss and Wulbert in [193]).

No general classification of L_1-predual spaces in known. The relationships between the classes of concrete L_1-predual spaces discussed in this text are developed in the exercises of this section. The reader can see [193] also.

Problems. 1. Let X be a Banach space and U^* the closed unit sphere of X^*. Then the following statements are equivalent:

a) X is an L_1-predual space,

b) for each $x^*\in U^*$ there is a unique $\mu\in C(U^*,\mathbb{R})^*$ such that $\|\mu\|\leqslant 1$, $|\mu|$ is maximal, $\mu=-\mu\circ\sigma$, and for each $x\in X$, $x^*(x)=\int y^*(x)\,d\mu(y^*)$.

2. Let T be a compact Hausdorff space and $\rho: T\to C(T,\mathbb{R})^*$ a non-negative barycentric mapping. Show that A_ρ is a simplex space. Show that if ρ is weak* continuous, then $A_\rho=C_0(T,\mathbb{R})$ for some locally compact Hausdorff space T.

3. Let T be a compact Hausdorff space, F a closed set in T, $\rho: T\to C(T,\mathbb{R})^*$ a mapping such that $\|\rho(t)\|\leqslant 1$ for all $t\in T$. Suppose further that $\rho(t)=\varepsilon_t$ for $t\notin F$ and that $\rho|F$ is weak* continuous. Let $F_1=\{t\in F:\rho(t)=\varepsilon_t\}$ and suppose that $|\rho(t)|(F\setminus F_1)=0$ for all $t\in F\setminus F_1$ and that $F\setminus F_1\neq T$.

Show that ρ is a barycentric mapping.

4. Using theorem 4 show that if X is an L_1-predual space and the closed unit ball of X has an extreme point, then X is linearly isometric to a closed subspace A (containing 1) of $C(T,\mathbb{R})$ for some compact Hausdorff space T, where A has the decomposition property. That is, $X=A(K,\mathbb{R})$ for some compact Hausdorff space K.

5. Prove that if K is a compact Choquet simplex with closed extreme points, then $A(K,\mathbb{R})=C(T,\mathbb{R})$ where $T=\operatorname{ext} K$.

6. Prove that if X is linearly isometric to both $C_\sigma(T,\mathbb{R})$ and $A(K,\mathbb{R})$ for some σ, T, K, then X is linearly isometric to $C(S,\mathbb{R})$ for some compact Hausdorff space S.

7. A *G-space* is a Banach space which is linearly isometric to $\{f\in C(T,\mathbb{R}): f(t_i)=\lambda_i f(t_i')\}$ for some compact Hausdorff space T, some

$\lambda_i \in [-1,1]$, and some collection of elements $t_i, t_i' \in T$ (compare with abstract M spaces and theorem 5 of section 9). Let X be a G space. Show that there is an M space $Y \supset X$ and a contractive projection of Y onto X. (Hint: Consider $T_0 = T \times \{1,2\}$ and $Y = \{f \in C(T_0, \mathbb{R}): f(t_i, 1) = \lambda_i f(t_i', 1)$ if $\lambda_i \geqslant 0$ and $f(t_i, 2) = -\lambda_i f(t_i', 2)$ if $\lambda_i < 0\}$). Use this to prove that X is an L_1-predual space.

8. Let X be an L_1-predual space and T be the weak* closure of $\operatorname{ext}(b(X^*))$. Show that X is a G space if and only if each $x^* \in T$ is of the form $x^* = \lambda y^*$ where $\lambda \in [-1,1]$ and $y^* \in \operatorname{ext}(b(X^*))$.

9. Let X be an L_1-predual space.

a) Show that if X is linearly isometric to both a G space and an $A(K, \mathbb{R})$ space, then it is linearly isometric to $C(T, \mathbb{R})$ for some compact Hausdorff space T.

b) Show that if X is linearly isometric to both a $C_\Sigma(T, \mathbb{R})$ and $C_0(T', \mathbb{R})$ for some Σ, T, and locally compact T', then X is linearly isometric to $C(S, \mathbb{R})$ for some compact Hausdorff space S.

c) Show that if X is linearly isometric to both a G space and an $A_0(K, \mathbb{R})$ space, then X is linearly isometric to an abstract M space.

d) Show that if X is linearly isometric to both an $A_0(K, \mathbb{R})$ space and a $C_\sigma(T, \mathbb{R})$ space, then it is linearly isometric to $C_0(T', \mathbb{R})$ for some locally compact Hausdorff space T'.

§ 22. Some Selection and Embedding Theorems for Real L_1-Predual Spaces

We wish to show that if X is a separable L_1-predual space and X^* is nonseparable, then X contains a subspace Y which is both the range of a contractive projection on X and linearly isometric to $C(\Delta, \mathbb{R})$, where Δ is the Cantor set. To accomplish this we prove certain selection theorems about the closed unit ball of X^*. We then use this to investigate necessary and sufficient conditions for X^* to be linearly isometric to $l_1(\Gamma, \mathbb{R})$ for some index set Γ.

By a *face* F of the closed unit ball U^* of X^*, we shall mean that F is a face in the sense of section 18 (that is, F is a convex set and if $x^* = a y^* + (1-a) z^*$ where $x^* \in F$, $0 \leqslant a \leqslant 1$, and $y^*, z^* \in U^*$, then $y^*, z^* \in F$) and each element of F has norm one. Clearly by Zorn's lemma, each face F is contained in a maximal face G. However, G is not necessarily weak* closed.

Let X be a Banach space and U^* the closed unit ball of X^*. A face F of U^* is said to be *essentially weak* closed* if the convex hull $\operatorname{co}(F \cup -F)$ of $F \cup -F$ is weak* closed. We shall use in the lemma below the well known fact that a subspace Y of X^* is weak* closed if and only if $Y \cap U^*$ is weak* closed (see [91]).

Lemma 1. *Let X be an L_1-predual space and F be an essentially weak* closed face of U^*. Then the span, Z, of F is weak* closed and $Z \cap V^* = \mathrm{co}(F \cup -F)$.*

Proof. It is clearly sufficient to prove that $Z \cap U^* = \mathrm{co}(F \cup -F)$ and to do this we need only to show that $Z \cap U^* \subset \mathrm{co}(F \cup -F)$.

Let G be a maximal proper face of U^* containing F and order X^* relative to G, that is, the positive cone is $(a x^*: a \geqslant 0, x^* \in G\}$. By theorem 2 of section 18, X^* is an abstract L_1 space under this ordering. Without loss of generality we assume that $X^* = L_1(\mu, \mathbb{R})$ for some regular measure μ on some locally compact Hausdorff space T (linearly isometric and order isomorphic with respect to the order generated by G). Now, let $x^* \in U^* \cap Z$ be nonzero. Then there are x_1^*, x_2^* in F and scalars a_1, $a_2 \geqslant 0$ such that $x^* = a_1 x_1^* - a_2 x_2^*$. Clearly the case $a_1 a_2 = 0$ is trivial and we may suppose that $a_1 \neq 0 \neq a_2$. Thus $(x^*)^+ = x^* \vee 0 \leqslant a_1 x_1^*$ and hence, $a_1 x_1^* = (x^*)^+ + a y^*$ for some $y^* \in G$ and $a \geqslant 0$. Suppose that $(x^*)^+ \neq 0$. Then $\dfrac{(x^*)^+}{\|(x^*)^+\|} \in G$ and $x_1^* = \dfrac{\|(x^*)^+\|}{a_1} \dfrac{(x^*)^+}{\|(x^*)^+\|} + \dfrac{a}{a_1} y^*$. Since $X^* = L_1(\mu, \mathbb{R})$ $a_1 = a + \|(x^*)^+\|$ and since F is a face of U^*, $\dfrac{(x^*)^+}{\|(x^*)^+\|} \in F$. Similarly, if $(x^*)^- \neq 0$, then $\dfrac{(x^*)^-}{\|(x^*)^-\|} \in F$. Thus, if $(x^*)^+ \neq 0 \neq (x^*)^-$, then $x^* = \|(x^*)^+\| \left(\dfrac{(x^*)^+}{\|(x^*)^+\|}\right) + \|(x^*)^-\| \left(\dfrac{(x^*)^-}{\|(x^*)^-\|}\right)$ is in $\mathrm{co}(F \cup -F)$. The case $(x^*)^+ = 0$ or $(x^*)^- = 0$ is trivial. ▮

If $f: U^* \to \mathbb{R}$ is a function, we say f is *odd* if $f(-x^*) = -f(x^*)$ for all $x^* \in U^*$.

Theorem 1. *Let X be an L_1-predual space and U^* be the closed unit ball of X^*. Suppose $g: U^* \to (-\infty, \infty]$ is a concave weak* lower semicontinuous function satisfying $g(x^*) + g(-x^*) \geqslant 0$ for all $x^* \in U^*$. Furthermore, suppose F is an essentially weak* closed face of U^* and $f: H \to \mathbb{R}$ is a weak* continuous odd affine function (where $H = \mathrm{co}(F \cup -F)$) such that $f \leqslant g$ on H. Then there is a weak* continuous odd affine extension h of f to U^* such that $h \leqslant g$.*

Proof. Clearly we may assume that g is finite and bounded from above on U^*. We first suppose that for some $\varepsilon > 0$,

(1) $$g(x^*) + g(-x^*) \geqslant 2\varepsilon$$

for all $x^* \in U^*$ and $g(x^*) \geqslant f(x^*) + \varepsilon$ for all $x^* \in H$. Moreover, we may assume that g is continuous. For, let f' be any continuous extension of f to U^* with $g(x^*) > f'(x^*) > -g(-x^*)$ for $x^* \in U^*$ and put

$f_1(x^*)=\frac{1}{2}(f'(x^*)-f'(-x^*))$ for all $x^*\in U^*$. Then f_1 is odd and $f_1(x^*)<g(x^*)$ for $x^*\in U^*$. By theorem 8 in the appendix and a standard compactness argument there is a continuous concave function g_1 with $g_1\leqslant g$ and $f_1(x^*)<g_1(x^*)$ for all $x^*\in U^*$. Clearly g_1 satisfies (1) for some $\varepsilon'>0$.

Let G be a maximal proper face on X^* containing F and order X^* by taking the cone generated by G as the positive cone and let Z be the span of F. Since f is odd and affine, f admits a weak* continuous linear extension to Z which shall also be denoted by f. (Note that $Z=(X/{}^{\perp}Z)^*$ and, hence, f can be thought of as a member of $X/{}^{\perp}Z$, where ${}^{\perp}Z=\{x\in X:x^*(x)=0$ for all $x^*\in Z\}$.) Consider the following sets in $X^*\times\mathbb{R}:A_1=\{(x^*,g(x^*)):x^*\in S^*\}$ and $A_2=\{(x^*,f(x^*)):x^*\in Z\}$. It will be shown that the weak* closed convex hull of A_1 is disjoint from A_2. From lemma 5 of section 20 it follows that for any $x^*\in U^*$, $\inf\{r:(x^*,r)\in\overline{\mathrm{co}}(A_1)\}=\inf\{r:(x^*,r)\in\mathrm{co}(A_1)\}$. Hence we need only show that if $(x^*,r)\in\mathrm{co}(A_1)$ and $x^*\in H$, then $r\geqslant\varepsilon+f(x^*)$.

Let $x^*\in H$ and $(x^*,r)\in\mathrm{co}(A_1)$. Then there are $x_i^*\in U^*$ and $a_i\in\mathbb{R}$ such that $a_i\geqslant 0$, $\sum_{i=0}^{n}a_i=1$, $x^*=\sum_{i=1}^{n}a_ix_i^*$, and $r=\sum_{i=1}^{n}a_ig(x_i^*)$. Since $U^*=\mathrm{co}(G\cup -G)$ and $H=\mathrm{co}(F\cup -F)$, there are $y_i^*,z_i^*\in G$, $y^*,z^*\in F$, and $1\geqslant b_i$, $b\geqslant 0$ such that $x_i^*=b_iy_i^*+(1-b_i)(-z_i^*)$ for $i=1,\dots,n$ and $x^*=by^*+(1-b)(-z^*)$. Hence $\sum_{i=1}^{n}a_ib_iy_i^*+(1-b)z^*=\sum_{i=1}^{n}b_i(1-a_i)z_i^*+by^*$ and since g is concave, $\sum_{i=1}^{n}b_i(1-a_i)g(-z_i^*)+\sum_{i=1}^{n}a_ib_ig(y_i^*)\leqslant r$. By the decomposition theorem for vector lattices, there are $u_{ij}^*\in G$ and $a_{ij}\geqslant 0$ for $i,j=1,\dots,n+1$ such that $a_ib_iy_i^*=\sum_{j=1}^{n+1}a_{ij}u_{ij}^*$ for $1\leqslant i\leqslant n$ and $b_j(1-a_j)z_j^*=\sum_{i=1}^{n+1}a_{ij}u_{i,j}^*$ for $1\leqslant j\leqslant n$, and $by^*=\sum_{i=1}^{n+1}a_{i,n+1}u_{i,n+1}^*$, $(1-b)z^*=\sum_{j=1}^{n+1}a_{n+1,j}u_{n+1,j}^*$. From these equations and the additivity of the norm on the positive elements of X^*, it follows that

$$a_ib_i=\sum_{j=1}^{n}a_{ij},\qquad a_j(1-b_j)=\sum_{i=1}^{n}a_{ij}$$

and

$$b=\sum_{i=1}^{n+1}a_{i,n+1},\qquad 1-b=\sum_{j=1}^{n}a_{n+1,j}.$$

Thus $a_ib_ig(y_i^*)\geqslant\sum_{j=1}^{n+1}a_{ij}g(u_{ij}^*)$ for $1\leqslant i\leqslant n$ and $a_j(1-b_j)g(-z_j^*)\geqslant\sum_{i=1}^{n+1}a_{ij}g(-u_{ij}^*)$ for $1\leqslant j\leqslant n$.

Now since F is a face of U^*, $u^*_{i,n+1}, u^*_{n+1,j}$ are in F when $a_{i,n+1} \neq 0$ and $a_{n+1,j} \neq 0$ respectively. Thus

$$r \geqslant \sum_{i=1}^{n} \sum_{j=1}^{n+1} a_{ij} g(u^*_{ij}) + \sum_{j=1}^{n} \sum_{i=1}^{n+1} a_{ij} g(-u^*_{ij}) \geqslant 2\varepsilon \sum_{i=1}^{n} \sum_{j=1}^{n} a_{ij}$$

$$= \sum_{i=1}^{n} a_{i,n+1} g(u^*_{i,n+1}) + \sum_{j=1}^{n} a_{n+1,j} g(u^*_{n+1,j}) \geqslant 2\varepsilon \sum_{i=1}^{n} \sum_{j=1}^{n} a_{ij}$$

$$+ \sum_{i=1}^{n+1} a_{i,n+1} f(u^*_{i,n+1}) + \varepsilon \sum_{i=1}^{n} a_{i,n+1} - \sum_{j=1}^{n+1} a_{n+1,j} f(u^*_{n+1,j}) + \varepsilon \sum_{j=1}^{n} a_{n+1,j}.$$

Thus $r \geqslant b f(y^*) - (1-b) f(z^*) + \varepsilon \left(\sum_{i=1}^{n} a_i b_i + \sum_{i=1}^{n} (1-b_i) a_i \right) = f(x^*) + \varepsilon.$

This proves the assertion concerning $\overline{\text{co}\, A_1}$. Now A_2 is closed by Lemma 1 and $\overline{\text{co}\, A_1}$ is compact. Thus there is a closed hyperplane V in $X^* \times \mathbb{R}$ such that $V \cap \overline{\text{co}\, A_1} = \emptyset$ and $A_2 \subset V$. Let h be defined on U^* by $h(x^*)$ is the unique number such that $(x^*, h(x^*)) \in V$ for $x^* \in U^*$. Then h is the required function.

We now pass to the general case. By the above argument there is a weak* continuous function h_1 on such that $h_1 | U^* \leqslant g + 2/3$ and $h_1 | H = f$. Assume that we have obtained $h_1, \ldots, h_n$ all weak* continuous linear extensions of f so that $h_i | U^* \leqslant g + (2/3)^i$ for $i = 1, \ldots, n$ and $\|h_i - h_{i-1}\| \leqslant (2/3)^{i+1}$ for $i = 2, \ldots, n$.

Let $g_{n+1} = \min[h_n + (2/3)^{n+1}, \; g + (2/3)^{n+1}]$. Then g_{n+1} is concave, weak* lower semicontinuous, and $g_{n+1} \geqslant f + (2/3)^{n+1}$ on H. Furthermore, $g_{n+1}(x^*) + g_{n+1}(-x^*) \geqslant \frac{1}{3}(2/3)^n$ for all $x^* \in U^*$. For, if $h_n(x^*) \leqslant g(x^*)$ and $h_n(-x^*) \leqslant g(-x^*)$, then $g_{n+1}(x^*) + g_{n+1}(-x^*) = (2/3)^{n+1}$ while if $h_n(x^*) > g(x^*)$, then $g(-x^*) \geqslant -g(x^*)$ and $|h_n(x^*) - g(x^*)| \leqslant (2/3)^n$, hence $g_{n+1}(x^*) + g_{n+1}(-x^*) \geqslant \min[h_n(x^*), g(x^*)] - \max[h_n(x^*), g_n(z^*)] + 2(2/3)^{n+1} = 2(2/3)^{n+1} - |h_n(x^*) - g(x^*)| \geqslant 1/3(2/3)^{n+1}$. A similar argument works if $h_n(-x^*) > g(-x^*)$. Thus g_{n+1} satisfies (1) for some $\varepsilon > 0$ and hence there is a weak* continuous linear extension h_{n+1} of f such that $h_{n+1} | U^* \leqslant g + (2/3)^n$ for $x^* \in U^*$ so that $\|h_{n+1} - h_n\| \leqslant (2/3)^{n+1}$. The limit of the sequence $\{h_n\}$ is the required function. ∎

We now have to introduce the notion of a selection. Let X be a locally convex hausdorff topological vector space, $k(X)$ denotes the set of all nonempty convex subsets of X, and $\bar{k}(X)$ denotes the set of all closed sets in $k(X)$. A mapping ϕ from a convex set K in Y to $k(X)$ (where Y is a locally convex Hausdorff topological linear space) is said to be *convex* if $a\phi(y_1) + (1-a)\phi(y_2) \subset \phi(a y_1 + (1-a) y_2)$ for y_1, y_2 in K and $0 \leqslant a \leqslant 1$. The mapping ϕ is said to be *lower semicontinuous* if for each open set $U \subset X$, $\{y \in K: \phi(y) \cap U \neq \emptyset\}$ is open in K. We say ϕ is odd if $\phi(-y) = -\phi(y)$ whenever y and $-y$ are both in K. A *selection for* ϕ is a mapping h of K into X such that $h(y) \in \phi(y)$ for all $y \in K$. We preserve the same notation as above.

Lemma 2. *Let X be an L_1-predual space, Y be a locally convex Hausdorff topological linear space and, $\phi: U^* \to k(Y)$ be a convex odd weak* continuous mapping (U^* is the closed unit ball of X^*). Let F, H, f be as in theorem 1. Then for each absolutely convex neighborhood U of $0 \in Y$ there is a weak* continuous odd affine $h: U^* \to Y$ such that $h(x^*) \in \phi(x^*) + U$ for all $x^* \in {}^*$ and $h(x^*) - f(x^*) \in U$ for all $x^* \in H$.*

Proof. Suppose $Y = \mathbb{R}$ and let U be a symmetric open interval about $0 \in \mathbb{R}$. Let g be defined by $g(x^*) = \sup\{\phi(x^*) + U: x^* \in U^*\}$. Then g is a weak* lower semicontinuous concave function which satisfies (1) of theorem 1 for some $\varepsilon > 0$. The existence of h follows from theorem 1.

Now suppose the lemma is valid for $\mathbb{R}^n$ and let $Y = \mathbb{R}^{n+1} = \mathbb{R} \times \mathbb{R}^n$. Let P, Q be the natural projections of Y onto $\mathbb{R}$ and $\mathbb{R}^n$ respectively. Let V, W be symmetric neighborhoods of 0 in $\mathbb{R}$ and $\mathbb{R}^n$ respectively so that $V \times W \subset U$. By theorem 1 and the above argument there is a weak* continuous odd affine function $g: U^* \to \mathbb{R}$ which extends $P \circ f$ and $g(x^*) \in P \circ \phi(x^*) + \frac{1}{2}V$ for all $x^* \in U^*$. The map $\psi': X^* \to k(Y)$ defined by $\psi'(x^*) = P^{-1}(g(x^*) + \frac{1}{2}V)$ is convex odd and weak* lower semicontinuous. Moreover, its graph is open in $U^* \times R^{n+1}$. The map $\psi: B^* \to k(Y)$ defined by $\psi(x^*) = P^{-1}(g(x^*) + \frac{1}{2}V) \cap \phi(x^*)$ for all $x^* \in U^*$ is convex and odd and weak* lower semicontinuous. The same properties hold for $Q \circ \psi$ and $Q \circ f$ is a selection for $Q \circ \psi | H$. From the induction hypothesis it follows that there is a weak* continuous odd affine selection τ of $x^* \to Q \circ \psi(x^*) + V$ such that $\tau(x^*) \in (Q \circ f(x^*) + W)$ for all $x^* \in H$. Then $h: S^* \to Y$ defined by $h(x^*) = (g(x^*), \tau(x^*))$ has the required properties.

Now suppose Y is infinite dimensional and for $y \in Y$ let $G_y = \{x^* \in U^*: y \in (\phi(x^*) + \frac{1}{2}V)\}$ and $G'_y = \{x^* \in H: y \in (f(x^*) + \frac{1}{2}U)\}$. Since ϕ is weak* lower semicontinuous and f is weak* continuous, $\{G_y\}_{y \in Y}$ and $\{G'_y\}_{y \in Y}$ are open covers of U^* and H respectively. Hence there are $y_1, \ldots, y_n$ in Y such that $U^* = \bigcup_{i=1}^{n} G_{y_i}$ and $H = \bigcup_{i=1}^{n} G'_{y_i}$. Consider the maps $\psi_1: U^* \to k(\mathbb{R}^n)$ and $\psi_2: H \to k(\mathbb{R}^n)$ defined by $\psi_1(x^*) = \left\{(a_1, \ldots, a_n) \in \mathbb{R}^n: \sum_{i=1}^{n} a_i y_i \in (\phi(x^*) + \frac{1}{2}U)\right\}$ and $\psi_2(x^*) = \left\{(a_1, \ldots, a_n) \in \mathbb{R}^n: \sum_{i=1}^{n} a_i y_i \in (f(x^*) + \frac{1}{4}U)\right\}$. Then ψ_1, ψ_2 are convex odd and weak* lower semicontinuous. Let W be an absolutely convex neighborhood of $0 \in \mathbb{R}^n$ such that $(a_1, \ldots, a_n) \in W$ implies $\sum_{i=1}^{n} a_i y_i \in \frac{1}{4} U$. By the first part of the proof the map defined by $x^* \to \psi_2(x^*) + W$ for $x^* \in H$ admits a weak* continuous odd selection g. Clearly $g(x^*) \in \psi_1(x^*)$ for $x^* \in H$. Moreover, there is a weak* continuous odd affine selection $h' = (h'_1, \ldots, h'_n): U^* \to \mathbb{R}^n$ such that $h'(x^*) \in (\psi_1(x^*) + W)$ for $x^* \in U^*$ and $h'(x^*) - g(x^*) \in W$ for $x^* \in H$. Let $h = \sum_{i=1}^{n} h'_i g_i$. Then h has the desired properties. ∎

We now come to the main theorem on selections due to Lazar and Lindenstrauss [175].

Theorem 2. *Let X be an L_1-predual space and Y be a complete metrizable locally convex Hausdorff topological vector space. Let $\phi: U^* \to \overline{k}(Y)$ be a convex odd weak* lower semicontinuous mapping. Then ϕ admits a weak* continuous odd affine section h. Moreover, if F is an essentially weak* closed face of U^*, $H = \mathrm{co}(F \cup -F)$, and $f: H \to Y$ is a weak* continuous odd affine selection of $\phi | H$, then the selection h can be chosen so that $h | H = f$.*

Proof. Clearly we need only prove the second assertion of the theorem. Let $\{P_n\}$ be an increasing sequence of seminorms on Y which determine the topology of Y. For $y \in Y$ and $r > 0$ let $B_n(y, r) = \{z \in Y: P_n(y, z) < r\}$. By repeated application of the lemma 2, there is a sequence $\{h_n\}$ of weak* continuous odd affine functions from U^* to Y such that $h_1(x^*) \in (\phi(x^*) + B_1(0, 1/2))$ for $x^* \in U^*$ and $(h_1(x^*) - f(x^*)) \in B_1(0, 1/2)$ for $x^* \in H$ and for $n > 1$, $h_n(x^*) \in (\phi(x^*) \cap B_{n-1}(h_{n-1}(x^*), 2^{-n+1}) + B_n(0, 2^{-n})$ for $x^* \in U^*$ and $(h_n(x^*) - f(x^*)) \in B_n(0, 2^{-n})$ for $x^* \in H$. The sequence $\{h_n\}$ converges to an h with the desired properties. ∎

Corollary. *Let X be an L_1-predual space and F be an essentially weak* closed face of U^* such that $H = \mathrm{co}(F \cup -F)$ is weak* metrizible. Then there is a weak* continuous odd affine map h of U^* onto H such that $h | H$ is the identity.*

Proof. The span, Z, of H is metrizable and complete in a topology which agrees with the weak* topology on the unit sphere H of Z. Let $\phi: U^* \to \overline{k}(Z)$ be defined by $\phi(x^*) = \{x^*\}$ if $x^* \in H$ and $\phi(x^*) = H$ if $x^* \in U^* \setminus H$. Then ϕ is a convex odd weak* lower semicontinuous mapping. ∎

What this corollary says in other terms is that for the natural map $\pi: X \to X/{}^{\perp}Z$, there is a linear isometry $A: X/{}^{\perp}Z \to X$ so that $\pi \circ A$ is the identity. Thus $A \circ \pi$ is a contractive projection of X onto $A(X/{}^{\perp}Z)$.

We now wish to investigate separable L_1-predual spaces and apply the above theorem to obtain an embedding theorem for these spaces. We first have the following result.

Theorem 3. *Let X be an L_1-predual space and suppose that $\operatorname{ext} U^*$ is infinite and countable. Then $X^* = l_1(\mathbb{N}, \mathbb{R})$. In particular X is separable.*

Proof. Since $\operatorname{ext} U^* = \{e_n^*\}$ is countable, by theorem 6 of section 20, for each $x^* \in U^*$ there is a probability measure μ on $\operatorname{ext} U^*$ such that $\mu(\operatorname{ext} U^*) = 1$ and $x^*(x) = \int y^*(x)\, d\mu(y^*) = \sum_{n=1}^{\infty} e_n^*(x)\mu(x_n^*)$ for all $x \in X$. Clearly $\mu(e_n^*) \geqslant 0$ and $\sum_{n=1}^{\infty} \mu(x_n^*) = 1$. Thus there is a $y^* \in U^*$ such that $y^* = \sum_{n=1}^{\infty} \mu(x_n^*) e_n^*$. Clearly $y^* = x^*$ and U^* is the norm closed convex hull of its extreme points. This clearly implies that $X^* = l_1(\mathbb{N}, \mathbb{R})$. ∎

We shall also need the following lemmas.

Lemma 3. *Let X, Y be Banach spaces with $X \subset Y$ and suppose X is an L_1-predual space. Then there is a contractive projection $P: Y^* \to Y^*$ whose kernel is $X^\perp$ and whose range is linearly isometric to X^* under the restriction mapping $y^* \to y^* | X$ for $y^* \in Y^*$.*

Proof. Let $J: X^* \to X^{***}$, $K: Y \to Y^{**}$, $L: X \to X^{**}$, and $i: X \to Y$ be the natural embeddings. Since X^* is an L_1 space, X^{**} is a $\mathscr{P}_1(\mathbb{R})$ space. Hence there is a contractive projection $Q: Y^{**} \to Y^{**}$ whose range is $i^{**}(X^{**})$. Set $P' = [((i^{**})^{-1} Q K)^* J]$. Then $\|P'\| \leqslant 1$ and $(P' x^*)(i x) = x^*(x)$ for all $x^* \in X^*$ and $x \in X$. For, $\langle [(i^{**})^{-1} Q K]^* J x^*, i x \rangle = \langle J x^*, (i^{**})^{-1} Q K i x \rangle = \langle (i^{**})^{-1} Q K i x, x^* \rangle = \langle (i^{**})^{-1} Q i^{**} L x, x^* \rangle = \langle L x, x^* \rangle = x^*(x)$. Hence P' is a linear isometry and $P = P' \circ i^*$ is the required projection. ∎

Lemma 4. *Let X, Y be L_1-predual spaces and let $X \subset Y$. Then there is an order preserving linear isometry A of X^* into Y^*.*

Proof. By lemma 3, there is a linear isometry $A: X^* \to Y^*$ such that $(A x^*)(x) = x^*(x)$ for all $x^* \in X^*$ and $x \in X$. Let F be a maximal proper face of the unit sphere of X^* and G a maximal proper face of the unit sphere of Y^* containing $A(F)$. Then as before, F and G generate orders and X^* and Y^* respectively under which they are abstract L spaces (and linearly order isometric to any other L_1 structures on X^*, Y^* respectively). Clearly A is order preserving.

Moreover, for any $0 \leqslant x^* \in X^*$, $A((x^*)^{\perp\perp}) \subset (A x^*)^{\perp\perp}$. For, $\{z^* : 0 \leqslant z^* \leqslant x^*\}$ is a set whose linear span is dense in $(x^*)^{\perp\perp}$. ∎

We now have the main embedding theorem due to Lazar and Lindenstrauss [175].

Theorem 4. *Let X be a separable L_1-predual space and suppose that* ext U^* *is uncountable. Then X contains a subspace Y linearly isometric to $C(\Delta, \mathbb{R})$ where Δ is the Cantor set. Moreover, there is an contractive projection of X onto Y. In particular, X^* is nonseparable.*

Proof. Since U^* is metrizable in the weak* topology, $\operatorname{ext} U^*$ is a G_δ set in U^*. Clearly there is a relatively closed uncountable set $K_1 \subset \operatorname{ext} U^*$ such that $K_1 \cap (-K_1) = \emptyset$. By theorem 2 of section 6 there is a homeomorphic copy Δ of the Cantor set contained in K_1. Thus $F = \overline{\operatorname{co} K}$ is a weak* closed face in U^*. By the corollary to theorem 2, Y is linearly isometric to $X/{}^{\perp}Z$, where Z is the span of F. It is easy to see that $X/{}^{\perp}Z$ is linearly isometric to $C(\Delta, \mathbb{R})$. ∎

We can now completely characterize the duals of separable L_1-predual spaces.

Theorem 5. *Let X be a separable L_1-predual space. Then*

(1) $X^* = l_1(\mathbb{N}, \mathbb{R})$ *if and only if* $\operatorname{ext} U^*$ *is infinite and countable.*

(2) $X^* = C([0,1], \mathbb{R})^*$ *if and only if* $\operatorname{ext} U^*$ *is uncountable.*

Proof. Statement (1) has already been established and the necessity of statement (2) is clear.

Now, let $Y \subset X$ be linearly isometric to $C(\Delta, \mathbb{R})$ where Δ is the Cantor set. By theorem 3 of section 15, $X^* = \left[l_1(\mathbb{R}, \mathbb{R}) \oplus \sum_{i \in I} L_1([0,1]^{m_i}, \mathbb{R})\right]_1$ for some cardinal numbers $m_i \geqslant \aleph_0$. By lemma 4, $m_i = \aleph_0$ since X can be linearly isometrically embedded in $C([0,1], \mathbb{R})$ and we know that if $L_1([0,1]^m, \mathbb{R}) \subset C([0,1], \mathbb{R})^*$ then $m = \aleph_0$. Since we know that $C([0,1], \mathbb{R})^* = \left[l_1(\mathbb{R}, \mathbb{R}) \oplus \sum_{c} L_1([0,1], \mathbb{R})\right]_1$ it only remains to show that $\operatorname{card} I = c$. Now, let $y_t^* \in Y^*$ with $0 \leqslant y_t^*$ (we assume the order set up as described in lemma 4 and that $A: Y^* \to X^*$ is an order preserving linear isometry) and $(y_t^*)^{\perp\perp} = L_1([0,1], \mathbb{R})$ for all $t \in [0,1]$ and $y_t^* \wedge y_s^* = 0$ for $s \neq t$. Recall that $y_t^* \wedge y_s^* = 0$ is equivalent to $\|y_t^* + y_s^*\| = \|y_t^* - y_s^*\|$. Since A is an isometry, $\|A(y_t^*) + A(y_s^*)\| = \|A(y_t^*) - A(y_s^*)\|$ and, thus, $A(y_t^*) \wedge A(y_s^*) = 0$ for $t \neq s$, since $A((y_t^{**})^{\perp\perp}) \subset A(y_t^*)^{\perp\perp}$, it follows that $(A\, y_t^*)^{\perp\perp} = [l_1(\Gamma_t, \mathbb{R}) \oplus B_t]_1$ where $0 \leqslant \operatorname{card} \Gamma_t \leqslant \aleph_0$ and $B_t = L_1([0,1], \mathbb{R})$ for all t. Thus, clearly $\operatorname{card} I \geqslant c$. Since $\operatorname{card} X^* \leqslant c$ since $X \subset C([0,1], \mathbb{R})$), it follows that $\operatorname{card} I = c$. ∎

The theorem has an isomorphic version due to Lewis and Stegall [180] and Stegall [265].

Theorem. *Let X be a separable $\mathscr{L}_\infty$ space. Then X^* is linearly isomorphic to $l_1(\mathbb{N},\mathbb{R})$ or $C([0,1],\mathbb{R})^*$ depending on whether or not X^* is separable.* ∎

We now wish to investigate when $X^* = l_1(\Gamma,\mathbb{R})$ for some index set Γ.

Lemma 5 (Lindenstrauss-Phelps). *Let X be an infinite dimensional reflexive Banach space. Then the set $\operatorname{ext} U$ of extreme points of the closed unit ball U of X is uncountable.*

Proof. Suppose that $\operatorname{ext} U = \{x_n\}$ and for each n let $F_n = \{x^* \in X^*: \|x^*\| \leq 1$ and $|x^*(x_n)| = \|x^*\|\}$. Then each F_n is weakly closed. By the Krein Milman theorem applied to U, the closed unit ball U^* of X^* is equal to $\bigcup_{n=1}^{\infty} F_n$. Thus, F_{n_1} has a weak nonempty interior relative to U^*. Let x_0^* be an interior point with $\|x_0^*\| < 1$. Thus there are $y_1, \ldots, y_n$ in X such that if $\|x^*\| < 1$ and $|x^*(y_i) - x_0^*(y_i)| < 1$ for $i = 1, \ldots, n$, then $x^* \in F_{n_1}$. Let $M = \{x^* \in X^* : x^*(y_i) = x^*(y_i)$ for $i = 1, \ldots, n$ and $x^*(x_{n_1}) = x_0^*(x_{n_1})\}$. Since X is infinite dimensional, M contains a line through x_0^* which contains a y_0^* with $\|y_0^*\| = 1$. Thus $y_0^* \in F_{n_1}$ which is a contradiction. ∎

For the proof of the following lemma see lemma 1 of section 23.

Lemma 6. *Let X be an L_1-predual space and $Y \subset X$ a separable subspace. Then there is a separable L_1-predual space Z with $Y \subset Z \subset X$.* ∎

The following theorem is due to Lacey and Morris [166] and traces back to Pełczyński and Semadeni [235]. An isomorphic version has recently been proved by Stegall [265] where X is a $\mathscr{L}_\infty$ space.

Theorem 6. *Let X be an L_1-predual space. Then the following are equivalent.*

(1) $X^* = l_1(\Gamma,\mathbb{R})$ *for some index set* Γ,

(2) $X^{**} = l_\infty(\Delta,\mathbb{R})$ *for some index set* Δ,

(3) $l_2(\mathbb{N},\mathbb{R})$ is not a continuous linear image of X,

(4) for each Banach space Y and each weakly compact linear operator $A: X \to Y$, A is compact.

(5) if $Y \subset X$ is an L_1-predual space, then $Y^* = l_1(\Gamma,\mathbb{R})$ for some index set Γ,

(6) if $Y \subset X$ is a separable infinite dimensional L_1-predual space, then $Y^* = l_1(\mathbb{N},\mathbb{R})$,

(7) if $Y \subset X$ is a separable subspace, then Y^* is separable,

(8) if $Y \subset X$ is reflexive, then Y is finite dimensional.

Proof. Clearly (1) and (2) are equivalent by theorem 11 of section 11. (1) implies (4). Let $A: X \to Y$ be a weakly compact operator. Then A^* is weakly compact and hence compact since each weakly compact set in X^* is compact. (4) implies (3) is clear. (1) implies (5). Let $Y \subset X$ be an L_1-predual space. Then by lemma 4, there is a linear isometry of Y^* into X^*. Thus $Y^* = l_1(\Delta, \mathbb{R})$ for some index set Δ. (5) implies (6). This is immediate from theorem 5. (6) implies (7). This is immediate from lemma 6.

(7) implies (6) is clear.

(6) implies (3). Let $A: X \to l_2(\mathbb{N}, \mathbb{R})$ be a bounded linear operator and $\{x_n\}$ be a bounded sequence in X. Choose a separable L_1-predual space $Z \subset X$ which contains $\{x_n\}$. By (6), $Z^* = l_1(\mathbb{N}, \mathbb{R})$. Let $A_0 = A|Z$. Then $A_0^*: l_2(\mathbb{N}, \mathbb{R}) \to l_1(\mathbb{N}, \mathbb{R})$ and, hence, is compact (see theorem 13 of section 14). Thus A_0 is compact and it follows immediately that A is compact. Hence A is not onto $l_2(\mathbb{N}, \mathbb{R})$. (3) implies (1). Suppose X^* is not of the type $l_1(\Gamma, \mathbb{R})$. Then by theorem 12 of section 14 X^* contains a copy of $L_1([0,1], \mathbb{R})$ which, in turn, contains a copy of $l_2(\mathbb{N}, \mathbb{R})$. Let A be a linear isomorphism of $l_2(\mathbb{N}, \mathbb{R})$ into X^*. Then $A^*|JX$ is onto since $(A^*|JX)^* = A$ and A has a bounded inverse, where $J: X \to X^{**}$ is the natural map. (6) implies (8). Suppose $Y \subset X$ is an infinite dimensional reflexive subspace. Then without loss of generality Y is separable. By lemma 6, $Y \subset Z \subset X, Z$ is separable, and $Z^* = l_1(\mathbb{N}, \mathbb{R})$ (from 6). By lemma 5 this is impossible. (8) implies (6). Suppose there is a separable infinite dimensional L_1-predual space Y such that $Y^* \neq l_1(\mathbb{N}, \mathbb{R})$. Then by theorem 4, Y contains a copy of $C(\Delta, \mathbb{R})$ where Δ is the Cantor set and hence a copy of any separable Banach space. ∎

Zippin [306] has shown that if X is an infinite dimensional L_1-predual space, then X contains a subspace linearly isometric to $c_0(\mathbb{N}, \mathbb{R})$. It also follows from [175] and [207] that a separable infinite dimensional Banach space X is an L_1-predual space if and only if there is a normalized Schauder basis $\{x_n\}$ for X such that for each n, $\operatorname{span}(x_1, \ldots, x_n) = l_\infty(n, \mathbb{R})$.

In the isomorphic theory the following embedding question is still open.

Question. *Does every infinite dimensional (separable) $\mathscr{L}_\infty$ space contain a subspace linearly isomorphic to $c_0(\mathbb{N}, \mathbb{R})$?*

For some embedding theorems for $\mathscr{L}_\infty$ spaces see [127]. Also, Rosenthal [248] has proved the following isomorphism theorem.

Theorem (Rosenthal). *Let X be a complemented subspace of $C([0,1], \mathbb{R})$ such that X^* is nonseparable. Then X is isomorphic to $C([0,1], \mathbb{R})$.* ∎

It is still open as to whether or not a complemented subspace of $C(T, \mathbb{R})$ is linearly isomorphic to $C(T, \mathbb{R})$ for some compact Hausdorff space S.

Finally, we have the following open question.

Question. *Is every $\mathscr{L}_\infty$ space linearly isomorphic to an L_1-predual space?*

It has recently been proved by Benjamini and Lindenstrauss [34] that there are separable L_1-predual spaces X and Y such that X^* is separable and Y^* is nonseparable and neither X nor Y are linearly isomorphic to a complemented subspace of a space $C(T, \mathbb{R})$.

Exercises. 1. Find a nonseparable Banach space X such that X^* is linearly isometric to $C([0,1], \mathbb{R})^*$.

2. Let X be an L_1-predual space. Prove that $X^* = l_1(\Gamma, \mathbb{R})$ for some index set Γ if and only if each bounded sequence in X has a weakly Cauchy subsequence.

3. Let X be an L_1-predual space. Then $X^* = l_1(\Gamma, \mathbb{R})$ for some index set Γ if and only if each maximal measure on $b(X^*)$ is purely atomic.

4. Let X and Y be infinite dimensional L_1-predual spaces with $X \subset Y$ and $\dim Y/X < \infty$. Show that X^* and Y^* are linearly isometric.

5. Let X and Y be L_1-predual spaces. Show that if X can be linearly isomorphically embedded in Y and Y can be linearly isomorphically embedded in X, then X^* and Y^* are linearly isomorphic.

6. Let X and Y be L_1-predual spaces. Suppose that every principle ideal in X^* is separable, i.e., $X^* = L_1(\mu, \mathbb{R})$ and if $v \in L_1(\mu, \mathbb{R})$ then $L_1(v, \mathbb{R})$ is separable. Show that if X can be linearly isomorphically embedded in Y and Y can be linearly isomorphically embedded in X, then X^* and Y^* are linearly isometrically isomorphic.

7. Let T be a compact Hausdorff space of weight m and suppose that T can be mapped continuously onto $[0,1]^m$. Then $C(T, \mathbb{R})^*$ is linearly isomorphic to $C([0,1]^m, \mathbb{R})^*$.

8. Let K be a compact Choquet simplex. Prove that $A(K, R) = X^*$ for some Banach space X if and only if $\operatorname{ext} K$ is closed and extremally disconnected and the union of the supports of the normal regular Borel measures on $\operatorname{ext} K$ is dense in $\operatorname{ext} K$.

§ 23. Characterizations of Complex L_1-Predual Spaces

In this section we shall primarily be concerned with complex L_1-predual spaces. However, the characterizations in theorems 2 and 3 are valid for both the real and complex case (we only give the argument in the complex case as it is readily adaptable to the real case). In particular, a Banach space is an L_1-predual space if and only if it is an $\mathscr{L}_{\infty,\lambda}$ space for all $\lambda > 1$ (the reader should compare this with theorem 7 of section 17). We also present the complex version of theorem 7 of section 21 and study complex L_1-predual spaces whose closed unit ball contains an extreme point.

The structure of separable L_1-predual spaces in also given in terms of finite dimensional subspaces (valid for both the real and complex cases).

We shall use the fact a Banach space X is linearly isometric to an abstract L_1 space if and only if X^* is a $\mathscr{P}_1(\mathbb{C})$ space (see theorem 11 and exercise 6 of section 11). Thus, we get immediately that X is an L_1-predual space if and only if X^{**} is a $\mathscr{P}_1(\mathbb{C})$ space.

We shall need the following theorem which has useful applications in several situations (see [190], [265], and [296] for example). The version presented here is due to Lindenstrauss and Rosenthal. There are other versions to be found in the above references.

Before stating the theorem, we make the following observation. Let X be a Banach space, K an open convex set in X, Y a finite dimensional subspace of X, and $L: X \to Y$ an onto bounded linear operator. Then $L^{**}(K^0) = L(K)$, where K^0 is the norm interior of the weak*-closure of K in X^{**} (we consider $X \subset X^{**}$ without reference to the natural embedding of X into X^{**}). For, since Y is reflexive and L^{**} is weak* continuous, $\overline{L(K)} \supset L^{**}(K^0)$, where the closure is with respect to the weak* topology. Since L is onto, $L(K)$ and $L^{**}(K^0)$ are open convex sets of Y with $L(K) \subset L^{**}(K^0)$. Hence $L(K) = L^{**}(K^0)$.

Theorem 1. (Principle of local reflexivity). *Let X be a Banach space, Y a finite dimensional subspace of X^{**}, and $\varepsilon > 0$. Then there is a one to one (bounded) linear operator $B: Y \to X$ such that $Bx = x$ for all x $x \in Y \cap X$ and $\|B\|\,\|B^{-1}\| \leqslant 1 + \varepsilon$.*

Proof. Let $0 < \delta < 1$ be such that $\dfrac{1+\delta}{(1-\delta)(1-2\delta) - \delta(1+\delta)} < 1 + \varepsilon$. Since the unit sphere of Y is compact there are $y_1, \ldots, y_m$ in Y such that $\|y_i\| = 1$ for all $i = 1, \ldots, m$ and if $y \in Y$ and $\|y\| = 1$, then $\|y - y_i\| < \delta$ for some $1 \leqslant i \leqslant m$.

Let $k = \dim Y/Y \cap X$ and let $z_1^{**}, \ldots, z_k^{**}$ be linearly independent and in $Y\,X$. Then for each $i = 1, \ldots, m$ there are scalars a_{ij} and $w_i^{**} \in X \cap Y$ so that $y_i = \sum_{j=1}^{k} a_{ij} z_j^{**} + w_i^{**}$. Also, we may choose $x_i^* \in X^*$ such that $|z_i^{**}(x_i^*)| \geqslant 1 - \delta$ and $\|x_i^*\| = 1$ for $i = 1, \ldots, m$.

Now let $K_i = \left\{(u_1, \ldots, u_k): u_j \in X \text{ and } \left\|\sum_{j=1}^{k} a_{ij} u_j + w_i^{**}\right\| < 1 + \delta\right\}$ and

$$C_i = \left\{(u_1, \ldots, u_k): u_j \in X \text{ and } \left|x_i^*\left(\sum_{j=1}^{k} a_{ij} u_j\right) - \left(\sum_{j=1}^{k} a_{ij} z_j^{**}\right)(x_i^*)\right| < \delta\right\}.$$

Suppose $\bigcap_{j=1}^{m} K_i \cap C_i = \emptyset$. Then since K_i and C_i are open convex sets in $X^k = X \times \cdots \times X$ (for k factors), we have by Klee's theorem (see the appendix) a finite dimensional subspace Z of X^k with $\dim Z \leqslant 2m-1$ and an onto bounded linear operator $L: X^k \to Z$ with $\bigcap_{i=1}^m (L(K_i) \cap L(C_i)) = \emptyset$. Now, for each i, let K_i^{**} and C_i^{**} be the subsets of $(X^{**})^k$ defined by $K_i^{**} = \left\{(u_1^{**}, \ldots, u_k^{**}): \left\| \sum_{j=1}^n a_{ij} u_j^{**} + w_i^{**} \right\| < 1 + \delta \right\}$ and $C_i^{**} = \left\{(u_1^{**}, \ldots, u_k^{**}): \left| \sum_{j=1}^n a_{ij} u_j^{**}(x_i^*) - \sum_{j=1}^k a_{ij} z_j^{**}(x_i^*) \right| < \delta \right\}$. Then clearly $(z_1^{**}, \ldots, z_k^{**}) \in K_i^{**} \cap C_i^{**}$ for all i. But, K_i^{**} and C_i^{**} are contained in the weak* closure of K_i and C_i respectively. For the case C_i^{**} it is easy to see. Let $(u_1^{**}, \ldots, u_k^{**}) \in K_i^{**}$ and suppose $a_{ij} \neq 0$ for some j (the case $a_{ij} = 0$ for all j is trivial). Then there are nets $\{v_\alpha\}$ and $\{x_\alpha^r\}$ in X for $1 \leqslant r \leqslant k$, $r \neq j$, such that $\|v_\alpha\| < 1 + \delta$ and $v_\alpha \to \sum_{i=1}^k a_{ij} u_j^{**} + w_i^{**}$ in the weak* topology and $x_\alpha^r \to u_r^{**}$ in the weak* topology. Now, let $y_\alpha^j = \frac{1}{\alpha_{ij}} \left(v_\alpha - \sum_{r \neq j} a_{ir} x_\alpha^r - w_i^{**} \right)$. Then $(y_\alpha^1, \ldots, y_\alpha^k) \to (u_1^{**}, \ldots, u_k^{**})$ in the weak* topology and $(y_\alpha^1, \ldots, y_\alpha^k) \in K_i$ for all α.

Let K_i^0, C_i^0 be the norm interior of the weak* closure of K_i and C_i respectively. Then by the observation $L(K_i) = L^{**}(K_i^0)$ and $L(C_i) = L^{**}(C_i^0)$. But, $L^{**}(z_1^{**}, \ldots, z_k^{**})$ is in $\bigcap_{i=1}^m (LK_i) \cap L(C_i)$ which is a contradiction. Now, let $B: Y \to X$ by $B\left(\sum_{i=1}^k a_i z_i^{**} + x \right) = x + \sum_{i=1}^k a_i x_i$, where $(x_1, \ldots, x_k)$ is fixed and belongs to $\bigcap_{i=1}^m K_i \cap C_i$. Then $Bx = x$ for $x \in X \cap Y$ and $|x_i^*(B y_i) - y_i(x_i^*)| < \delta$ and $1 - 2\delta < \|B y_i\| < 1 + \delta$ for $i = 1, \ldots, m$.

From the above inequalities it follows that $\|B\| \leqslant \frac{1+\delta}{1-\delta}$ and $\|B^{-1}\| \leqslant \frac{1-\delta}{(1-\delta)(1-2\delta) - \delta(1+\delta)}$. Thus $\|B\| \, \|B^{-1}\| < 1 + \varepsilon$ by the choice of δ. ∎

Corollary. *Let X, Y, Z be Banach spaces with $Y \subset Z$, and X and Z finite dimensional. Suppose $L: Y \to X$ is a linear operator and $\lambda \geqslant \|L\|$. Then the following statements are equivalent.*

(1) *For every $\varepsilon > 0$ there is a linear operator $B: Z \to X$ such that $\|B\| \leqslant \lambda + \varepsilon$ and $B \mid Y = L$.*

(2) *For every finite dimensional Banach space W and every bounded linear operator $B: X \to W$ there is a linear operator $C: Z \to W$ such that $C \mid Y = BL$ and $\|C\| \leqslant \lambda \|B\|$.* ∎

We can now state the following characterization theorem for L_1-predual spaces which is valid both in the real and complex case.

Theorem 2. *Let X be a Banach space. Then the following are equivalent.*

(a) *X is an L_1-predual space,*

(b) *X is an N_λ space for all $\lambda > 1$,*

(c) *X is an $\mathscr{L}_{\infty,\lambda}$ space for all $\lambda > 1$,*

(d) *for each $\varepsilon > 0$ and each finite set $A \subset X$ there is an integer n and a linear operator $B: l_\infty(n, \mathbb{C}) \to X$ such that $(1+\varepsilon)^{-1}\|y\| \leqslant \|By\| \leqslant (1+\varepsilon)\|y\|$ for all $y \in l_\infty(n, \mathbb{C})$ and $d(x, B(l_\infty(n, \mathbb{C}))) < \varepsilon$ for all $x \in A$.*

Proof. We leave the equivalence of (b), (c), and (d) to the reader (see theorem 7 of section 17). From theorems 1 and 2 of section 21 it follows that (b) implies (a). For, theorems 1 and 2 are valid for both the real and complex case and for from these we get that (b) implies that X^{**} is a $\mathscr{P}_\lambda(\mathbb{C})$ space for all $\lambda > 1$ (the M.A.P. is not used in this part of the proof of theorem 1 of section 21). We leave it to the reader to show that this implies that X^{**} is a $\mathscr{P}_1(\mathbb{C})$ space. Thus it is only necessary to show that (a) implies (c). Now if X is an L_1-predual space, then X^{**} is a $\mathscr{P}_1(\mathbb{C})$ space, that is, $X^{**} = C(T, \mathbb{C})$ where T is an extremally disconnected compact Hausdorff space. Now by considering continuous simple functions on T it is easy to see that $C(T, \mathbb{C})$ is an $\mathscr{L}_{\infty,\lambda}$ space for all $\lambda > 1$. Let Y be a finite dimensional subspace of X and $J: X \to X^{**}$ be the natural mapping. Then for any $\lambda > 1$, by the above, there is a finite dimensional subspace Z of X^{**} containing $J(Y)$ and with the property that $d(Z, l_\infty(n, \mathbb{C})) < \lambda$ where $n = \dim Z$. By the principle of local reflexivity there is a linear operator $B: Z \to X$ such that $B(Jy) = y$ for all $y \in Y$ and $\|B\|\,\|B^{-1}\| \leqslant \lambda$. Thus for $W = B(Z)$ we have that $W \supset Y$ and $d(W, l_\infty(n, \mathbb{C})) \leqslant \lambda^2$. ∎

We can now give an easy proof of lemma 6 of section 21 valid for both the real and complex case.

Lemma 1. *Let X be an L_1-predual space and Y be a separable subspace of X. Then there is a separable subspace Z of X such that $Z \supset Y$ and Z is an L_1-predual space.*

Proof. Let $\{y_n\}$ be a dense set in Y and choose a sequence $\{Z_n\}$ of finite dimensional subspaces of X such that $y_n \in Z_n$, $Z_n \subset Z_{n+1}$ for all n, and $d(Z_n, l_\infty(\dim Z_n, \mathbb{C})) < 1 + 1/n$. Then $Z = \overline{\bigcup Z_n}$ is such a space. ∎

We now wish to prove that a separable infinite dimensional Banach space X is an L_1-predual space if and only if there is a sequence $\{Y_n\}$ of finite dimensional subspaces of X such that Y_n is linearly isometric to $l_\infty(n, \mathbb{C})$, $Y_n \subset Y_{n+1}$ for all n, and $\bigcup_{n=1}^{\infty} Y_n$ is dense in X. Note that by theorem 2 the sufficiency is obvious. For the necessity we shall need several lemmas on isometric embeddings of $l_\infty(n,\mathbb{C})$ into $l_\infty(m,\mathbb{C})$. The development here is due to Michael and Pełczyński [205].

For $x \in l_\infty(n,\mathbb{C})$, $N(x)=\{i: 1 \leqslant i \leqslant n, |x(i)|=\|x\|\}$ and e_i^n denotes the ith coordinate vector in $l_\infty(n,\mathbb{C})$, that is $e_i^n(j)=1$ if $i=j$ and 0 otherwise. In what follows m and n stand for positive integers.

Lemma 1. *Let* $L: l_\infty(m,\mathbb{C}) \to l_\infty(n,\mathbb{C})$ *be a linear operator. Then* $\|L\| = \max\left\{\left|\sum_{j=1}^m L(e_j^n)(i)\right| : 1 \leqslant i \leqslant n\right\}$. ∎

The following lemma characterizes linear isometric embeddings of $l_\infty(m,\mathbb{C})$ into $l_\infty(n,\mathbb{C})$.

Lemma 2. *Let* $L: l_\infty(m,\mathbb{C}) \to l_\infty(n,\mathbb{C})$ *be a linear operator. Then L is a linear isometry if and only if (a), (b), and (c) below are valid.*

(a) $\sum_{i=1}^m |L(e_j^m)(i)| \leqslant 1$ *for all* $1 \leqslant i \leqslant n$.

(b) $\|L(e_j^m)\| = 1$ *for all* $1 \leqslant j \leqslant m$.

(c) if $i \in N(L(e_j^m))$ *and* $k \neq j$, *then* $L(e_k^m)(i)=0$.

Proof. Clearly if L is a linear isometry, then (a) and (b) are valid. To see (c) let $|\lambda|=1$ be chosen so that $|L(e_j^m)(i)| + \lambda L(e_k^m)(i) = |L(e_j^m)(i)| + |L(e_k^m)(i)| = 1 + |L(e_k^m)(i)|$. Then $1 = \|L(e_j^m) + \lambda L(e_k^m)\| \geqslant 1 + |L(e_k^m)(i)|$ which implies that $L(e_k^m)(i)=0$.

Now suppose that (a), (b), and (c) are valid. For $1 \leqslant k \leqslant m$ let $i_k \in N(L(e_k^m))$ and let $x = \sum_{j=1}^m t_j e_j^m$ be in $l_\infty(m,\mathbb{C})$. Then by (a) and lemma 1, $\|L(x)\| \leqslant \|x\|$. On the other hand, $\|L(x)\| = \max\left\{\left|\sum_{j=1}^m t_j e_j^m(i)\right| : 1 \leqslant i \leqslant n\right\}$ $\geqslant \max\left\{\left|\sum_{j=1}^m t_j e_j^m(i_k)\right| : 1 \leqslant k \leqslant m\right\} = \max\{|t_k| : 1 \leqslant k \leqslant m\} = \|x\|$. ∎

We now prove that we can approximate operators which are almost isometries by ones which are isometries.

Lemma 3. *Let* $0 < \varepsilon < 1$. *Then for any linear operator* $L: l_\infty(m,\mathbb{C}) \to l_\infty(n,\mathbb{C})$ *such that* (*) $\|x\| \leqslant \|L(x)\| \leqslant (1+\varepsilon)\|x\|$ *for all* $x \in l_\infty(m,\mathbb{C})$, *there is a linear isometry* $L': l_\infty(m,\mathbb{C}) \to l_\infty(n,\mathbb{C})$ *such that* $\|L - L'\| < \varepsilon$.

Proof. Let $f_j = L(e_j^m)$ for $1 \leqslant j \leqslant m$ and put $N_j = \{i: 1 \leqslant i \leqslant n, |f_j(i)| \geqslant 1\}$. From the left side of (*) above we see that $1 = \|e_j^m\| \leqslant \|f_j\|$ and thus for each $1 \leqslant j \leqslant m$ there is an i_j with $1 \leqslant i_j \leqslant n$ and $|f_j(i_j)| = \|f_j\| \geqslant 1$. Hence N_j is nonempty for each j and, moreover, the N_j's are pairwise disjoint. For, by the right side of (*) and lemma 1 we have that $1+\varepsilon \geqslant \|L\| \geqslant \sum_{j=1}^{m} |f_j(i)|$ for $i \leqslant i \leqslant n$. Thus if $k \neq j$ and $i \in N_j$, then $|f_k(i)| \leqslant 1+\varepsilon - |f_j(i)| \leqslant \varepsilon < 1$. Hence $i \notin N_k$. Now for $1 \leqslant j \leqslant m$ and $1 \leqslant i \leqslant n$

$$\text{let } y_j(i) = \begin{cases} f_j(i)/|f_j(i)| & \text{if } i \in N_j, \\ 0 \quad \text{if } i \in N_k & \text{for } k \neq j, \\ f_j(i)/(1+\varepsilon) & \text{otherwise.} \end{cases}$$

Then $\sum_{j=1}^{m} |y_j(i)| \leqslant 1$, $\|y_j\| = 1$ and $N(y_j) = N_j$. Let us define $L' : l_\infty(n, \mathbb{C})$ by $L'(x) = \sum_{j=1}^{m} x(j) y_j$. Then it is easy to see that L' is a linear isometry since it satisfies (a), (b), and (c) of lemma 2. Moreover, $\|L - L'\| < \varepsilon$. To see this consider the following two cases. If $i \notin \bigcup_{k=1}^{m} N_k$, then $\sum_{j=1}^{m} |f_j(i) - y_j(i)| = \sum_{j=1}^{m} |f_j(i)| \dfrac{\varepsilon}{1+\varepsilon} < \varepsilon$. If $i \in N_k$ for some k, then $\sum_{j=1}^{m} |f_j(i) - y_j(i)| = \sum_{\substack{j=1 \\ j \neq k}}^{m} |f_j(i)| + |f_k(i)| - 1 \leqslant \|L\| - 1 < \varepsilon$. ∎

We next show that we can fit the spaces $l_\infty(n, \mathbb{C})$ together in the correct fashion.

Lemma 4. *Let X be a linear subspace of $l_\infty(n, \mathbb{C})$ which is linearly isometric to $l_\infty(m, \mathbb{C})$ where $m < n$. Then there is a linear subspace X_1 of $l_\infty(n, \mathbb{C})$ containing X such that X_1 is linearly isometric to $l_\infty(m+1, \mathbb{C})$.*

Proof. Let e_j be the image of e_j^m under some fixed linear isometry of $l_\infty(m, \mathbb{C})$ onto X. Then clearly $\{e_j\}$ satisfies conditions (a), (b) and (c) of lemma 2. Since $m < n$, by (c) of lemma 2 we get that either one of the sets $N(e_j)$ contains at least two indicies or there is an index $i (1 \leqslant i \leqslant n)$ which does not belong to any $N(e_j)$. In both cases we may choose an i_0 $(1 \leqslant i_0 \leqslant n)$ such that $N(e_j) \setminus \{i_0\}$ is nonempty for $1 \leqslant j \leqslant n$. Let $f_j = e_j - e_j(i_0) e_{i_0}^n$ for $1 \leqslant j \leqslant m$ and $f_{m+1} = e_{i_0}^n$. Then for X_1 we take the linear span of the f_j's. Clearly $X_1 \supset X$ and from lemma 2 it follows that $L(x) = \sum_{j=1}^{m+1} x(j) f_j$ for $x \in l_\infty(m+1, \mathbb{C})$ defines a linear isometry of $l_\infty(m+1, \mathbb{C})$ onto X_1. ∎

We need one final lemma on approximation by linear isometries.

Lemma 5. *Let X be a Banach space and suppose $0<\varepsilon<1/6$ is fixed. Then for $0<\delta<\varepsilon/8$, we have that if $L_1: l_\infty(m,\mathbb{C})\to X$ and $L_2: l_\infty(n,\mathbb{C})\to X$ are linear operators satisfying*

$$(1+\varepsilon)^{-1}\|x\|\leqslant\|L_1(x)\|\leqslant(1+\varepsilon)\|x\| \quad \textit{for all } x\in l_\infty(m,\mathbb{C}),$$

$$(1+\delta)^{-1}\|y\|\leqslant\|L_2(y)\|\leqslant(1+\delta)\|y\| \quad \textit{for all } y\in l_\infty(n,\mathbb{C})$$

and for all nonzero x in $l_\infty(m,\mathbb{C})$, $d(L_1(x), L_2(l_\infty(n,\mathbb{C})))<\delta\|x\|$, then there is a linear isometry $S: l_\infty(m,\mathbb{C})\to l_\infty(n,\mathbb{C})$ such that $\|L_2S-L_1\|<12\varepsilon$.

Proof. Since $l_\infty(n,\mathbb{C})$ is a $\mathscr{P}_1(\mathbb{C})$ space, there is a projection P of X onto $L_2(l_\infty(n,\mathbb{C}))$ such that $\|P\|\leqslant\|L_2\|\,\|L_2^{-1}\|\leqslant(1+\varepsilon)^2<1+3\varepsilon$. Let $x\in l_\infty(m,\mathbb{C})$ and choose $y\in L_2(l_\infty(n,\mathbb{C}))$ such that $\|L_1(x)-y\|<\delta\|x\|$. Since $P(y)=y$, $\|L_1(x)-P(L_2x)\|\leqslant\|L_1(x)-y\|+\|Py-P(L_2x)\|\leqslant(1+\|P\|)\delta\|x\|\leqslant4\delta\|x\|$. Thus $\|P(L_1(x))\|\geqslant\|L_1(x)\|-4\delta\|x\|\geqslant[(1+\varepsilon)^{-1}-\varepsilon/2]\|x\|\geqslant(1-3\varepsilon/2)\|x\|$. Hence using the inequalities $8\delta<\varepsilon$ and $0<\varepsilon<1/6$, we get that $\|L_2^{-1}(PL_1(x))\|\geqslant(1+\delta)^{-1}\|PL_1(x)\|\geqslant(1-2\varepsilon)\|x\|$. Let $L=(1-2\varepsilon)^{-1}L_2^{-1}PL_1$. Then since $\|L_2^{-1}\|\,\|P\|\,\|L_1\|\leqslant(1+\delta)^3(1+\varepsilon)<1+2\varepsilon$, $\|x\|\leqslant\|Lx\|\leqslant[(1+2\varepsilon)/(1-2\varepsilon)]\|x\|\leqslant(1+6\varepsilon)\|x\|$. Thus by lemma 3 there is a linear isometry $S: l_\infty(m,\mathbb{C})\to l_\infty(n,\mathbb{C})$ such that $\|S-L\|<6\varepsilon$. Hence $\|L_2S-L_1\|\leqslant\|L_2\|\,\|S-L\|+\|L_2L-L_1\|<(1+\varepsilon)6\varepsilon+\|PL_1-L_1\|+\|PL_1\|[|1-(1-2\varepsilon)^{-1}|]<7\varepsilon+4\delta+(1+3\delta)(1+\varepsilon)(2\varepsilon/(1-2\varepsilon))\leqslant12\varepsilon$. ∎

We are now ready to prove the theorem. First let us note that if X is an L_1-predual space, then for any finite dimensional subspace Y of X and any $\varepsilon>0$, there is an n and a linear operator $L: l_\infty(n,\mathbb{C})\to X$ such that $(1+\varepsilon)^{-1}\leqslant\|x\|\leqslant\|Lx\|\leqslant(1+\varepsilon)\|x\|$ for all $x\in l_\infty(n,\mathbb{C})$ and $d(y, L(l_\infty(n,\mathbb{C})))<\varepsilon\|x\|$ for all nonzero y in Y.

Theorem 3 (Michael-Pełczyński). *Let X be a separable infinite dimensional Banach space. Then X is an L_1-predual space if and only if there is a sequence $\{Y_n\}$ of finite dimensional subspaces of X such that $Y_n\subset Y_{n+1}$ for all n, each Y_n is linearly isometric to $l_\infty(n,\mathbb{C})$, and $\bigcup_{n=1}^{\infty} Y_n$ is dense in X.*

Proof. Let $\{x_n\}$ be a dense set in X. By lemma 4 and the remark above we can inductively define a sequence of linear operators L_k, and a sequence of positive integers n_k, and a sequence of linear isometries S_k such that $L_k: l_\infty(n_k,\mathbb{C})\to X$, $S_k: l_\infty(n_k,\mathbb{C})\to l_\infty(n_{k+1},\mathbb{C})$, $\|L_{k+1}S_k-L_k\|<12\cdot2^{-3k}$, $\max(\|L_k\|,\|L_k^{-1}\|)<1+2^{-3k}$, and $d(f, L_{k+1}(l_\infty(n_{k+1},\mathbb{C})))<2^{-3k}\|f\|$ for all nonzero f in $F_k=\operatorname{span}(L_k(l_\infty(n_k,\mathbb{C})), x_1,\dots,x_k)$. For $j=1,2,\dots$ and $k=j+1,j+2,\dots$ let $V_{k,j}=(L_kS_{k-1}\dots S_j): l_\infty(n_k,\mathbb{C})\to X$. Since each S_k is an isometry, $\|V_{k+1,j}-V_{k,j}\|\leqslant\|L_{k+1}S_k-L_k\|<12\cdot2^{-3k}$. Hence for all j, $V_{\infty,j}=\lim_k V_{k,j}$ exists. Put $Z_j=V_{\infty,j}(l_\infty(n_j,\mathbb{C}))$. Then since $\lim\|L_k\|=\lim\|L_k^{-1}\|=1$, $\lim_k\|V_{k,j}\|=\lim_k\|V_{k,j}^{-1}\|=\|V_{\infty,j}\|=\|V_{\infty,j}^{-1}\|=1$. Thus $V_{\infty,j}$

is an isometric embedding (since its domain is finite dimensional). Now clearly $Z_j \subset Z_{j+1}$ since $V_{k,j+1}(l_\infty(n_{j+1}, \mathbb{C})) \supset V_{k,j}(l_\infty(n_j, \mathbb{C}))$ for all $j=1,2,\dots;$ $k=j+1, j+2, \dots$.

Since we can fill in the gaps by lemma 4, all we need to do to finish the proof is show that $\bigcup_{j=1}^{\infty} Z_j$ is dense in X.

Let x_m and $j>m$ be fixed. Choose $x \in l_\infty(n_j, \mathbb{C})$ such that $\|L_j(x)-x_j\| < 2^{-3j}\|x_m\|$. Now clearly $\|L_j(x)-V_{\infty,j}(x)\| = \lim_k \|L_j(x)-V_{k,j}(x)\|$. Furthermore, $\|L_j(x)-V_{k,j}(x)\| \leqslant \|L_j(x)-V_{j=1,j}(x)\| + \sum_{i=1}^{k-j-1} \|(V_{j+i+1,j} - V_{j+i,j})(x)\| \leqslant \sum_{i=j}^{k-1} 12\cdot 2^{-3i}\|x\| \leqslant \sum_{i=j}^{\infty} 12\cdot 2^{-3i}\|x\|$. By choice of x, $\|x\| \leqslant \|L_j^{-1}\|\,\|L_j(x)\| \leqslant \|L_j^{-1}\|(1+2^{-3j})\|x_m\|$. Thus $\|x_m - V_{\infty,j}(x)\| \leqslant \|L_j(x) - x_m\| + \|L_j(x)-V_{\infty,j}(x)\| \leqslant 2^{-3j}\|x_m\| + \|L_j^{-1}\|(2^{-3j}+1)\sum_{i=j}^{\infty} 12\cdot 3^{-i}\|x_m\|$. Since $\lim \|L_j^{-1}\|=1$, we have $\lim_j d(x_m, Z_j)=0$ and $\bigcup_{j=1}^{\infty} Z_j$ is dense in X. ∎

Suppose Y is linearly isometric to $l_\infty(n,\mathbb{C})$, we say a basis $y_1,\dots,y_n$ for Y is *admissible* if there is a linear isometry L of $l_\infty(n,\mathbb{C})$ onto Y such that $L(e_i^n)=y_i$ for $i=1,\dots,n$, clearly by lemma 2 above the admissible basis of $l_\infty(n,\mathbb{C})$ are of the form $\lambda_1 e^n_{\pi(1)},\dots,\lambda_n e^n_{\pi(n)}$ where $|\lambda_i|=1$ and π is a permutation of $\{1,\dots,n\}$. Thus, if $X=\overline{\bigcup_{n=1}^{\infty} Y_n}$ where each Y_n is linearly isometric to $l_\infty(n,\mathbb{C})$ and $Y_n \subset Y_{n+1}$ for all n, then it is possible to find admissible bases $\{y_i^n : 1\leqslant i\leqslant n\}$ and scalars $\{a_i^n : 1\leqslant i\leqslant n\}$ for $n=1,2,\dots$ such that $y_i^n = y_i^{n+1} + a_i^n y_{n+1}^{n+1}$ for $1\leqslant i\leqslant n$ and $\sum_{i=1}^{n} |a_i^n| \leqslant 1$ for $n=1,2,\dots$.

We have noted that if X is a real L_1-predual space such that the closed unit ball of X has an extreme point e, then for $K=\{x^*\in X^* : x^*(e) = 1 = \|x^*\|\}$, K is a (weak*) compact Choquet simplex and the mapping $x\to\hat{x}$ (where $\hat{x}(x^*)=x^*(x)$ for $x\in X$ and $x^*\in K$) is a linear isometry of X into $C(K,\mathbb{R})$ such that $\hat{e}=1$. We wish to prove a similar theorem for the complex case which is due to Hirsberg and Lazar [293]. It will require some lemmas and the use of theorem 3 above.

Lemma 6. *Let X and Y be Banach spaces with $Y\subset X$, X linearly isometric to $l_\infty(n,\mathbb{C})$, and Y is the range of a contractive projection P on X (so that Y is linearly isometric to $l_\infty(m,\mathbb{C})$ for some m). Suppose that $\{x_i : 1\leqslant i\leqslant n\}$ and $\{y_j : 1\leqslant j\leqslant m\}$ are admissible bases of X and Y respectively such that $P(x_j)=y_j$ for $1\leqslant j\leqslant m$ and $P(x_j)=0$ for $m+1\leqslant j\leqslant n$. Let $x\in X$ with $\|x\|\leqslant 1$ and suppose that $P(x)=1/2(u+v)$ with $\|u\|$, $\|v\|\leqslant 1$,*

$u, v \in Y$, and $u \neq P(x)$. Then there are y, z in X with $\|y\|, \|z\| \leqslant 1$ such that $x = 1/2(y+z)$, $Py = u$, $Pz = v$, and $\|y-u\|, \|z-v\| \leqslant \max\{4\|x-Px\|, 4\|x-Px\|^{1/4}\}$.

Proof. Clearly from lemma 2 and the fact that P is a contractive projection it follows that there are scalars a_{ij} $(1 \leqslant i \leqslant m; m+1 \leqslant j \leqslant n)$ such that $y_i = x_i + \sum_{j=m+1}^{n} a_{ij} x_j$ for $1 \leqslant i \leqslant m$, and $\sum_{i=1}^{m} |a_{i,j}| \leqslant 1$ for $m+1 \leqslant j \leqslant n$. Now $x = \sum_{j=1}^{n} b_j x_j$ for some set of set of scalars b_j and $P(x) = \sum_{i=1}^{m} b_i y_i$, $u = \sum_{i=1}^{m} (b_i + \delta_i) y_i$, and $v = \sum_{i=1}^{m} (b_i - \delta_i) y_i$, where (1) $\max_{1 \leqslant j \leqslant n} |b_j| \leqslant 1$ and $\max_{1 \leqslant i \leqslant m} |b_i \pm \delta_i| \leqslant 1$. Let $\varepsilon = \|x - P(x)\|$. Then since $P(x) = \sum_{i=1}^{m} b_i x_i + \sum_{j=m+1}^{n} \sum_{i=1}^{m} b_i a_{ij} x_j$, we have that $\left| b_j - \sum_{i=1}^{m} b_i a_{ij} \right| \leqslant \varepsilon$ for $m+1 \leqslant j \leqslant n$. If $\varepsilon \geqslant 1/3$, we put $y = \sum_{i=1}^{m} (\lambda_i + \delta_i) x_i + \sum_{j=m+1}^{n} b_j x_j$ and $z = \sum_{i=1}^{m} (\lambda_i - \delta_i) x_i + \sum_{j=m+1}^{n} b_j x_j$. By (1) we have that $|\delta_i| \leqslant 1$ so that $\|y-u\| = \max_{m+1 \leqslant j \leqslant n} \left| b_j - \sum_{i=1}^{m} (b_i + \delta_i) a_{ij} \right| \leqslant \varepsilon + 1 \leqslant 4\varepsilon$. By a similar argument we obtain that $\|z - v\| \leqslant 4\varepsilon$.

Now suppose that $0 \leqslant \varepsilon < 1/3$ and that $m+1 \leqslant j \leqslant n$ is fixed. If $|b_j| \geqslant 1 - \varepsilon^{\frac{1}{2}}$, let $\delta_j = 0$. If $|b_j| < 1 - \varepsilon^{\frac{1}{2}}$ and $\sum_{i=1}^{m} \delta_i a_{ij} = 0$, let $\delta_j = 0$ again. If $|b_j| < 1 - \varepsilon^{\frac{1}{2}}$ and $\sum_{i=1}^{m} \delta_i a_{ij} \neq 0$, we define δ_j as follows. We can find unique positive numbers t_1 and t_2 so that $1 = \left| b_j + t_1 \left(\sum_{i=1}^{m} \delta_i a_{ij} \right) \right| = \left| b_j - t_2 \left(\sum_{i=1}^{m} \delta_i a_{ij} \right) \right|$. Then for $t = \min(1, t_1, t_2)$, we put $\delta_j = t \left(\sum_{i=1}^{m} \delta_i a_{ij} \right)$. Let us note that $0 \leqslant 1 - t \leqslant \varepsilon^{\frac{1}{2}}$. If $t = 1$, then it is clear. If $t = t_1$ or $t = t_2$, then

$$1 = \left| (1-t) b_j + t \left(b_j \pm \sum_{i=1}^{m} \delta_i a_{ij} \right) \right|$$
$$\leqslant (1-t)|b_j| + t \left| b_j \pm \sum_{i=1}^{m} \delta_i a_{ij} \right| \leqslant (1-t)(1 - \varepsilon^{\frac{1}{2}}) + t(1+\varepsilon).$$

Hence $t \geqslant (\varepsilon^{\frac{1}{2}} + 1)^{-1}$ and $1 - t \leqslant \varepsilon^{\frac{1}{2}}$.

Now let $y=\sum_{j=1}^{n}(b_j+\delta_j)x_j$ and $z=\sum_{j=1}^{n}(\lambda_j-\delta_j)x_j$. Then $x=\frac{1}{2}(y+z)$, $P(y)=u$ and $P(z)=v$. Moreover, $|b_j\pm\delta_j|\leqslant 1$ for $1\leqslant j\leqslant n$ so that $\|y\|, \|z\|\leqslant 1$.

Now, $y-u=\sum_{j=m+1}^{n}\left[b_j+\delta_j-\sum_{i=1}^{m}(b_i+\delta_i)a_{ij}\right]x_j$ and, hence, $\|y-u\| = \max_{m+1\leqslant j\leqslant n}\left|b_j+\delta_j-\sum_{i=1}^{n}(b_i+\delta_i)a_{ij}\right|$. Suppose that $|b_j|\geqslant 1-\varepsilon^{\frac{1}{2}}$. Then $\left|\sum_{i=1}^{m}b_ia_{ij}\right|\geqslant 1-\varepsilon^{\frac{1}{2}}-\varepsilon>0$ and $\left|\sum_{i=1}^{m}(b_i\pm\delta_i)a_{ij}\right|\leqslant 1$ so that $\left|\sum_{i=1}^{m}\delta_ia_{ij}\right| \leqslant\left(1-\left|\sum_{i=1}^{m}b_ia_{ij}\right|^2\right)^{\frac{1}{2}}\leqslant 2\varepsilon^{\frac{1}{4}}$. Thus $\left|b_j+\delta_j-\sum_{i=1}^{m}(b_i+\delta_i)a_{ij}\right|\leqslant\left|b_j-\sum_{i=1}^{m}b_ia_{ij}\right| +\left|\sum_{i=1}^{m}\delta_ia_{ij}\right|\leqslant\varepsilon+2\varepsilon^{\frac{1}{4}}\leqslant 4\varepsilon^{\frac{1}{4}}$. If $|b_j|<1-\varepsilon^{\frac{1}{2}}$ and $\sum_{i=1}^{m}\delta_ia_{ij}=0$, then clearly $\left|b_j+\delta_j-\sum_{i=1}^{m}(b_i+\delta_i)a_{ij}\right|\leqslant 4\varepsilon^{\frac{1}{4}}$.

Let us now suppose that $|b_j|<1-\varepsilon^{\frac{1}{2}}$ and that $\sum_{i=1}^{m}b_ia_{ij}\neq 0$. Then

$$\left|b_j=\delta_j-\sum_{i=1}^{m}(b_i+\delta_i)a_{ij}\right|\leqslant\left|b_j-\sum_{i=1}^{m}b_ia_{ij}\right|+(1-t)\left|\sum_{i=1}^{m}\delta_ia_{ij}\right|\leqslant\varepsilon+\varepsilon^{\frac{1}{2}}\leqslant 4\varepsilon^{\frac{1}{4}}.$$

It follows readily that $\|y-u\|, \|z-v\|\leqslant 4\varepsilon^{\frac{1}{4}}$. ∎

Let X be a separable infinite dimensional L_1-predual space and $\{Y_n\}$ be a sequence of finite dimensional subspaces of X such that $Y_n\subset Y_{n+1}$ for all n, each Y_n is linearly isometric to $l_\infty(n,\mathbb{C})$, and $\bigcup_{n=1}^{\infty}Y_n$ is dense in X. Let $\{x_i^n: 1\leqslant i\leqslant n, n+1, 2, \ldots\}$ be chosen so that $x_1^n,\ldots,x_n^n$ is an admissible basis for Y_n and $x_i^n=x_i^{n+1}+a_i^n x_{n+1}^{n+1}$ for $1\leqslant i\leqslant n$ and $n=1,2,\ldots$. We define a sequence $\{x_i^*\}$ in X^* as follows: if $i\leqslant n$ and $x=\sum_{j=1}^{n}a_jx_j^n$, let $x_i^*(x)=a_i$. Then x_i^* is well defined on $\bigcup_{n=1}^{\infty}Y_n$ (and hence is uniquely extendable to X) and $\|x_i^*\|=1$.

We use the above notation in the following lemma.

Lemma 7. *Let X be a separable infinite dimensional L_1-predual space and suppose that u is an extreme point of the closed unit ball of X. Then there is a sequence $\{P_n\}$ of contractive projections on X such that the range of each P_n is Y_n, $\lim P_n(x)=x$ for all $x\in X$, and $P_n(u)$ is an extreme point of the closed unit ball of Y_n for each n.*

Proof. Let $P_n = \sum_{i=1}^{n} x_i^*(u)x_i^n$. Then clearly P_n satisfies the first three of the stated conditions. So, we only need show that $P_n(u)$ is an extreme point of the closed unit ball of Y_n for each n. Assume that this is not true for some n_0. Then for some distinct $x_0, y_0 \in Y_{n_0}$ with $\|x_0\|, \|y_0\| \leqslant 1$, we have that $P_{n_0}(u) = \frac{1}{2}(x_0 + y_0)$. Choose an increasing sequence $\{n_k\}$ of positive integers such that $n_1 > n_0$ and $\|P_{n_{k+1}}(u) - P_{n_k}(u)\| < 2^{-4k}$ for all k. Since for $m \geqslant n$ we have that $P_n P_m = P_n$, it follows from lemma 6 that there are x_1, y_1 in Y_n, with $\|x_1\|, \|y_1\| \leqslant 1$ and $P_{n_1}(u) = \frac{1}{2}(x_1 + y_1)$, $P_{n_0}(x_1) = x_0$, and $P_{n_0}(y_1) = y_0$. By using lemma 6 we can construct inductively two sequences $\{x_k\}, \{y_k\}$ such that $x_k, y_k \in Y_{n_k}$, $\|x_k\|, \|y_k\| \leqslant 1$, $P_{n_k}(u) = \frac{1}{2}(x_k + y_k)$, $P_{n_k}(x_{k+1}) = x_k$, $P_{n_k}(y_{k+1}) = y_k$, and $\|y_{k+1} - y_k\|, \|x_{k+1} - x_k\| \leqslant 4 \cdot 2^{-k}$. Hence $\{x_k\}, \{y_k\}$ are Cauchy sequences and converge to some x and y respectively in X and $u = \frac{1}{2}(x + y)$. Clearly $\|u - x\| \geqslant \|P_{n_0}(u) - x_0\| > 0$ which is a contradiction. ∎

Lemma 8. *Let X be an L_1-predual space and u be an extreme point of the closed unit ball of X. Then for any $x \in X$ with $\|x\| = 1$ and $\varepsilon > 0$ there is a functional $x^* \in X^*$ such that $\|x^*\| = x^*(u) = 1$ and $|x^*(x)| > 1 - \varepsilon$.*

Proof. It is clear if X is finite dimensional. If X is infinite dimensional, let $Y = \overline{\bigcup_{n=1}^{\infty} Y_n}$ be a subspace of X containing u and x such that $Y_n \subset Y_{n+1}$ for all n and each Y_n is linearly isometric to $l_\infty(n, C)$. Let $\{P_n\}$ be the sequence of projections defined in lemma 7 and choose n so that $\|x - P_n(x)\| < \varepsilon$. Since $P_n(u)$ is an extreme point of the closed unit ball of Y_n, $|x_i^*(P_n u)| = 1$ for all $1 \leqslant i \leqslant n$ and since $\|P_n(x)\| > 1 - \varepsilon$, $|x_i^*(P_n(x))| > 1 - \varepsilon$ for some $1 \leqslant i \leqslant n$. Thus for this index i and a suitable scalar a with $|a| = 1$ we can put $x^* = a x_i^* \circ P_n$. ∎

Theorem 4 (Hirsberg-Lazar). *Let X a complex L_1-predual space and suppose that u is an extreme point of the closed unit ball of X and that $K = \{x^* \in X^* : x^*(u) = 1 = \|x^*\|\}$. Then the mapping $x \to \hat{x}$ where $\hat{x}(x^*) = x^*(x)$ for all $x \in X$ and $x^* \in K$ is a linear isometric mapping of X into $C(K, \mathbb{C})$ such that $\hat{u} = 1$ (where K has the weak* topology).*

Proof. Clearly $\|\hat{x}\| \leqslant \|x\|$ and by lemma 8 we get that, indeed, $\|\hat{x}\| = \|x\|$. ∎

We now wish to give the complex version of theorem of section and use it to further investigate L_1-predual spaces (in particular when the closed unit ball contains an extreme point). This will require additional terminology and the use of theorem 11 in section 20. As in this section, we let $\Gamma = \{\alpha \in C : |\alpha| = 1\}$ and we shall denote by $d\alpha$ the normalized are measure on Γ, (see [130]).

Let X be a complex Banach space and V^* denote the closed unit ball of X^* with the weak* topology. For each $\alpha\in\Gamma$ let $\sigma_\alpha: V^*\to V^*$ be defined by $\sigma_\alpha(x^*)=\alpha x^*$ for all $x^*\in V^*$. Then clearly σ_α is an affine homomorphism of V^* onto itself and $\sigma_\alpha^{-1}=\sigma_{\alpha-1}$ for all $\alpha\in\Gamma$. We call a function f on V^* *Γ-invariant* if $f(\alpha x^*)=f(x^*)$ for all $x^*\in V^*$ and $\alpha\in\Gamma$. Similarly, we call f *Γ-homogeneous* if $f(\alpha x^*)=\alpha f(x^*)$ for all $x^*\in V^*$ and $\alpha\in\Gamma$. If $\mu\in M(V^*,\mathbb{C})$, we say that μ is *Γ-invariant* if $\mu\circ\sigma_\alpha^{-1}=\mu$ for all $\alpha\in\Gamma$ and we say that μ is *Γ-homogeneous* if $\mu\circ\sigma_\alpha^{-1}=\mu$ for all $\alpha\in\Gamma$. Let $C_I(V^*,\mathbb{C})=\{f\in C(V^*,\mathbb{C}):f$ is Γ-invariant$\}$, $C_H(V^*,\mathbb{C})$ $=\{f\in C(V^*,C):f$ is Γ-homogeneous$\}$, $M_I(V^*,\mathbb{C})=\{\mu\in M(V^*,\mathbb{C}):\mu$ is Γ-invariant$\}$, and $M_H(V^*,\mathbb{C})=\{\mu\in M(V^*,\mathbb{C}):\mu$ is Γ-homogeneous$\}$.

Furthermore, let $Q:C(V^*,\mathbb{C})\to C(V^*,\mathbb{C})$ and $H:C(V^*,\mathbb{C})\to C(V^*,\mathbb{C})$ be defined by $(Qf)(x^*)=\int_\Gamma f(\alpha x^*)d\alpha$ and $(Hf)(x^*)=\int_\Gamma \alpha^{-1}f(\alpha x^*)d\alpha$ for $f\in C(V^*,\mathbb{C})$. Then Q is a contractive projection of $C(V^*,\mathbb{C})$ onto $C_I(V^*,\mathbb{C})$ and H is a contractive projection of $C(V^*,\mathbb{C})$ onto $C_H(V^*,\mathbb{C})$. Clearly Q^* is a (weak* continuous) contractive projection of $M(V^*,\mathbb{C})$ onto $M_I(V^*,\mathbb{C})$ and H^* is a (weak* continuous) contractive projection of $M(V^*,\mathbb{C})$ onto $M_H(V^*,\mathbb{C})$. All of the above remarks are easily verified and are left to the reader.

Now we have two continuous mappings of Γ into V^* which we wish to consider. Namely, $\alpha\to\mu\circ\sigma_\alpha^{-1}$ and $\alpha\to\alpha^{-1}(\mu\circ\sigma_\alpha^{-1})$. Note that if $f\in C(V^*,\mathbb{C})$ and $\mu\in M(V^*,\mathbb{C})$ then $(Q^*\mu)(f)=\int\int f(\alpha x^*)d\alpha\, d\mu(x^*)$ $=\int\int f(\alpha x^*)d\mu(x^*)d\alpha=\int(\mu\circ\sigma_\alpha^{-1})f\,d\alpha$. That is, $\int(\mu\circ\sigma_\alpha^{-1})d\alpha$ in the weak* sense. A similar argument shows that for all $f\in C(V^*,\mathbb{C})$, $(H^*\mu)(f)$ $=\int\alpha^{-1}(\mu\circ\sigma_\alpha^{-1})f\,d\alpha$ or, $H^*\mu=\int\alpha^{-1}(\mu\circ\sigma_\alpha^{-1})d\alpha$ in the weak* sense. In particular, $|H^*\mu|\leqslant Q^*|\mu|$. Finally, it is easy to see that $H^*(\mu\circ\sigma_\alpha^{-1})$ $=\alpha H^*\mu$.

Now let $\mu\in M(V^*,\mathbb{C})$ and h be a Borel measurable function such that $|h|=1$ and $\mu=h|\mu|$ (see section 8). Then $\omega: V^*\to V^*$ defined by $\omega(x^*)=h(x^*)x^*$ for all $x^*\in V^*$ is also Borel measurable. If $f\in C_H(V^*,\mathbb{C})$, then $(|\mu|\circ\omega^{-1})(f)=\int f(h(x^*)x^*)d|\mu|(x^*)=\int h(x^*)f(x^*)d|\mu|(x^*)=\mu(f)$. Hence $H^*(|\mu|\circ\omega^{-1})=H^*\mu$. On the other hand, if $f\in C_I(V^*,\mathbb{C})$, then $(|\mu|\circ\omega^{-1})(f)=|\mu|(f)$ so that $Q^*(|\mu|\circ\omega^{-1})=Q^*(|\mu|)$.

Recall that for $x\in X$, $\hat{x}\in C(V^*,\mathbb{C})$ is defined by $\hat{x}(x^*)=x^*(x)$ for $x^*\in V^*$ and that $x\to\hat{x}$ is a linear isometry of X into $C(V^*,\mathbb{C})$ (actually it is easy to see that the range of this mapping is $A(V^*,\mathbb{C})\cap C_H(V^*,\mathbb{C})$, i.e., the space of affine Γ-homogeneous continuous complex valued functions on V^*). Thus for $\mu\in M(V^*,\mathbb{C})$, we get an element $r(\mu)\in X^*$ defined by $r(\mu)(x)=\int\hat{x}\,d\mu$ for all $x\in X$. Moreover, (μ) is the resultant of μ if $\mu\in M^*(V^*,\mathbb{R})$ and $\|\mu\|=1$. That is, for each $f\in A(V^*,\mathbb{R})$, $f(r(\mu))$ $=\int f\,d\mu$. For, by the approximation lemma for affine continuous functions (see the appendix), it suffices to assume that $f=g+a$ where $a\in\mathbb{R}$ and g is the restriction to V^* of a real linear weak* continuous functional on X^*. Moreover, $g=\operatorname{Re}\hat{x}$ where $\hat{x}$ is defined by $\hat{x}(x^*)=g(x^*)-ig(ix^*)$ for $x^*\in V^*$ and $\hat{x}$ is affine and Γ-homogeneous. Thus $f(r(\mu))=g(r(\mu))+a=\operatorname{Re}\hat{x}(r(\mu))\neq a=\operatorname{Re}\int\hat{x}\,d\mu+a=\int\operatorname{Re}\hat{x}\,d\mu+a=\int f\,d\mu$. (We shall call positive normalized regular Borel measures *probability measures* for short.)

It is easily seen that $r: M(V^*,\mathbb{C})\to X^*$ is a weak* continuous, norm decreasing, linear mapping onto X^* and that $r(\mu\circ\sigma_\alpha^{-1})=\alpha r(\mu)$, $r(H\mu)=r(\mu)$, and $r(|\mu|\circ\omega^{-1})=r(H(|\mu|\circ\omega^{-1}))=r(\mu)$ for all $\mu\in M(V^*,\mathbb{C})$.

As in section 20, we let $\bar{f}$ denote the upper envelope of a real valued function for V^* and recall that for $f\in C(V^*,\mathbb{R})$, $\bar{f}(x^*)=\sup\{v(f): v$ is a probability measure on V^* and $r(v)=x^*\}$. As before, $B(f)=\{x^*\in V^*: \bar{f}(x^*)=f(x^*)\}$ and $B(f)$ is a G_δ set. Moreover, a positive measure μ on V^* is maximal (in the sense of Choquet) if and only if $\mu(V^*\setminus B(f))=0$ for all $f\in C(V^*,\mathbb{R})$ (or, all continuous real-valued convex functions on V^*). Recall also from section 20 that $\mu\in M(V^*,\mathbb{C})$ is said to be a *boundary measure* if $|\mu|$ is maximal.

Lemma 9. *If $f\in C(V^*,\mathbb{R})$ is convex, then $B(Qf)\subset B(f)$.*

Proof. If $x^*\notin B(f)$, then $\bar{f}(x^*)>f(x^*)$ and there is a probability measure μ on V^* with $r(\mu)=x^*$ and $\mu(f)>f(x^*)$. For all $\alpha\in\Gamma$, $r(\mu\circ\sigma_\alpha^{-1})=\alpha x^*$ and since f is convex, $(\mu\circ\sigma_\alpha^{-1})(f)\geqslant f(\alpha x^*)$. Since $\alpha\to f\circ\sigma_\alpha$ is norm continuous, the function $\alpha\to(\mu\circ\sigma_\alpha^{-1})(f)-f(\alpha x^*)$ is continuous, positive, and strictly positive at $\alpha=1$. Thus

$$\begin{aligned}0&<\int[(\mu\circ\sigma_\alpha^{-1})(f)-f(\alpha x^*)]\,d\alpha\\&=(Q^*\mu)(f)-(Qf)(x^*)=\mu(Qf)-(Qf)(x^*)\\&\leqslant(Qf)(x^*)-(Qf)(x^*)\quad\text{and}\quad x^*\notin B(Qf).\ \blacksquare\end{aligned}$$

In the proof of the next lemma we need the fact that the mapping $\alpha\to(|\mu|\circ\sigma_\alpha^{-1})(V^*\setminus B(f))$ is integrable with respect to $d\alpha$ where $\mu\in M(V^*,\mathbb{C})$ and $f\in C(V^*,\mathbb{R})$ are fixed. This follows readily from the fact $B(f)$ is a G_δ set and we leave it as an exercise for the reader (for a slightly more general result see [289]).

Lemma 10. *If $\mu\in M^+(V^*,\mathbb{C})$ is a boundary measure, then so are $H\mu$ and $|\mu|\circ\omega^{-1}$.*

Proof. Let $f\in C(V^*,\mathbb{R})$ be convex and $g=Qf$. Since $V^*\setminus B(g)$ is invariant under σ_α for all $\alpha\in\Gamma$,

$$\begin{aligned}(Q^*|\mu|)(V^*\setminus B(g))&=\int|\mu|\circ\sigma_\alpha^{-1}(V^*\setminus B(g))\,d\alpha\\&=\int|\mu|(V^*\setminus B(g))\,d\alpha=|\mu|(V^*\setminus B(g))=0\,.\end{aligned}$$

Hence $Q^*|\mu|$ is maximal and since $|H\mu|\leqslant Q^*|\mu|$, it follows that $H\mu$ is a boundary measure.

Now $|\mu|\circ\omega^{-1}(V^*\setminus B(g))=Q^*|\mu|(V^*\setminus B(g))=0$ so that $|\mu|\circ\omega^{-1}$ is also maximal. ∎

In order to prove the theorem we need an approximation lemma in $L_1(T,\Sigma,\mu,\mathbb{C})$. We say a countable set $\{A_n\}$ in Σ is a *partition* of T with respect to $f\in L_1(T,\Sigma,\mu,\mathbb{C})$ if

(a) $0<\mu(A_n)<\infty$ for all n,

(b) $A_n\cap A_m=\phi$ for $n\neq m$,

(c) $S(f)\subset\bigcup_{n=1}^{\infty} A_n$ where $S(f)=\{t\in T: f(t)\neq 0\}$.

If $\{f_k\}$ is a sequence in $L_1(T,\Sigma,\mu,\mathbb{C})$, then using the sets $C_{kj}=\{t\in T: |f_k(t)|\geqslant 1/j\}$ one can construct a partition which is common to each f_k.

Given $f\in L_1(T,\Sigma,\mu,\mathbb{C})$ and a partition $\pi=\{A_n\}$ with respect to f, we define $E(f,\pi)=\sum_{n=1}^{\infty}[\mu(A_n)^{-1}\int_{A_n} f\,d_\mu]f_{A_n}$ where f_{A_n} is the characteristic function of A_n.

It is easy to see that the following relations hold where $f,g\in L_1(T,\Sigma,\mu,\mathbb{C})$ and π is a partition common to both f and g.

(1) $E(f+g,\pi)=E(f,\pi)+E(g,\pi)$,

(2) $E(af,\pi)=aE(f,\pi)$,

(3) $|E(f,\pi)|\leqslant E(|f|,\pi)$,

(4) $\int E(f,\pi)d\ =\int f\,d\mu$,

(5) $\|E(f,\pi)\|_1\leqslant\|f\|_1$.

Lemma 10 (Effros). *Let $f_1,\ldots,f_m$ be in $L_1(T,\Sigma,\mu,\mathbb{C})$ and $\varepsilon>0$ be given. Then there is a partition π common to $f_1,\ldots,f_m$ such that*

$$\|f_k-E(f_k,\pi)\|_1\leqslant\varepsilon \quad \textit{for}\ k=1,\ldots,m\,.$$

Proof. Let $C_{n,k}=\{t\in T: 1/n\leqslant|f_k(t)|\leqslant n\}$. Then $\{\int_{T\setminus C_{n,k}}|f_k|\,d\mu\}$ decreases to zero. Thus there is an n_0 and a $B\in\Sigma$ such that for all k,

(a) $|f_k(t)|\leqslant n_0$ for all $t\in B$,

(b) $\int_{T\setminus B}|f_k|\,d\mu\leqslant\varepsilon/3$,

(c) $\mu(B)<\infty$.

Let D be the closed disk in the complex plane of radius n_0 and center 0 and let $D_j(j=1,\ldots,n)$ be closed disks of diameter less than $\varepsilon/3\mu(B)$ with $D\subset\cup D_j$. By taking intersections of the sets $S_{jk}=f_k^{-1}(D_j)\cap B$ for $j=1,\ldots,$ and $k=1,\ldots,m$, we obtain pairwise disjoint sets $B_1,\ldots,B_s$ in Σ such that for all h,k, $|f_k(t)-f_k(t')|<\varepsilon/3\mu(B)$ for all $t,t'\in B_h$, and $B=\cup B_h$. We also select $B_{s+1},B_{s+2},\ldots,$ in Σ such that $B_{s+h}\subset T\setminus B$, $\mu(B_{s+h})<\infty$, and $\bigcup_k S(f_k)B\subset\bigcup_k B_{s+h}$. By deleting null sets, we get a partition $\pi=\{B_1,B_2,\ldots\}$ which is common to each $f_1,\ldots,f_m$.

Given k and $t \in B_h$ for $h \leqslant s$, we have that

$$\begin{aligned}|f_k(t)-E(f_k,\pi)| &= |f_k(t)-\mu(B_h)^{-1}\int_{B_h} f_k d_\mu| \\ &= \mu(B_h)^{-1}\int_{B_h}[f_k(t)-f_k(t')]\,d_\mu(t') \leqslant \varepsilon/3\,\mu(B).\end{aligned}$$

If $t \in B_h$ and $h > s$, then

$$|f_k(t)-E(f_k,\pi)(t)| \leqslant |f_k(t)| + \mu(B_h)^{-1}\int_{B_h}|f_k|\,d\mu.$$

Thus $\|f_k - E(f_k,\pi)\|_1 \leqslant \sum_{h=1}^{s} \varepsilon\mu(B_h)/3\,\mu(B) + 2\int_{T\setminus B}|f_k|\,d\mu \leqslant \varepsilon.$ ∎

We shall use the above lemmas and notation in the following theorem (Compare with theorem 7 of section 21).

Theorem 5 (Effros). *Let X be a complex Banach space and V^* be the closed unit ball of X^*. Then X is an L_1-predual space if and only if whenever v_1 and v_2 are two maximal normalized positive measures on V^* with $r(v_1)=r(v_2)$, then $H^* v_1 = H^* v_2$.*

Proof. Suppose X is an L_1-predual space and v_1, v_2 are maximal probability measures on V^* with $r(v_1)=r(v_2)$ and let $v=\frac{1}{2}(v_1+v_2\circ\sigma^{-1})$. Then v is a maximal probability measure with $(v)=0$ and it suffices to show that $H^* v = 0$.

Clearly we may choose a net of purely atomic measures $v_\gamma = \sum_{k=1}^{n(\gamma)} c_{\gamma k}\varepsilon_{x^*_{\gamma k}}$ where $\sum_{k=1}^{n(\gamma)} c_{\gamma k}=1$, $c_{\gamma k} \geqslant 0$, and $\{v_\gamma\}$ converges to v in the weak* topology so that $r(v_\gamma)=\sum_{k=1}^{n(\gamma)} c_{\gamma k} x^*_{\gamma k}=0$.

Given $\varepsilon > 0$ and γ, we obtain by lemma 10, a common partition $\pi^\varepsilon_\gamma = \{B_1, B_2, \ldots\}$ for $x^*_{\gamma 1}, \ldots, x^*_{\gamma n(\gamma)}$ such that for all k

$$\|x^*_{\gamma k} - E(x^*_{\gamma k}, \pi^\varepsilon_\gamma)\| < \varepsilon.$$

Letting $f_{\gamma k} = E(x^*_{\gamma k}, \pi^\varepsilon_\gamma)$, the probability measures

$$v^\varepsilon_\gamma = \sum_k c_{\gamma k}\,\varepsilon_{f_{\gamma k}}$$

also converge weak* to v as $\gamma \to \infty$ and $\varepsilon \to 0$, and

$$\begin{aligned} r(v^\varepsilon_\gamma) &= \Sigma_k\, E(c_{\gamma k} x^*_{\gamma k}, \pi^\varepsilon_\gamma) \\ &= E(\Sigma_k c_{\gamma k} x^*_{\gamma k}, \pi^\varepsilon_\gamma) \\ &= 0.\end{aligned}$$

It suffices to show that for each γ and $\varepsilon > 0$, there is a probability measure $\lambda^\varepsilon_\gamma$ on V^* with $v^\varepsilon_\gamma \prec \lambda^\varepsilon_\gamma$ (in the sense of Choquet), and $H^*(\lambda^\varepsilon_\gamma)=0$, since if this is the case, let $\{\lambda_\delta\}$ be a weak* convergent subnet of $\lambda^\varepsilon_\gamma$. Letting $\lambda=\lim_\delta \lambda_\delta$, it is evident that $v\prec\lambda$, hence $\lambda=v$. Thus

$$H^* v=\lim_\delta H^*\lambda_\delta=0\,.$$

We have

$$f_{\gamma k}=\Sigma_{j=1}^\infty a_{jk} f_{B_j}\,,$$

where

$$\Sigma_j |a_{jk}|\,\mu(B_j)=\|f_{\gamma k}\|\leqslant\|x^*_{\gamma k}\|\leqslant 1\,.$$

Letting $\alpha_{jk}=\alpha_{jk}|a_{jk}|$, $\alpha_{jk}\in\Gamma$, and $q_j=\mu(B_j)^{-1}f_{B_j}$, the probability measure

$$\lambda^\varepsilon_{\gamma k}=\Sigma_j |a_{jk}|\,\mu(B_j)\,\varepsilon_{\alpha_{jk}q_j}+(1-\|p^\varepsilon_{\gamma k}\|)\,\varepsilon_0$$

has resultant $f_{\gamma k}$. It follows that

$$\lambda^\varepsilon_\gamma=\Sigma_k c_{\gamma k}\lambda^\varepsilon_{\gamma k}$$

is such that $v^\varepsilon_\gamma\prec\lambda^\varepsilon_\gamma$. On the other hand, since $r(v^\varepsilon_\gamma)=0$,

$$\Sigma_k c_{\gamma k} f_{\gamma k}=0\,,$$

hence if we multiply by f_{B_j}

$$\Sigma_k c_{\gamma k} a_{jk}=0\,.$$

It follows that

$$\begin{aligned}H^*\lambda^\varepsilon_\gamma&=\Sigma_k c_{\gamma k} H^*\lambda^\varepsilon_k\\&=\Sigma_{j,k} c_{\gamma k}|a_{jk}|\,\mu(B_j)\,H^*\varepsilon_{\alpha_{jk}q_j}\\&=\Sigma_j[\Sigma_k c_{\gamma k}a_{jk}]\,\mu(B_j)\,H^*\varepsilon_{\alpha_{jk}q_i}\\&=0\,.\end{aligned}$$

For the converse, let $M_{Hb}(V^*,\mathbb{C})=\{\mu\in M_H(V^*,\mathbb{C}):\mu$ is a boundary measure$\}$. We shall show that $M_{Hb}(V^*,\mathbb{C})$ is linearly isometric to X^* and that $M_{Hb}(V^*,\mathbb{C})$ is also linearly isometric to an abstract L_1 space.

For $x^*\in V^*$, let v be any maximal probability measure on V^* with $r(v)=x^*$ and put $L(x^*)=H^*\mu$. By assumption, $L(x^*)$ is well defined. It is easy to see that L is affine since if μ, v are maximal probability measures with resultants x^* and y^* respectively, then for any $0<t<1$, $t\mu+(1-t)v$ is a maximal probability measure having resultant $tx^*+(1-t)y^*$. Thus

$$\begin{aligned}L(tx^*+(1-t)y^*)&=H^*(t\mu+(-t)v)\\&=tH^*(\mu)+(1-t)H^*(v)=tL(x^*)+(1-t)L(y^*)\,.\end{aligned}$$

If $x^* \in V^*$ and $\alpha \in \Gamma$ and μ is a maximal probability measure with resultant x^* then $\mu \circ \sigma_\alpha^{-1}$ is maximal and has resultant αx^*. Thus $L(\alpha x^*) = H^*(\mu \circ \sigma_\alpha^{-1}) = \alpha H^*(\mu) = \alpha L(x^*)$. Thus it follows that L has unique linear extension of X^* to $M_{Hb}(V^*, \mathbb{C})$.

If $\|x^*\| = 1$, let μ be a maximal probability measure with resultant x^*. Then $L(x^*) = H^*(\mu)$ implies that $\|L(x^*)\| \leqslant 1$. But, since $L(x^*)$ is an extension of x^* (i.e. $L(x^*)(\hat{x}) = x^*(x)$ for all $x \in X$), $\|L(x^*)\| \geqslant \|x^*\|$. Thus we get that L is an isometry.

If $v \in M_{Hb}(V^*, \mathbb{C})$ and $\|v\| \leqslant 1$, choose $x^* \in \operatorname{ext} V^*$ and let $\mu = |v| \circ \omega^{-1} + (1 - \||v| \circ \omega^{-1}\|/2)(\varepsilon_{x^*} + \varepsilon_{-x^*})$. By lemma 10, μ is a maximal probability measure and $H^* \mu = H^* v = v$. Thus for $x^* = r(\mu)$, $L(x^*) = v$ and L is onto.

Now note that $M_b(V^*, \mathbb{C}) = \{\mu \in M(V^*, \mathbb{C}) : \mu$ is a boundary measure$\}$ $= \{\mu \in M(V^*, \mathbb{C}) : |\mu|(V^* \setminus B(f)) = 0$ for all $f \in C(V^*, \mathbb{C})\}$ is an ideal in $M(V^*, \mathbb{C})$. In particular, it is an abstract L_1 space.

From lemma 10 it follows that H^* restricted to $M_b(V^*, \mathbb{C})$ is a contractive projection of $M_b(V^*, \mathbb{C})$ onto $M_{Hb}(V^*, \mathbb{C})$. Hence by theorem 3 of section 17, $M_{Hb}(V^*, \mathbb{C})$ is linearly isometric to an abstract L_1 space and, hence, so is X^*. ∎

We wish to investigate further complex L_1-predual spaces such that the closed unit ball contains an extreme point. By theorem 4 we may restrict ourselves to spaces of the form $A \subset C(T, \mathbb{C})$ where T is a compact Hausdorff space and A is a closed linear subspace of $C(T, \mathbb{C})$ containing the constants and separating the points of T. For such an A, we again let $K = \{x^* \in A^* : x^*(1) = 1 = \|x^*\|\}$ and use the fact that K is a weak* compact set. We also introduce a new set $Z = \operatorname{co}(K \cup -iK)$. Clearly Z is also a weak* compact comvex set and K is a face of Z. Moreover, for each $z^* \in Z \setminus (K \cup -iK)$ there are $x^*, y^* \in K$ and a unique a such that $0 < a < 1$ $z^* = a x^* + (1-a) y^*$.

Let $\theta : A \to A(Z, \mathbb{R})$ be defined by $(\theta x)(z^*) = \operatorname{Re} z^*(x)$ for all $x \in A$ and $z^* \in Z$. Then θ is an isomorphism of A onto $A(Z, \mathbb{R})$. For, by the approximation theorem for affine continuous functions, we have that the range of θ is dense in $A(Z, \mathbb{R})$. Thus, we need only show that θ is an isomorphism. Clearly $\|\theta(x)\| \leqslant \|x\|$ for all $x \in A$. On the other hand, if $x \in A$, then there is an $x^* \in \operatorname{ext}(K)$ such that $|x^*(x)| = \|x\|$ and without loss of generality, we take $\|x\| = 1$. Thus, either $|\operatorname{Re} x^*(x)| \geqslant \frac{1}{2}$ or $|\operatorname{Im} x^*(x)| \geqslant \frac{1}{2}$. Hence, either $|\theta(x)(x^*)| \geqslant \frac{1}{2}$ or $|\theta(x)(-ix^*)| \geqslant \frac{1}{2}$ and we have that $\|x\| \leqslant 2\|\theta(x)\|$. Therefore θ is an isomorphism and is onto $A(Z, \mathbb{R})$.

Let $\varphi : T \to K$ be defined as usual by $\varphi(t)(x) = x(t)$ for all $x \in A$ and recall that φ is a homeomorphic embedding and that $\varphi(T) \supset \operatorname{ext}(K)$. As in section 20 we let $L : C(T, \mathbb{C}) \to C(\Gamma \times T, \mathbb{C})$ be defined by $(Lf)(a, t) = \alpha f(t)$ for $f \in C(T, \mathbb{C})$ and $(\alpha, t) \in \Gamma \times T$ and we let $\Phi : \Gamma \times T \to V^*$ be defined by $\Phi(\alpha,t)=\alpha\varphi(t)$ for $(\alpha,t)\in\Gamma\times T$ (V^* is the closed unit ball of A^*). Recall that L is a linear isometric embedding, Φ is a homeomorphic embedding, and $\Phi(\Gamma\times T)\supset \operatorname{ext} V^*$. Moreover, by theorem 11 of section 20, if μ is a maximal measure on V^*, then $L^*(\mu\circ\Phi)$ is a boundary measure on T (with respect to A).

Lemma 11. *Let μ be a maximal probability measure on V^*, then for $\nu=L^*(\mu\circ\Phi)$ we have that $H^*(\mu)=H^*(\nu\circ\varphi^{-1})$.*

Proof. Let $f\in C(V^*,\mathbb{C})$. Then $L((Hf)\circ\varphi)(\alpha,t)=H(f)(\alpha\varphi(t))$ and, thus,

$$\begin{aligned} H^*(\nu\circ\varphi^{-1})(f) &= \int_T (Hf)\circ\varphi\, d\nu = \int_T (Hf)\circ\varphi\, dL(\mu\circ\Phi)\\ &= \int_{\Gamma\times T} L(H(f)\circ\varphi)\; d\mu\circ\Phi = \int_{\Phi(\Gamma\times T)} L(H(f)\circ\Phi^{-1})\; d\mu\\ &= \int_{V^*} H(f)\,d\mu = (H^*\mu)(f)\,. \end{aligned}$$ ∎

We now characterize when K and $-iK$ are complimentary faces of Z (see section 18).

Lemma 12. *The faces K and $-iK$ are complimentary in Z if and only if A is self-adjoint.*

Proof. Suppose K and $-iK$ are complimentary faces in Z. Let $f\in A$ and write $f=f_1+if_2$ where $f_1,f_2\in C(T,\mathbb{R})$. We define $g_1\in A(K,\mathbb{R})$ and $g_2\in A(-iK,\mathbb{R})$ by $g_1(x^*)=(\theta f)(x^*)$ and $g_2(-ix^*)=-(\theta f)(-ix^*)$ for all $x^*\in K$. Since K and $-iK$ are closed complimentary faces, there is an $h\in A$ such that $\theta h|K=g_1$ and $\theta h|(-iK)=g_2$ (see exercise 6). Thus for $t\in T$, $f_1(t)-if_2(t)=(\theta f)(\varphi(t))-i(\theta f)(-i\varphi(t))=(\theta h)(\varphi(t))+i(\theta h)(-i\varphi(t))=h(t)$. That is, $\overline{f}=h\in A$.

Now suppose that A is self-adjoint and suppose that $ax_1^*+(1-a)(-ix_2^*)=ay_1^*+(1-a)(-iy_2^*)$ for x_1^*,x_2^*,y_1^*,y_2^* in K and $0<a<1$. If $x_1^*+y_1^*$, then by the Hahn-Banach theorem there is an $f\in A$ (and we can take $f=\overline{f}$ since A is self-adjoint) such that $(\theta f)(x_1^*)\neq(\theta f)(y_1^*)$. Moreover, $a(\theta f)(x_1^*)+(1-a)(\theta f)(-ix_2^*)=a(\theta f)(y_1^*)+(1-a)(\theta f)(-iy_2^*)$. Since $(\theta f)|(-iK)=0$, we have that $a(\theta f)(x_1^*)=a(\theta f)(y_1^*)$ which is a contradiction. ∎

We are now ready to prove the characterization theorem due to Hirsberg and Lazar [293].

By $M_b(T,\mathbb{C})$ we mean the space of boundary measures on T with respect to A.

Theorem 6 (Hirsberg-Lazar). *The following are equivalent.*

(1) *A is an L_1-predual space,*

(2) $A^\perp\cap M_b(T,\mathbb{C})=\{0\}$,

(3) *$Z=\operatorname{co}(K\cup -iK)$ is a Choquet simplex,*

(4) *A is self-adjoint and $\operatorname{Re}A$ is a (real) L_1-predual space.*

Proof. (1) implies (2). Let $\mu \in A^\perp \cap M_b(T,C)$ and write $\mu = a_1\mu_1 - a_2\mu_2 + i(a_3\mu_3 - a_4\mu_4)$ where $a_i \geqslant 0$ and μ_i are probability measures on T for $i=1,2,3,4$. Let $x_i^* = r(\mu_i \circ \varphi^{-1})$ be the resultant of $\mu_i \circ \varphi^{-1}$ for $i=1,2,3,4$. Then $0 = a_1 x^* - a_2 x_2^* + i(a_3 x_3^* - a_4 x_4^*)$, that is, $a_1 x_1^* + a_4(-i x_4^*) = a_2 x_2^* + a_3(-i x_3^*) = z^*$. Since $1 \in A$, $a_1 = a_2$ and $a_3 = a_4$. Hence we may assume that $a_1 + a_4 = a_2 + a_3 = 1$. That is, $z^* \in Z$.

If $\psi: V^* \to V^*$ is defined by $\psi(x^*) = -ix^*$ for $x^* \in V^*$, then $v_1 = a_1(\mu_1 \circ \varphi^{-1}) + a_4 \mu_4 \circ \varphi^{-1} \circ \psi^{-1}$ and $v_2 = a_2(\mu_2 \circ \varphi^{-1}) + a_3(\mu_3 \circ \varphi^{-1} \circ \psi^{-1})$ are maximal probability measures representing z^*.

Since A is a complex L_1-predual space, by theorem 5 we get that $H^* v_1 = H^* v_2$. Let $f \in C(T,\mathbb{C})$ and define g on $\Phi(\Gamma \times T)$ by $g(\alpha\varphi(t)) = \alpha f(t)$ for $(\alpha,t) \in \Gamma \times T$ and assume without loss of generality that $g \in C(V^*,\mathbb{C})$ (i.e. extend g to V^* by the Tietze extension theorem). Then $H(g)(\varphi(t)) = f(t)$ for all $t \in T$ and $(H^* v_1)(g) = a_1 \int_T H(g) \circ \varphi \, d\mu_1 + a_4 \int_T H(g) \circ \psi \circ \varphi) \, d\mu_4 = a_1 \int_T f \, d\mu_1 - a_4 i \int_T f \, d\mu_4$.

Similarly, $(H v_2)(g) = a_2 \int_T f \, d\mu_2 - a_3 i \int_T f \, d\mu_3$ and, hence, $0 = a_1\mu_1(f) - a_2\mu_2(f) + i(a_3\mu_3(f) - a_4\mu_4(f)) = \mu(f)$.

That (2) implies (1) can be seen by using theorem 5.

(2) implies (3). First note that by (2) K is a Choquet simplex since it admits no nonzero real annihilating boundary measures. If A is self-adjoint, then by lemma 12, K and $-iK$ are complimentary faces in Z and it follows immediately that Z is also a Choquet simplex.

Let $f \in A$ and suppose $\overline{f} \notin A$. Then there is a $\mu \in A^\perp$ such that $\mu(\overline{f}) \neq 0$. Let $\mu = M_1 + i\mu_2$ where $\mu_i \in M(T,\mathbb{R})$ for $i=1,2$ and choose boundary measures $v_i \in M(T,\mathbb{R})$ such that $\mu_i - v_i \in A^\perp$ for $i=1,2$. If $v = v_1 + i v_2$, then $v \in A^\perp \cap M_b(T,\mathbb{C})$ and by (2), $v=0$. That is, $v_1 = v_2 = 0$. Hence $\mu_i \in A^\perp$ for $i=1,2$ which contradicts $\mu(\overline{f}) = 0$. (3) implies (2). Let $\mu \in A^\perp \cap M_b(T,\mathbb{C})$ and write $\mu = \mu_1 + i\mu_2$ with $\mu_i \in M(T,\mathbb{R})$ for $i=1,2$. Since A is self-adjoint, $\mu_i \in A^\perp$ and since K is a Choquet simplex, $\mu_i = 0$ for $i=1,2$. Hence $\mu = 0$.

We leave the equivalence of (3) and (4) to the reader. ∎

Corollary. *Let X be a complex Banach space. Then X is linearly isometric to $C(T,\mathbb{C})$ for some compact Hausdorff space T if and only if*

(1) *X is an L_1-predual space,*

(2) $\operatorname{ext} V^*$ *is weak* closed (V^* is the closed unit ball of X^*),*

(3) $\operatorname{ext} V \neq \emptyset$ *(V is the closed unit ball of X).* ∎

Exercises. 1. Let A be a linear subspace of $C(T,\mathbb{R})$ which contains constants and separates the points of T. Prove that A is an L_1-predual space if and only if $A^\perp \cap M_b(T,\mathbb{R}) = \{0\}$ where $M_b(T,\mathbb{R}) = \{\mu \in M(T,\mathbb{R}):$ μ is a boundary measure with respect to $A\}$.

2. Let A be a linear subspace of $C(T,\mathbb{C})$ which contains the constants and separates the points of T. Prove that $M_b(T,\mathbb{C})$ is an abstract

L_1 space and that A is an L_1-predual space if and only if A^* is linearly isometric to $M_b(T,\mathbb{C})$.

3. Suppose that X is a Banach space such that for each $\lambda > 1$, X^* is a $\mathscr{P}_\lambda(\mathbb{C})$ space. Prove that X^* is a $\mathscr{P}_1(\mathbb{C})$ space.

4. Prove that every separable L_1-predual space has a (monotone) Schauder basis.

5. Let S and T be compact Hausdorff spaces and suppose that $\sigma: S \to M(T,\mathbb{C})$ is a mapping such that $\sigma(s) > 0$ and $\|\sigma(s)\| \leqslant 1$ for all $s \in S$. Suppose further that σ is continuous with respect to the weak* topology on $M(T,\mathbb{C})$ and that $\mu \in M(S,\mathbb{C})$ is a positive measure. Show that if $f \geqslant 0$ is lower semicontinuous on T, then $s \to \int f\, d\sigma(s)$ is measurable on S and $\int [\int f\, d\sigma(s)]\, d\mu(s) = \int f\, dv$ where $v \in M(T,\mathbb{C})$ is defined by $\int g\, dv = \int [\int g\, d\sigma(s)]\, d\mu(s)$ for all $g \in C(T,\mathbb{C})$. What if f is Borel measurable?

6. Let K be a compact convex set and suppose that K_1 and K_2 are complementary faces in K. Let $g_1 \in A(K_1,\mathbb{C})$ and $g_2 \in A(K_2,\mathbb{C})$. Show that there is a (unique) $g \in A(K,\mathbb{C})$ such that $g|K_1 = g_1$ and $g|K_2 = g_2$.

Appendix

We list some of the theorems which we use throughout the text and prove some theorems from functional analysis whose proofs are not readily accessible.

We shall need the most general form of the Hahn-Banach theorem. Consequently, we shall state it in this manner.

Definition 1. Let X be a real vector space. A *gauge* g on X is a mapping from X to $\mathbb{R}$ such that $g(x+y) \leqslant g(x)+g(y)$ and $g(ax)=ag(x)$ for all $x, y \in X$ and $a \in \mathbb{R}^+$.

If X is a real or complex vector space, a *seminorm* on X is a mapping p from X to $\mathbb{R}^+$ such that $p(x+y) \leqslant p(x)+p(y)$ and $p(ax)=|a|\,p(x)$ for all $x, y \in X$ and all $a \in \mathbb{R}$ (or $a \in \mathbb{C}$). A *norm* is a seminorm p with the additional property that $p(x)=0$ implies $x=0$.

Theorem 1 (Hahn-Banach). *Real Form.* Let X be a real vector space, g a gauge on X, Y a linear subspace of X, and $f: Y \to \mathbb{R}$ a linear functional on X such that $f \leqslant g$ (that is $f(y) \leqslant g(y)$ for all $y \in Y$). Then there is a linear functional $f': X \to \mathbb{R}$ such that $f=f'$ on Y and $f' \leqslant g$.

Complex Form. Let X be a complex vector space, p a seminorm on X, Y a linear subspace of X and $f: Y \to \mathbb{C}$ a linear functional such that $|f| \leqslant p$. Then there is a linear functional $f': X \to \mathbb{C}$ such that $f'=f$ on Y and $|f'| \leqslant p$. ∎

We shall also make use of two basic separation theorems which are consequences of the Hahn-Banach theorem (see [91]).

Let X be a (real or complex) vector space. A set $K \subset X$ is said to be *convex* if whenever $x, y \in K$ and $0 \leqslant a \leqslant 1$, $ax+(1-a)y \in K$. If $A \subset X$ is an nonempty set, the *convex hull* of A, $\mathrm{co}(A)$, is the smallest convex set containing A. Clearly $x \in \mathrm{co}(A)$ if and only if $x = \sum_{i=1}^{n} a_i x_i$ where $a_i \geqslant 0$, $\sum_{i=1}^{n} a_i = 1$, and $x_i \in A$.

Let A and B be disjoint subsets of X. A linear functional x^* on X is said to *separate* A and B if $\mathrm{Re}(x^*(x)) \leqslant \mathrm{Re}(x^*(y))$ for all $x \in A$ and

$y \in B$. It is said to *strictly separate* A and B if $\sup\{\mathrm{Re}(x^*(x)): x \in A\} < \inf\{\mathrm{Re}(x^*(y)): y \in B\}$.

Theorem 2 (First Basic Separation Theorem). *Let X be a locally convex Hausdorff topological vector space and A and B be two disjoint convex sets in X. If A has nonempty interior, then there is a continuous linear functional x^* on X which separates A and B.* ∎

Theorem 3 (Second Basic Separation Theorem). *Let X be a locally convex Hausdorff topological vector space and suppose that A and B are disjoint closed convex sets in X. If A is compact, then there is a continuous linear functional x^* on X which strictly separates A and B.* ∎

We shall need the following theorem due to Klee [153]. The proof here is from [190].

Theorem 4 (Klee's Separation Theorem). *Let X be a locally convex Hausdorff topological vector space and suppose that $K_0, \ldots, K_m$ are nonempty open convex sets such that $K_0 \cap \cdots \cap K_m = \emptyset$. Then there is a bounded linear operator $L: X \to F^m$ (where $F = \mathbb{R}$ in the real case and $F = \mathbb{C}$ in the complex case) such that $L(K_0) \cap \cdots \cap L(K_m) = \emptyset$.*

Proof. Let $K = \{(x_0 - x_1, \ldots, x_0 - x_m): x_i \in K_i,\ i = 0, \ldots, m\}$. Then $K \subset F^m$ is an open convex set which does not contain 0. Thus by the first basic separation theorem there are linear functionals $x_1^*, \ldots, x_m^*$ in X^* such that for all $x_i \in K_i$ $(i = 0, \ldots, m)$, we have that $\sum_{i=1}^{m} \mathrm{Re}(x_i^*(x - x_i)) > 0$. Let L be defined by $L(x) = (x_1^*(x), \ldots, x_m^*(x))$. Then L has the desired properties. ∎

Recall in a vector space X, a point x of a convex set K is said to be an *extreme point* of K if whenever $x = ty + (1-t)z$ with y, z in K and $0 < t < 1$, it follows that $x = y = z$. The set of all extreme points of K is denoted by $\mathrm{ext}\, K$.

A function $f: K \to [-\infty, \infty]$ is said to be *convex* if $f(tx + (1-t)y) \leqslant tf(x) + (1-t)f(y)$ for all x, y in K and $0 \leqslant t \leqslant 1$. If $-f$ is convex, f is said to be *concave*. If h is real valued, convex and concave, it is said to be *affine*.

The following theorem is proved because it has as corollaries some important results.

Theorem 5. Let X be a locally convex Hausdorff topological vector space and K a nonempty compact convex set in X. Then for every open convex set U in K with $\mathrm{ext}\, K \subset U$, $U = K$.

Proof. Let $\mathcal{U}$ be the family of all open convex sets in K which are not equal to K. Let $U_0 \in \mathcal{U}$ and $\mathcal{C}$ be a chain in $\mathcal{U}$ with each element of $\mathscr{C}$ containing U_0. Then $U=\bigcup\{V: V\in\mathscr{C}\}$ is an open convex set in K and since $\{K\setminus V: V\in\mathscr{C}\}$ is a family of closed sets with the finite intersection property, $U\neq K$. Thus by Zorn's lemma there is a maximal element of $\mathscr{U}$ containing U_0.

Let V be a maximal element of $\mathscr{U}$. For each $x\in K$ and $0\leqslant t\leqslant 1$ let $f_{x,t}: K\to K$ be defined by $f_{x,t}(y)=ty+(1-t)x$. Then $f_{x,t}$ is continuous for all $0\leqslant t\leqslant 1$ and a homeomorphism for $0<t<1$.

Suppose $x\in V$ and $0\leqslant t<1$. Then $f_{x,t}^{-1}(V)$ is an open convex set K which contains V properly (if $t=0$, then $f_{x,0}^{-1}(V)=K$ and $0<t<1$ and $y\in\overline{V}\setminus V$, then $y\in f_{x,t}^{-1}(V)$), and by the maximality of V, it follows that $f_{x,t}^{-1}(V)=K$. Hence, $f_{x,t}(K)\subset V$. Therefore for any open convex set U in K, $V\cap U$ is convex (and open). Hence $V\cup U=V$ or $V\cup U=K$. In particular, $K\setminus V$ contains exactly one point e and $e\in\operatorname{ext}K$. ∎

Corollary 1. (Krein-Milman theorem). *Let X be a locally convex Hausdorff topological linear space and K a nonempty compact convex set in X. Then K is the closed convex hull of* $\operatorname{ext}K$.

Proof. Let L be the closed convex hull of $\operatorname{ext}K$ (that is, the closure of the convex hull of $\operatorname{ext}K$). By the second basic separation theorem, L is the intersection of all open convex sets containing L. But, if U is open in K and $U\supset L$, then $U\supset\operatorname{ext}K$. Hence $U=K$ and $L=K$. ∎

Corollary 2. (Bauer Maximum Principle). *Let X be a locally convex Hausdorff topological linear space and f a convex upper semicontinuous function from X to $[-\infty,\infty)$. Then f attains its supremum on an extreme point of K.*

Proof. Let $M=\sup\{f(x): x\in K\}$ and suppose f does not attain its supremum at a point of $\operatorname{ext}K$. Then $U=\{x\in K: f(x)<M\}$ is a convex open set containing $\operatorname{ext}K$. Hence $U=K$ which is a contradiction. ∎

Some important theorems concerning convex, concave, and affine functions are now established. Let X be a real locally convex Hausdorff topological linear space and K a nonempty compact convex set. The symbol $\boldsymbol{K}$ will denote the set of all continuous concave real valued functions on K and $A(K,\mathbb{R})$ denotes the set of all continuous affine real valued functions on K. The above setting will hold for the next few theorems.

Theorem 6. *Let f be a real valued upper semicontinuous concave function on K. Then $f=\inf\{h: h\in A(K,\mathbb{R}), f<h\}$.*

Proof. Clearly $f\leqslant\inf\{h: h\in A(K,\mathbb{R}), f<h\}$. Let $x_0\in K$ and $a_0\in R$ such that $f(x_0)<a_0$. Then $L=\{(x,a): x\in K, a\leqslant f(x)\}$ is a closed con-

vex set in $X \times \mathbb{R}$. By the second basic separation theorem there is a continuous linear functional y^* on $X \times \mathbb{R}$ and a $b \in \mathbb{R}$ such that $y^*(x,a) < b < y^*(x_0, a_0)$ for $(x,a) \in L$. Now, y^* has the form $y^*(x,a) = x^*(x) + ac$ where $c = y^*(0,1)$ and x^* is a continuous linear functional on X. Thus, $(x^*(x) + ac) < b < (x^*(x_0) + a_0 c)$ for all $(x,a) \in L$. Since Since $(x_0, f(x_0)) \in L$, it follows that $c > 0$. Now $(x, f(x)) \in L$ implies $x^*(x) + f(x)c < b$ and, hence, $f(x) < \dfrac{-1}{c}(x^*(x) - b)$. Thus for $h = \dfrac{-1}{c}(x^*(x) - b)$, $h \in A(K, \mathbb{R})$, $h > f$ and $h(x_0) < a_0$. The result follows. ∎

Theorem 7. *Let f be an affine upper semicontinuous real valued function on K. Then $\{h \in A(K,\mathbb{R}) : h > f\}$ is directed downwards and $f = \inf\{h \in A(K,\mathbb{R}) : h > f\}$.*

Proof. The fact that $f = \inf\{h \in A(K,\mathbb{R}) : h > f\}$ follows by theorem 6 above.

Let $h_1, h_2 \in A(K,R)$ with $h_1 > f$, $h_2 > f$ and let $b \in \mathbb{R}$ with $b > h_1(x)$, $h_2(x)$ for all $x \in K$. Let $V_i = \{(x,a) : h_i(x) \leqslant a \leqslant b\}$ for $i = 1, 2$ and $F = \{(x,a) : a \leqslant f(x)\}$. Then F is convex and closed in $X \times \mathbb{R}$ and $\mathrm{co}(V_1 \cup V_2)$ is disjoint from F since $V_1 \cup V_2 \subset \{(x,a) : f(x) < a\}$. Since V_1, V_2 are convex and compact, $\mathrm{co}(V_1 \cup V_2) = Y$ is compact. Hence by the second basic separation theorem there is a continuous linear functional y^* on $X \times \mathbb{R}$ and a $c \in \mathbb{R}$ such that $y^*(x,a) < c < y^*(y,b)$ for $(x,a) \in F$ and $(y,b) \in Y$. As above, $y^*(x,a) = x^*(x) + ad$ for some $d \in \mathbb{R}$ and a continuous linear functional x^* on X. Then $d > 0$ and for $h = \dfrac{-1}{d}(x^* - c)$, $h \in A(K,\mathbb{R})$ and $h > f$ with $h < h_1$, $h < h_2$. ∎

Theorem 8. *Let f be a real valued lower semi continuous function on K. Then $f = \sup\{u : u \in \boldsymbol{K}, u < f\}$.*

Proof. Clearly $f \geqslant \sup\{u : u \in \boldsymbol{K}, u < f\}$. Let $x_0 \in K$ and $a_0 \in \mathbb{R}$ such that $a_0 < f(x_0)$. Since f is lower semicontinuous there is a closed absolutely convex neighborhood V of 0 in X such that $f(x) > a_0$ for all x in $K \cap (x_0 + V)$. Let p be Minkowski support functional for V (see [91]). Then p is continuous and convex and $V = \{x \in X : p(x) \leqslant 1\}$. Choose $b < \inf\{f(x) - a_0 : x \in K\}$ and u be defined by $u(x) = (a_0 - b)(1 - p(x - x_0)) + b$. Then u is continuous and concave and $u(x_0) = a_0$. If x is in $K \cap (x_0 + V)$, then $u(x) < f(x)$. If x is not in $K \cap (x_0 + V)$, then $p(x - x_0) > 1$ and $u(x) < b \leqslant f(x)$. ∎

Theorem 9. *Let f be a real valued lower semicontinuous concave function on K. Then $\{u \in \boldsymbol{K} : u < f\}$ is directed upwards and $f = \sup\{u : u \in \boldsymbol{K}, u < f\}$.*

Proof. The second statement follows from theorem 8. Let u_1, u_2 be in K with $u_1 < f$, $u_2 < f$. Since f is bounded from below, it can be assumed that $0 \leqslant u_1(x)$, $u_2(x)$ for all $x \in K$. Let $Y = \{(x,a): a \geqslant f(x)\}$ and $V_i = \{(x,a): 0 \leqslant a \leqslant u_i(x)\}$ for $i = 1, 2$. Now Y is closed in $X \times \mathbb{R}$ and $L = (X \times \mathbb{R}) \setminus Y$ is convex. Also, V_1, V_2 are convex and compact. Hence there are open absolutely convex neighborhoods W_1, W_2 of 0 in $X \times \mathbb{R}$ such that $(W_1 + Y) \cap (W_1 + V_1) = \emptyset = (W_2 + Y) \cap (W_2 + V_2)$. Thus, if $W = W_1 \cup W_2$, then $\emptyset = [(V_1 \cup V_2) + W] \cap (Y + W)$. Moreover, $M = \mathrm{co}(V_1 \cup V_2)$ is compact and convex and disjoint from Y. Hence there is an open absolutely convex neighborhood W_0 of 0 in $X \times \mathbb{R}$ such that $(M + W_0) \cap (Y + W_0) = \emptyset$. Let u be defined by $u(x) = \sup\{a: (x,a) \in (M + W_0)\}$ for $x \in K$. Since $(x, u_i(x))$ is in $(M + W_0)$ for $x \in K$, $u \geqslant u_i$ for $i = 1, 2$. Since $\overline{M + W_0} \subset L$, $u < f$. Hence it remains to show that u is concave and continuous.

Let x, y be in K, $0 < t < 1$ and $a, b \geqslant 0$ with (x,a) and (y,b) in $(M + W_0)$. Then $u(tx + (1-t)y) \geqslant ta + (1-t)b$ since $(M + W_0)$ is convex. Hence u is concave.

Let $x_0 \in K$ with $u(x_0) > a_0$. Then $S = (M + W_0) \cap \{(x,a): a > a_0, x \in X\}$ is open and non-empty. Thus its coordinate projection S_0 in X is an open neighborhood of x_0. For x in $K \setminus S_0$, there is an $a > a_0$ such that (x,a) is in $(M + W_0)$. Thus $u(x) > a_0$ for all x in $K \setminus S_0$ and u its lower semicontinuous.

Let $x_0 \in K$ and $u(x_0) < a_0$. Then (x_0, a_0) is not in $(M + W_0)$ and there is an open absolutely convex neighborhood W' of 0 in $X \times \mathbb{R}$ such that $[(x_0, a_0) + W'] \cap (M + W_0) = \emptyset$. Without loss of generality, $W' = V' \times (-\varepsilon, \varepsilon)$ for some open absolutely convex neighborhood of 0 in X and some $\varepsilon > 0$. If for x in $(x_0 + V')$, $u(x) \geqslant a_0$, then (x, a_0) is in $(M + W_0)$ since $(x, 0)$ is.

Thus $u(x) < a_0$ for all x in $(x_0 + V')$ and u is upper semicontinuous. ∎

The following result is an immediate consequence of the above theorems.

Theorem 10 (Comparison Principle). *Let f_1, f_2 be two real valued semicontinuous functions on a compact convex set K with f_1 convex and f_2 concave. If $f_1 \leqslant f_2$ on $\mathrm{ext}\, K$, then $f_1 \leqslant f_2$. In particular, if h_1, h_2 are two affine semicontinuous functions on K and $h_1 = h_2$ on $\mathrm{ext}\, K$, then $h_1 = h_2$.* ∎

Theorem 11 (Approximation theorem for affine continuous functions). *Let X be a locally convex Hausdorff topological linear space and K a nonempty convex compact set in X. Then for each $h \in A(K, \mathbb{R})$ and $\varepsilon > 0$ three is an $x^* \in X^*$ and a $c \in \mathbb{R}$ with $\|h - (x^* + c)|K\| < \varepsilon$.*

Proof. Let $L_1 = \{(x,a): x \in K, a = h(x)\}$ and $L_2 = \{(x,a): x \in K, a = h(x) + \varepsilon\}$. Then L_1, L_2 are compact, convex, nonempty, and disjoint. In particular, $0 \notin (L_2 - L_1)$ and there is a continuous linear functional g on $X \times \mathbb{R}$ such that $g(0) = 0 < \inf g(L_2 - L_1) \leqslant \inf g(L_2) - \inf g(-L_1) = \inf g(L_2) - \sup g(L_1)$. Hence $\sup g(L_1) < \inf g(L_2)$. Choose t so that $\sup g(L_1) < t < \inf g(L_2)$ and define y^* by $g(x, y^*(x)) = t$. Then y^* is continuous and affine on X and for $x \in K$, $h(x) < y^*(x) < h(x) + \varepsilon$. ∎

Bibliography

1. Alfsen, E.: Compact Convex Sets and Boundary Integrals. Berlin-Heidelberg-New York: Springer 1971.
2. Alfsen, E.: On the decomposition of a Choquet simplex into a direct convex sum of complementary faces. Math. Scand. **17**, 169—176 (1965).
3. Alfsen, E., Effros, E. G.: Structure in real Banach spaces. Ann. of Math. **96**, 98—173 (1972).
4. Amir, D.: Continuous function spaces with the bounded extension property. Bull. Res. Counc. Israel FIO, 133—138 (1962).
5. Amir, D.: On projections and simultaneous extensions. Israel J. Math. **2**, 245—248 (1964).
6. Amir, D.: On Isomorphisms of Continuous Function Spaces. Israel J. Math. **3**, 205—210 (1969).
7. Amir, D.: Projections onto continuous function spaces. Proc. Amer. Math. Soc. **15**, 396—402 (1964).
8. Amir, D., Lindenstrauss, J.: The structure of weakly compact sets in Banach spaces. Ann. of Math. **88**, 35—46 (1968).
9. Anderson, F. W.: Approximation in systems of real-valued continuous functions. Trans. Amer. Math. Soc. **103**, 249—271 (1962).
10. Anderson, F. W., Blair, R. L.: Characterizations of the algebra of all real-valued continuous functions on a completely regular space. Illinois J. Math. **3**, 121—133 (1959).
11. Anderson, F. W., Blair, R. L.: Characterizations of certain lattices of functions. Pacific J. Math. **9**, 335—364 (1959).
12. Ando, T.: Banachverbande und positive projection. Math. Z. **109**, 121—130 (1969).
13. Ando, T.: Contractive projections in L_p spaces. Pacific J. Math. **17**, 391—405 (1966).
14. Ando, T.: Invariant Masse positiver Kontraktionen in $C(X)$. Studia Math. **31**, 173—187 (1968).
15. Ando, T.: On fundamental properties of a Banach space with a cone. Pacific J. Math. **12**, 1163—1169 (1962).
16. Arens, R. F.: Projections on Continuous Function Spaces. Duke Math. J. **32**, 469—478 (1965).
17. Arens, R. F.: Representation of *-algebras. Duke Math. J. **14**, 269—282 (1947).
18. Arens, R. F., Kelley, J. L.: Characterizations of the space of continuous functions over a compact Hausdorff space. Trans. Amer. Math. Soc. **62**, 499—508 (1947).
19. Aronszajn, N., Panitchpaki, P.: Extensions of uniformly continuous transformations and hyperconvex metric spaces. Pacific J. Math. **6**, 405—439 (1956).
20. Asimow, L., Ellis, A. J.: Facial decomposition of linearly compact simplexes and separation of functions on cones. Pacific J. Math. **34**, 301—310 (1970).
21. Bade, W. G.: The Banach Space $C(S)$. Lecture Notes. Aarhus Universitet, **26** (1971).

22. Bade, W.G.: The space of continuous functions on a compact Hausdorff space. Lecture Notes. Berkeley: University of California 1960.
23. Baker, J.W.: Compact spaces homeomorphic to a ray of ordinals. Fund. Mat. **65**, 1—9 (1972).
24. Baker, J.W.: Dispersed images of topological spaces and uncomplemented subspaces of $C(X)$. (to appear).
25. Baker, J.W.: Ordinal subspaces of topological spaces. General Top. and its Appl. **3**, 85—91 (1973).
26. Baker, J.W.: Projection constants for $C(S)$ spaces with the separable projection property. (to appear).
27. Baker, J.W.: Some uncomplemented subspaces of $C(X)$ of the type $C(Y)$. Studia Mat. **36**, 85—103 (1970).
28. Banach, S.: Théorie des Opérations Linéaires. Warsaw: Monografje Math. 1932. Reprinted New York: Chelsa 1955.
29. Bauer, H.: Konvexitat in Topologischen Vektorräumen. Lecture Notes. University of Hamburg 1963/64.
30. Bauer, H.: Silovscher rand und Dirichtletsches problem. Ann. Inst. Fourier (Grenoble) **11**, 89—136 (1961).
31. Bednar, J.B.: Facial characterizations of simplexes. J. Functional Analysis **8**, 422—431 (1971).
32. Bednar, J.B., Lacey, H.E.: Concerning Banach spaces whose duals are L spaces. Pacific J. Math. **41**, 13—24 (1972).
33. Bear, H.S.: The Silov boundary for a linear space of continuous functions. Amer. Math. Monthly **68**, 483—485 (1961).
34. Benyamini, Y., Lindenstrauss, J.: A predual of l_1 which is not isomorphic to a $C(K)$ space. Israel J. Math. **13**, 246—254 (1973).
35. Bernard, Alain: Une characterization de $C(X)$ parmi les algebras de Banach. C. R. Acad. Sci. Paris Sér. A—B **267**, A 634—635 (1968).
36. Bessaga, C., Pełczyński, A.: On bases and unconditional convergence of series of Banach spaces. Studia Math. **27**, 151—164 (1958).
37. Bessaga, C., Pełczyński, A.: On extreme points in separable conjugate spaces. Israel J. Math. **4**, 262—264 (1966).
38. Bessaga, C., Pełczyński, A.: Some remarks on conjugate spaces containing subspaces isomorphic to the space c_0. Bull. Acad. Polon. Sci. Sér. Sci. Math. Astronom. Phys. **6**, 249—250 (1958).
39. Bessaga, C., Pełczyński, A.: Spaces of continuous functions IV. Studia Math. **19**, 53—60 (1960).
40. Bernau, S.J.: A note on L_p Spaces. Math. Ann. **200**, 281—286 (1973).
41. Bernau, S.J.: A note on M spaces. J. London Math. Soc. **39**, 541—543 (1964).
42. Bernau, S.J., Lacey, H.E.: The range of a contractive projection on an L_p Space. (to appear).
43. Birkhoff, G.: Lattice Theory. Third Ed. Amer. Math. Soc. Colloq. Publ., vol. **25**. Providence, R.I.: American Mathematical Society 1967.
44. Bishop, E., De Leeuw, K.: The representation of linear functionals by measures on sets of extreme points. Ann. Inst. Fourier (Grenoble) **9**, 305—331 (1959).
45. Bishop, E., Phelps, R.R.: A proof that every Banach space is subreflexive. Bull. Amer. Math. Soc. **67**, 97—98 (1961).
46. Blumenthal, R.M., Lindenstrauss, J., Phelps, R.R.: Extreme operators in $C(K)$. Pacific J. Math. **15**, 747—756 (1965).
47. Boboc, N., Cornea, A.: Convex cones of lower semicontinuous functions on compact spaces. Rev. Roumaine Math. Pures Appl. **12**, 471—525 (1967).

48. Bohnenblust, F.: On axiomatic characterization of L_p spaces. Duke Math. J. **6**, 627—640 (1940).
49. Bohnenblust, F.: A Characterization of complex Hilbert space. Portugal. Math. **3**, 103—109 (1942).
50. Bohnenblust, F., Kakutani, S.: Concrete representation of *M* spaces. Ann. of Math. **42**, 1025—1028 (1941).
51. Bonsall, F.F.: Semialgebras of continuous functions. Proc. Internat. Sym. on Linear Spaces. Oxford: Jerusalem Pergamon, 1961.
52. Bretagnolle, J., Dacunha-Castelle, D.: Application de L'etude de Certaines Formes Linéaires Aleatoires Au Plongment d'Espaces de Banach dan les Espaces L_p. Ann. Sci. École Norm. Sup. **2**, 437—480 (1969).
53. Bretagnolle, J., Dacunha-Castelle, D., Krivine, J.: Lois stables et espaces L_p. Ann. Inst. H. Poincaré, vol. II, **3**, 231—259 (1966).
54. Braunschweiger, C.: A geometric construction of *L* spaces. Duke Math. J. **23**, 271—280 (1956).
55. Braunschweiger, C.: A geometric construction of the *M* space conjugate to an *L* space. Proc. Amer. Math. Soc. **10**, 77—82 (1959).
56. Buck, R.C.: Bounded continuous functions on a locally compact space. Michigan Math. J. **5**, 95—104 (1958).
57. Calder, J.R.: A property of l_p spaces. Proc. Amer. Math. Soc. **17**, 202—206 (1966).
58. Cambern, M.: A generalized Banach-Stone Theorem. Proc. Amer. Math. Soc. **17**, 396—400 (1966).
59. Cambern, M.: On isomorphisms with small bound. Proc. Amer. Math. Soc. **18**, 1062—1066 (1967).
60. Cambern, M.: On mappings of sequence spaces. Studia Math. **30**, 73—77 (1968).
61. Cambern, M.: Isomorphisms of $C_0(X)$ onto $C_0(X)$. Pacific J. Math.
62. Cater, S.: Algebras of bounded functions in L_p. Duke Math. J. **30**, 595—603 (1963).
63. Cherkas, Barry M.: Compactness in L_∞ spaces. Proc. Amer. Math. Soc. **25**, 347—350 (1970).
64. Choquet, G.: Lectures on Analysis. J. Marsden, T. Lance, S. Gelbart (Ed.): New York: Benjamin Inc. 1969.
65. Choquet, G.: Le théorème de représentation intégral dans les ensemble convexes compacts. Ann. Inst. Fourier (Grenoble) **10**, 333—344 (1960).
66. Choquet, G., Meyer, P.A.: Existence et unicité des représentations intégrals dan les convexes compacts quelconque. Ann. Inst. Fourier (Grenoble) **13**, 139—154 (1965).
67. Clarkson, J.A.: A characterization of *C* spaces. Ann. of Math. **48**, 845—850 (1947).
68. Collins, H.S.: Affine images of certain sets of measures. Czechoslovak Math. J. **18**, 57—65 (1968).
69. Cohen, H.B.: Injective envelopes of Banach spaces. Bull. Amer. Math. Soc. **70**, 723—726 (1964).
70. Corson, H.H.: Metrizability of compact convex sets. Trans. Amer. Math. Soc. **151**, 589—596 (1970).
71. Corson, H.H.: The weak topology of a Banach space. Trans. Amer. Math. Soc. **101**, 1—15 (1961).
72. Corson, H.H., Klee, V.: Exposed points of convex sets. Pacific J. Math. **17**, 33—43 (1966).
73. Corson, H.H., Lindenstrauss, J.: On simultaneous extensions of continuous functions. Bull. Amer. Math. Soc. **71**, 542—545 (1965).
74. Cunningham, F.: *L* structure in *L* spaces. Trans. Amer. Math. Soc. **95**, 274—299 (1960).
75. Cunningham, F.: *M* structure in *M* spaces. Proc. Cambridge Philos. Soc. **64**, 613—624 (1967).

76. Davies, E. B.: On the Banach space duals of certain spaces with the Riesz decomposition property. Quart. J. Math. Oxford Ser. **18**, 109—111 (1967).
77. Davies, E. B.: The Choquet theory and representation of ordered Banach spaces. Illinois J. Math. **13**, 176—187 (1969).
78. Davies, E. B.: The Structure and Ideal Theory of the Predual of Banach Lattice. Trans. Amer. Math. Soc. **131**, 544—555 (1968).
79. Davies, E. B., Vincent-Smith, G. F.: Tensor products, infinite products, and projective limits of Choquet simplexes. Math. Scand. **22**, 145—164 (1968).
80. Darst, R. B.: Perfect null sets in compact Hausdorff spaces. Proc. Amer. Math. Soc. **16**, 845 (1965).
81. Day, M. M.: Normed Linear Spaces. Ergebnisse Math. Berlin-Göttingen-Heidelberg: Springer 1958.
82. Dean, D. W.: Direct factors of AL spaces. Bull. Amer. Math. Soc. **71**, 368—371 (1965).
83. Dean, D. W.: Projections in certain continuous function spaces $C(H)$ and subspaces of $C(H)$ isomorphic to $C(H)$. Canad. J. Math. **14**, 385—401 (1962).
84. Dean, D. W.: Schauder decompositions in m. Proc. Amer. Math. Soc. **18**, 619—623 (1967).
85. Dean, D. W.: Subspaces of $C(H)$ which are direct factors of $C(H)$. Proc. Amer. Math. Soc. **16**, 237—242 (1965).
86. Dieudonné, J.: Sur les espaces L_1. Arch. Math. **10**, 151—152 (1959).
87. Ditor, S.: Linear operators of averaging and extension. Thesis. Berkeley: University of California 1968.
88. Ditor, S.: On a lemma of Milutin concerning averaging operators in continuous function spaces. Trans. Amer. Math. Soc. **149**, 443—452 (1970).
89. Dixmeir, J.: Sur certains espaces considérés par M. H. Stone. Summa Brasil. Math. **2**, 151—182 (1951).
90. Douglas, R. G.: Contractive projections on an L_1 space. Pacific J. Math. **15**, 443—462 (1965).
91. Dunford, N., Schwartz, J.: Linear Operators I. New York: Interscience 1958.
92. Dugundji, J.: An extension of Tietze's theorem. Pacific J. Math. **1**, 353—367 (1951).
93. Edwards, D. A.: Minimum stable wedges of semicontinuous functions. Math. Scand. **19**, 15—26 (1966).
94. Edwards, D. A.: Séparation des functions rélles définies sur un simplexe de Choquet. C. R. Acad. Sci. Paris Sér. A—B **261**, 2798—2800 (1965).
95. Edwards, D. A., Vincent-Smith, G.: A Weierstrass-Stone theorem for Choquet simplexes. Ann. Inst. Fourier (Grenoble) **18**, 261—282 (1968).
96. Edwards, R. E.: Fourier Series Vol. II. New York: Holt, Rinehart and Winston 1967.
97. Effros, E. G.: On a class of real Banach spaces. Israel J. Math. **9**, 430—458 (1971).
98. Effros, E. G.: Structure in simplex spaces. Acta Math. **117**, 103—121 (1967).
99. Eilenberg, S.: Banach spaces methods in topology. Ann. of Math. **43**, 568—579 (1942).
100. Ellis, A. J.: Extreme positive operators. Quart. J. Math. Oxford Ser. **15**, 342—344 (1964).
101. Ellis, A. J.: An order theoretic proof of D. A. Edwards' separation theorem for simplexes. J. London Math. Soc. **3**, 475—476 (1971).
102. Ellis, A. J.: The duality of partially ordered Banach spaces. J. London Math. Soc. **39**, 730—744 (1964).
103. Ellis, H. W., Snow, D. O.: On $(L^1)^*$ for general measure spaces. Canad. Math. Bull. **6**, 211—229 (1963).
104. Engleking, R.: Outlines of General Topology. New York: John Wiley 1968.
105. Fakhoury, H.: Espaces fortement reticules de functions affines. Séminaire Choquet 9e annee **7**, (1969/70) (mimeographed pp. 40).

106. Fakhoury, H.: Preduax de L espaces; proprites der G spaces et des C_σ espaces. C.R. Acad. Sci. Paris Sér. A **271**, 941—944 (1970).
107. Fakhoury, H.: Une characterisation des L espaces duax. Bull. Sci. Math. **96**, 129—144 (1972).
108. Fullerton, R.E.: A characterization of L spaces. Fund. Math. **38**, 127—136 (1951).
109. Fakhoury, H.: Geometrical characterizations of certain function spaces. Proc. Internat. Sympos. Linear Spaces (Jerusalem, 1960). Oxford: Jerusalem Pergamon 1961.
110. Geba, K., Semadeni, Z.: On the M subspace of the Banach space of continuous functions. Zeszyty Naukowe Universytetu Im. A. Mickiewicz Matematyka, Fizyka, Chenira, Zeszyt **2**, 54—68 (1960).
111. Gelfand, I.M., Raikov, D.A., Silov, G.E.: Commutative Normed Rings. New York: Chelsea 1964.
112. Gleason, A.: Projective topological spaces. Illinois J. Math. **2**, 482—489 (1958).
113. Gleit, A.: On the existence of simplex spaces. Israel J. Math. **9**, 199—209 (1971).
114. Goodner, D.B.: Projections in normed linear spaces. Trans. Amer. Math. Soc. **69**, 89—108 (1950).
115. Goodner, D.B.: Separable spaces with the extension property. J. London Math. Soc. **35**, 239—240 (1960).
116. Gordon, Hugh.: Measures defined by abstract L_p spaces. Pacific J. Math. **10**, 557—562 (1960).
117. Gorin, E.A.: Characterization of the ring of all continuous functions on a bicompactum. Dokl. Akad. Nauk/Tadžik. SSR **42**, 781—784 (1962).
118. Grossberg, J., Krein, M.: Sur la decomposition des fonctionnelles en composantes positives. C.R. (Doklady) Acad. Sci. USSR (N.S.) **24**, 723—726 (1939).
119. Grothendieck, A.: Sur les applications linéaires faiblement compactes d'espaces du type $C(K)$. Canad. J. Math. **5**, 129—173 (1953).
120. Grothendieck, A.: Une caractérisation vectorille-métrique des espaces L^1. Canad. J. Math. **7**, 552—561 (1955).
121. Grunbaum, B.: Projection Constants. Trans. Amer. Math. Soc. **95**, 451—465 (1960).
122. Gurarii, V.I.: Spaces of universal disposition, isotropic spaces and the Magur problem on rotations of Banach spaces. Sibirsk. Mat. Z. **7**, 1002—1013 (1966).
123. Guseman, L.F.: Spaces of affine continuous functions. Thesis. Austin, Texas: University of Texas 1968.
124. Hardy, J., Lacey, H.E.: Extension of Regular Borel Measures. Pacific J. Math. **24**, 277—282 (1968).
125. Hardy, J., Lacey, H.E.: Notes on perfectness and total disconnectedness. Amer. Math. Monthly **75**, 602—606 (1968).
126. Hagler, James: Some More Banach Spaces which Contain l^1. Studia Math. **46**, 35—42 (1973).
127. Hagler, James, Stegall, C.: Banach spaces whose duals contain complemented subspaces isomorphic to $C[0,1]^*$. Preprint. Binghampton: State University of New York 1972. J. Functional Analysis **13**, 233—251 (1973).
128. Hasumi, M.: The extension property of complex Banach spaces. Tôhoku Math. J. **10**, 135—142 (1958).
129. Hebert, D.J., Lacey, H.E.: On supports of regular Borel measures. Pacific J. Math. **27**, 101—118 (1968).
130. Hewitt, E., Ross, K.: Abstract Harmonic Analysis I. Berlin-Göttingen-Heidelberg: Springer 1963.
131. Hewitt, E., Stromberg, K.: Real and Abstract Analysis. Berlin-Heidelberg-New York: Springer 1965.
132. Honda, H., Yamamuro, S.: A characteristic property of L_p spaces ($p>1$). Proc. Japan Acad. **35**, 446—448 (1959).

133. Honda, H.: A characteristic property of L_p spaces ($p>1$) II. Proc. Japan Acad. **39**, 348—351 (1963).
134. Holsztyński, W.: Continuous mappings induced by isometries of spaces of continuous functions. Studia Math. **26**, 133—136 (1966).
135. Isbell, J. R.: Injective envelopes of Banach spaces are rigidly attached. Bull. Amer. Math. Soc. **70**, 727—729 (1964).
136. Isbell, J. R.: Three Remarks on Injective Envelopes of Banach Spaces. J. Math. Anal. Appl. **27**, 516—518 (1969).
137. Isbell, J. R., Semadeni, Z.: Projections constants and spaces of continuous functions. Trans. Amer. Math. Soc. **107**, 38—48 (1963).
138. James, R. C.: Projections in the space (M). Proc. Amer. Math. Soc. **6**, 899—902 (1955).
139. James, R. C.: Weakly compact sets. Trans. Amer. Math. Soc. **13**, 129—140 (1964).
140. Jameson, G. J. O.: Topological M spaces. Math. Z. **103**, 139—150 (1968).
141. Jellett, F.: Homomorphisms and inverse limits of Choquet simplexes. Math. Z. **103**, 219—226 (1968).
142. Jellet, F.: On the direct sum decomposition of the affine space of a Choquet simplex. Quart J. Math. Oxford Ser. **18**, 233—237 (1967).
143. Jerison, M.: Characterizations of certain spaces of continuous functions. Trans. Amer. Math. Soc. **70**, 103—113 (1951).
144. Kadec, M. I.: On linear dimension of the spaces L_p. Uspehi Mat. Nauk, **13**, 95—98 (1958).
145. Kadec, M., Pełczyński, A.: Bases, Lacunary sequences and complemented subspaces in the spaces L_p. Studia Math. **21**, 161—176 (1962).
146. Kadison, R. V.: Representation theory for commutative topological algebras. Mem. Amer. Math. Soc. **7**, 39 (1951).
147. Kakutani, S.: Concrete representation of abstract L spaces and the mean ergodic theorem. Ann. of Math. **42**, 523—537 (1941).
148. Kakutani, S.: Concrete representations of abstract M spaces. Ann. of Math. **42**, 994—1024 (1941).
149. Kakutani, S.: Some characterizations of Euclidean space. Japan J. Math. **16**, 93—97 (1939).
150. Kaplansky, I.: Lattices of continuous functions. Bull. Amer. Math. Soc. **53**, 617—623 (1947).
151. Kaufman, R.: A type of extension of Banach spaces. Acta Sci. Math. (Szeged) **27**, 163—166 (1964).
152. Kelley, J. L.: Banach spaces with the extension property. Trans. Amer. Math. Soc. **72**, 323—326 (1952).
153. Klee, V.: On certain intersection properties of convex sets. Canad. J. Math. **3**, 272—275 (1951).
154. Knowles, J. D.: Measures on Topological Spaces. Proc. London Math. Soc. **17**, 139—156 (1967).
155. Knowles, J. D.: On the existence of non atomic measures. Mathematika **14**, 62—67 (1967).
156. Köethe, G.: Hebbare lokalkonvexe räume. Math. Ann. **165**, 181—195 (1966).
157. Krein, M.: Sur la décomposition minimale d'une fonctionnelle linéaire en composantes positive. C. R. (Doklady) Acad. Sci. URSS (N. S.) **28**, 18—22 (1940).
158. Kuratowski, K.: Topology I. New York: Academic Press 1966.
159. Kuratowski, K.: Topology II. New York: Academic Press 1971.
160. Lacey, H. E.: A note concerning $A^* = L_1(\mu)$. Proc. Amer. Math. Soc. **29**, 525—528 (1971).
161. Lacey, H. E.: On the classification of Lindenstrauss spaces. J. Math. **47**, 139—145 (1973).

162. Lacey, H. E.: Separable quotients of Banach spaces. An. Acad. Brasil. Ci. **44**, 185—189 (1972).
163. Lacey, H. E., Bernau, S. J.: Characterizations and classifications of some classical Banach spaces. **12**, 1—35 (1974).
164. Lacey, H. E., Cohen, H. B.: On injective envelopes of Banach spaces. J. Functional Analysis **4**, 11—30 (1969).
165. Lacey, H. E., McPherson, R. V.: Classes of Banach spaces whose duals are abstract L spaces. J. Functional Analysis **11**, 425—435 (1972).
166. Lacey, H. E., Morris, P. D.: Continuous linear operators on spaces of continuous functions. Proc. Amer. Math. Soc. **17**, 848—853 (1966).
167. Lacey, H. E., Morris, P. D.: On spaces of type $A(K)$ and their duals. Proc. Amer. Math. Soc. **23**, 151—157 (1969).
168. Lacey, H. E., Morris, P. D.: On universal spaces of the type $C(X)$. Proc. Amer. Math. Soc. **19**, 350—353 (1968).
169. Lacey, H. E., Withley, R. J.: Conditions under which all the bounded linear operators are compact. Math. Ann. **158**, 1—5 (1965).
170. Lazar, A. J.: Affine functions on simplexes and extreme operators. Israel J. Math. **5**, 31—43 (1967).
171. Lazar, A. J.: Affine products of simplexes. Math. Scand. **22**, 165—175 (1968).
172. Lazar, A. J.: Spaces of affine continuous functions on simplexes. Trans. Amer. Math. Soc. **134**, 503—525 (1968).
173. Lazar, A. J.: The unit ball in conjugate L_1 spaces. Duke Math. J. **36**, 1—8 (1972).
174. Lazar, A. J.: Polyhedral Banach spaces and extensions compact operators. Israel J. Math. **7**, 357—364 (1969).
175. Lazar, A. J., Lindenstrauss, J.: Banach spaces whose duals are L_1 spaces and their representing matrices. Acta Math. **126**, 165—194 (1971).
176. Lazar, A. J., Lindenstrauss, J.: On Banach spaces whose duals are L_1 spaces. Israel J. Math. **4**, 205—207 (1966).
177. Lamperti, J.: On the isometries of certain function spaces. Pacific J. Math. **8**, 459—466 (1958).
178. Léger, C.: Convexes compacts et leurs point extrémaux. C. R. Acad. Sci. Paris Sér. A—B **267**, 92—93 (1968).
179. Lewis, D. R.: Spaces on which each absolutely summing map is nuclear. Proc. Amer. Math. Soc. **31**, 195—198 (1972).
180. Lewis, D. R., Stegal, C.: Banach spaces whose duals are isomorphic to $l_1(\Gamma)$. J. Functional Analysis **12**, 177—187 (1973).
181. Lindenstrauss, J.: A remark on $\mathscr{L}_1$ spaces. Israel J. Math. **8**, 80—82 (1970).
182. Lindenstrauss, J.: Extensions of compact operators. Mem. Amer. Math. Soc. **48** (1964).
183. Lindenstrauss, J.: Geometric theory of Banach spaces. Lecture Series. Florida State University, July 1969.
184. Lindenstrauss, J.: On a certain subspace of l_1. Bull. Acad. Polon. Sci. Ser. Sci. Math. Astronom. Phys. **12**, 539—542 (1964).
185. Lindenstrauss, J.: On some subspace of l_1 and c_0. Bull. Res. Counc. Israel, **10F**, 74—80 (1961).
186. Lindenstrauss, J.: On complemented subspaces of m. Israel J. Math. **5**, 153—156 (1967).
187. Lindenstrauss, J., Pełczyński, A.: Contributions to the theory of classical Banach spaces. J. Functional Analysis **8**, 225—249 (1971).
188. Lindenstrauss, J., Pełczyński, A.: Absolutely summing operators in L_p spaces and their applications. Studia Math. **29**, 275—326 (1968).
189. Lindenstrauss, J., Rosenthal, H. P.: Automorphisms in c_0, l_1, and m. Israel J. Math.

190. Lindenstrauss, J., Rosenthal, H.P.: The $\mathcal{L}_p$ spaces. Israel J. Math. **7**, 325—349 (1969).
191. Lindenstrauss, J., Phelps, R.R.: Extreme Point Properties of Convex Bodies in Reflexive Banach Spaces. Israel J. Math. **6**, 39—48 (1968).
192. Lindenstrauss, J., Tzafriri, L.: On the complemented subspace problem. Israel J. Math. **9**, 263—269 (1971).
193. Lindenstrauss, J., Wulbert, D.E.: On the classification of the Banach spaces whose duals are L_1 spaces. J. Functional Analysis **4**, 332—349 (1969).
194. Lindenstrauss, J., Zippin, M.: Banach spaces with a unique unconditional basis. J. Functional Analysis **3**, 115—125 (1969).
195. Lloyd, S.P.: On certain projections in spaces of continuous functions. Pacific J. Math. **13**, 171—175 (1963).
196. Lloyd, S.P.: On extreme averaging operators. Proc. Amer. Math. Soc. **14**, 305—310 (1963).
197. Luxemberg, W.A.J., Zannen, A.C.: Riesz Spaces. New York: American Elsevier 1971.
198. MacDowell, R.: Banach spaces and algebras of continuous functions. Proc. Amer. Math. Soc. **6**, 67—68 (1955).
199. Maharam, D.: On homogeneous measure algebras. Proc. Nat. Acad. Sci. **28**, 108—111 (1942).
200. Marti, J.T.: Introduction to the theory of bases. Berlin-Heidelberg-New York: Springer 1969.
201. Marti, J.T.: Topological representation of abstract L_p spaces. Math. Ann. **185**, 315—321 (1970).
202. Masurkiewicz, S., Sierpinski, W.: Contribution à la topologie des ensembles dénombrables. Fund. Math. **1**, 17—27 (1920).
203. McWilliams, R.D.: On projections of separable subspaces of m onto c_0. Proc. Amer. Math. Soc. **10**, 872—876 (1959).
204. Meyers, S.B.: Banach spaces of continuous functions. Ann. of Math. **49**, 132—140 (1948).
205. Michael, E., Pełczyński, A.: A linear extension theorem. Illinois J. Math. **11**, 563—579 (1967).
206. Michael, E., Pełczyński, A.: Peaked partitions subspaces of $C(X)$. Illinois J. Math. **11**, 555—562 (1967).
207. Michael, E., Pełczyński, A.: Separable Banach spaces which admit $l_\infty(n)$ approximations. Israel J. Math. **4**, 189—198 (1966).
208. Miliutin, A.A.: On spaces of continuous functions. Dissertation. Moscow State University 1952.
209. Miliutin, A.A.: Isomorphisms of the spaces of continuous functions over compact sets of cardinality of the continuum. Teor. Funkciĭ Funkcional. Anal. i Priložen. **2**, 150—156 (1966).
210. Mokobodzki, G.: Balayage défini par un cône convexe de fonctions numériques sur un espace compact. C. R. Acad. Sci. Paris Sér. A—B **254**, 803—805 (1962).
211. Morris, P.D.: Spaces of Continuous functions on dispersed spaces. Thesis. Austin: University of Texas 1966.
212. Morris, P.D., Phelps, R.R.: Theorems of Krein Milman type for certain convex sets of operators. Trans. Amer. Math. Soc. **150**, 183—200 (1970).
213. McPherson, R.V.: Classes of Banach spaces whose duals are L spaces. Thesis. Austin: University of Texas 1971.
214. Nachbin, Leopoldo: A characterization of the normed vector ordered spaces of continuous functions over a compact space. Amer. J. Math. **71**, 701—705 (1949).
215. Nachbin, Leopoldo: A theorem of the Hahn Banach type for linear transformations. Trans. Amer. Math. Soc. **68**, 28—46 (1950).

216. Nachbin, Leopoldo: On the Hahn Banach theorems. An. Acad. Brasil. Ci. **21**, 151—154 (1949).
217. Nakano, H.: Über normierte teilweise gerardte Modular. Proc. Imp. Acad. Tokoyo **17**, 301—307 (1941). (printed In: Semiordered linear spaces. Tokoyo 1955).
218. Nakano, H.: Riesz-Fischerecher Satz in normierten teilweise geordneten Modul. Proc. Imp. Acad. Tokoyo **180**, 350—353 (1942).
219. Namoika, I.: On certain actions of semigroups on L spaces. Studia Math. **29**, 63—77 (1963).
220. Namoika, L.: Partially ordered linear topological spaces. Mem. Amer. Math. Soc. **24** (1957).
221. Namoika, L., Phelps, R. R.: Tensor products of compact convex sets. Pacific J. Math. **31**, 469—480 (1969).
222. Nirenberg, R., Pamzone, P.: On the spaces L^1 which are isomorphic to a B^*. Rev. Un. Mat. Argentina **21**, 119—130 (1963).
223. Ng, Kung-Fu: A representation theorem for partially ordered Banach algebras. Proc. Cambridge Philos. Soc. **64**, 53—59 (1968).
224. Ng, Kung-Fu: A note on partially ordered Banach spaces. J. London Math. Soc. **2**, 520—524 (1969).
225. Ng, Kung-Fu: The duality of partially ordered Banach spaces. Proc. London Math. Soc. **19**, 269—288 (1969).
226. Olson, M. P.: The order continuous operators on L_p spaces I. Trans. Amer. Math. Soc. **133**, 477—504 (1968).
227. Peck, N. T.: Representation of functions in $C(X)$ by means of extreme points. Proc. Amer. Math. Soc. **18**, 133—135 (1967).
228. Pełczyński, A.: Linear extensions, linear averagings, and their applications to linear topological classification of spaces of continuous functions. Dissertationes Math. Rozprawy Mat. **58**, 1968.
229. Pełczyński, A.: On $C(S)$-subspaces of separable Banach spaces. Studia Math. **31**, 513—522 (1968).
230. Pełczyński, A.: On strictly singular and strictly cosingular operators I and II. Bull. Acad. Polon. Sci. Sér. Sci. Math. Astronom. Phys. **13**, 31—36, 37—41 (1965).
231. Pełczyński, A.: On the universality of certain Banach spaces. Vestnik Leningrad. Univ. **17**, 22—29 (1962).
232. Pełczyński, A.: On Banach spaces containing $L_1(\mu)$. Studia Math. **30**, 231—246 (1968).
233. Pełczyński, A.: Projections in certain Banach spaces. Studia Math. **19**, 209—228 (1960).
234. Pełczyński, A.: On the impossibility of embedding the space L in certain Banach spaces. Colloq. Math. **8**, 199—203 (1961).
235. Pełczyński, A., Semadeni, Z.: Spaces of continuous functions III. Studia Math. **18**, 211—222 (1959).
236. Phelps, R. R.: A Banach Space Characterization of Purely Atomic Measure Spaces. Proc. Amer. Math. Soc. **12**, 447—452 (1961).
237. Phelps, R. R.: Lectures on Choquet's theorem. Math. Studies. Princeton: Van Nostrand 1966.
238. Phelps, R. R.: Uniqueness of Hahn Banach extensions and unique best approximations. Trans. Amer. Math. Soc. **95**, 238—255 (1960).
239. Phillips, R. S.: A characterization of Euclidean space. Bull. Amer. Math. Soc. **46**, 930—933 (1940).
240. Pitt, H. R.: A note on bilinear forms. J. London Math. Soc. **11**, 174—180 (1936).
241. Peressini, A. L.: Ordered Topological Vector Spaces. New York: Harper and Row 1967.
242. Rainwater, J.: A note on projective resolutions. Proc. Amer. Math. Soc. **10**, 734—735 (1959).

243. Rickart, C.: General Theory of Banach Algebras. Princeton: Van Nostrand 1960.
244. Riesz, F.: Sur la décomposition des opérations fonctionelles linéaires. Atti. del congresso Bolonga **3**, 143—148 (1928).
245. Rosenthal, H.P.: On injective Banach spaces and the space $L_\infty(\mu)$ for finite measures. Acta Math. **124**, 205—248 (1970).
246. Rosenthal, H.P.: On relatively disjoint families of measures, with some applications to Banach space theory. Studia Math. **37**, 13—36 (1970).
247. Rosenthal, H.P.: On quasicomplemented subspaces of Banach spaces with an appendix on compactness of operators from $L_p(\ )$ to $L_r(\nu)$. J. Functional Analysis **4**, 176—214 (1969).
248. Rosenthal, H.P.: On factors of $C[0,1]$ with non-separable dual. Israel J. Math. **13**, 361—378 (1972).
249. Rosenthal, H.P.: On subspaces of L_p. Ann. of Math. **97**, 344—373 (1973).
250. Rogalski, M.: Operateurs de Lion, projecteurs boreliens et simplexes analytiques. J. Functional Analysis **2**, 458—488 (1968).
251. Rogasinski, W.W.: Continuous linear functionals on subspaces of L_p and C. Proc. Amer. Math. Soc. **6**, 175—190 (1956).
252. Royden, H.L.: On a paper of Rogasinski. J. London Math. Soc. **35**, 225—228 (1960).
253. Royden, H.L.: Real Analysis. New York: Macmillan 1968.
254. Schwarz, J.: A note on the space L_p^*. Proc. Amer. Math. Soc. **2**, 270—275 (1951).
255. Seever, G.L.: Non-negative projections on $C_0(X)$. Pacific J. Math. **17**, 159—166 (1968).
256. Semadeni, Z.: Banach spaces of continuous functions. Warsaw: PWN 1971.
257. Semadeni, Z.: Free compact convex sets. Bull. Acad. Polon. Sci. Sér. Sci. Math. Astronom. Phys. **13**, 141—146 (1965).
258. Semadeni, Z.: Isomorphic properties of Banach spaces of continuous functions. Studia Math. (ser specjalne) Zeszyt **1**, 93—102 (1963).
259. Semadeni, Z.: Selected topics related to Banach spaces of continuous functions. Lecture notes. University of Washington 1962.
260. Singer, I.: Bases in Banach spaces I. Berlin-Heidelberg-New York: Springer 1970.
261. Sine, Robert: On a paper of Phelps. Proc. Amer. Math. Soc. **18**, 484—486 (1967).
262. Shaeffer, H.H.: Topological Vector spaces. New York: Macmillan 1966.
263. Sobcyzk, A.: Projections of the space m onto its subspace c_0. Bull. Amer. Math. Soc. **47**, 938—947 (1941).
264. Stegall, C.: Characterizations of Banach spaces whose duals are L_1 spaces. Israel J. Math. **11**, 299—308 (1972).
265. Stegall, C.: Banach Spaces Whose Duals Contain $l_1(\Gamma)$ with Applications to the Study of Dual $L_1(\mu)$ Spaces. Trans. Amer. Math. Soc. **176**, 463—478 (1973).
266. Stegall, C., Reutherford, J.R.: "Fully nuclear and completely nuclear operators with applications to $\mathscr{L}_1$ and $\mathscr{L}_\infty$ spaces". Trans. Amer. Math. Soc. **163**, 457—492 (1972).
267. Stein, J.: Extreme points in Function Algebras. Curtis Seminar Notes, pp. 8. Berkeley 1968.
268. Sullivan, F.E.: Norm characterizations of real L_p spaces. Bull. Amer. Math. Soc. **74**, 153—154 (1968).
269. Thorp, E.O., Whitley, R.J.: Operator representation theorems. Illinois J. Math. **9**, 595—601 (1965).
270. Tong, H.: Some characterizations of normal and perfectly normal spaces.
271. Tzafriri, L.: An isomorphic characterization of L_p and c_0 spaces. Studia Math. **32**, 295—304 (1969).
272. Tzafriri, L.: Remarks on contractive projections in L_p spaces. Israel J. Math. **7**, 9—15 (1969).
273. Tzafriri, L.: An isomorphic characterization of L_p and c_0-spaces II. Michigan Math. J. **18**, 21—31 (1971).

274. Veech, W. A.: Short proof of Sobczyk's theorem. Proc. Amer. Math. Soc. **28**, 627—728 (1971).
275. Whitley, R. J.: Strictly singular operators and their conjugates. Trans. Amer. Math. Soc. **113**, 252—261 (1964).
276. Wulbert, D. E.: A note on the characterization of conditional expectation operators. Pacific J. Math. **24**, 285—288 (1970).
277. Wulbert, D. E.: Averaging projections. Illinois J. Math. **13**, 689—693 (1969).
278. Wulbert, D. E.: Some complemented function spaces in $C(X)$. Pacific J. Math. **24**, 589—602 (1968).
279. Wojtaszcyzk, P.: Existence of some special bases in Banach spaces. Studia Math. **47**, 83—93 (1973).
280. Wojtaszcyzk, P.: Some remarks on the Gurarii space. Studia Math. **41**, 207—210 (1972).
281. Zippin, M.: $\mathscr{L}_\infty$ subspaces of c_0.
282. Zippin, M.: On some subspaces of Banach spaces whose duals are L_1 spaces. Proc. Amer. Math. Soc. **23**, 378—385 (1969).
283. Zippin, M.: On perfectly homogeneous bases in Banach spaces. Israel J. Math. **4**, 265—272 (1966).
284. Zippin, M.: On bases in Banach spaces. Thesis. Jerusalem: Hebrew University 1968.

Supplemental Bibliography

285. Asimow, L.: Decomposable compact convex sets and peak sets for function spaces. Proc. Amer. Math. Soc. **25**, 75—79 (1970).
286. Bernau, S. J.: Theorem of Korovkin Type for L_p-Spaces. (to appear in Pacific J. Math.).
287. Bourbaki, N.: Integration Ch. V. Paris: Hermann 1967.
288. Dieudonne, J.: Foundations of Modern Analysis. New York: Academic Press 1960.
289. Effros, E. G.: On a class of complex Banach spaces. (to appear in Illinois J. Math.).
290. Enflo, Per.: A counterexample to the approximation problem in Banach spaces. Acta Math. **130**, 309—317 (1973).
291. Fuhr, R., Phelps, R. R.: Uniqueness of complex representing measures on the Choquet boundary. J. Functional Analysis **14**, 127 (1973).
292. Hirsberg, B.: Représentations intégrales des forms linéaires complexes. C. R. Acad. Sci. Paris Sér. A **274**, 1222—1224 (1972).
293. Hirsberg, B., Lazar, A. J.: Complex Lindenstrauss spaces with extreme points. (to appear).
294. Hustad, O.: A norm preserving complex Choquet theorem. Math. Scand. **29**, 272—278 (1971).
295. Johnson, W. B., Rosenthal, H. P.: On w^*-basic sequences and their applications to the study of Banach spaces. Studia Math. **43**, 77—92 (1972).
296. Johnson, W. B., Rosenthal, H. P., Zippin, M.: On bases, finite dimensional decompositions and weaker structures in Banach spaces. Israel J. Math. **9**, 488—506 (1971).
297. Lindenstrauss, J., Tzafriri: Classical Banach Spaces. Springer-Verlag Lecture Notes 338 (1973).
298. Ng, Kung-Fu: L_p-Conditions in Partially Ordered Banach Spaces. J. London Math. Soc. **5**, 387—394 (1972).
299. Paley, R. E. A. C.: Some theorems on Abstract Spaces. Bull. Amer. Math. Soc. **42**, 235—240 (1936).
300. Phillips, R. S.: On linear transformations. Trans. Amer. Math. Soc. **48**, 516—541 (1940).
301. Saki, S.: C^*-algebras and w^*-algebras. Berlin-Heidelberg-New York: Springer 1971.

302. Samuel, C.: Sur certains espaces $C_\sigma(S)$ et sur les sous-espaces complementes de $C(S)$. Bull. Sci. Math. **95**, 65—82 (1971).

303. Sikorski, R.: Boolean Algebras. Berlin-Göttingen-Heidelberg-New York, Springer 1964.

304. Singer, I.: Bases in Banach Spaces I. Berlin-Heidelberg-New York: Springer 1970.

305. Stone, M. H.: Applications of the theory of Boolean rings to general topology. Trans. Amer. Math. Soc. **41**, 375—481 (1937).

306. Zippin, M.: On some subspaces of Banach spaces whose duals are L_1 spaces. Proc. Amer. Math. Soc. **23**, 378—385 (1967).

Subject Index

Die Grundlehren der mathematischen Wissenschaften in Einzeldarstellungen mit besonderer Berücksichtigung der Anwendungsgebiete

Eine Auswahl

---

23. Pasch: Vorlesungen über neuere Geometrie. In Vorbereitung
41. Steinitz: Vorlesungen über die Theorie der Polyeder. In Vorbereitung
45. Alexandroff/Hopf: Topologie. Band 1
46. Nevanlinna: Eindeutige analytische Funktionen
63. Eichler: Quadratische Formen und orthogonale Gruppen
102. Nevanlinna/Nevanlinna: Absolute Analysis
114. Mac Lane: Homology
127. Hermes: Enumerability, Decidability, Computability
131. Hirzebruch: Topological Methods in Algebraic Geometry
135. Handbook for Automatic Computation. Vol. 1/Part a: Rutishauser: Description of ALGOL 60
136. Greub: Multilinear Algebra
137. Handbook for Automatic Computation. Vol. 1/Part b: Grau/Hill/Langmaack: Translation of ALGOL 60
138. Hahn: Stability of Motion
139. Mathematische Hilfsmittel des Ingenieurs. 1. Teil
140. Mathematische Hilfsmittel des Ingenieurs. 2. Teil
141. Mathematische Hilfsmittel des Ingenieurs. 3. Teil
142. Mathematische Hilfsmittel des Ingenieurs. 4. Teil
143. Schur/Grunsky: Vorlesungen über Invariantentheorie
144. Weil: Basic Number Theory
145. Butzer/Berens: Semi-Groups of Operators and Approximation
146. Treves: Locally Convex Spaces and Linear Partial Differential Equations
147. Lamotke: Semisimpliziale algebraische Topologie
148. Chandrasekharan: Introduction to Analytic Number Theory
149. Sario/Oikawa: Capacity Functions
150. Iosifescu/Theodorescu: Random Processes and Learning
151. Mandl: Analytical Treatment of One-dimensional Markov Processes
152. Hewitt/Ross: Abstract Harmonic Analysis. Vol. 2: Structure and Analysis for Compact Groups. Analysis on Locally Compact Abelian Groups
153. Federer: Geometric Measure Theory
154. Singer: Bases in Banach Spaces I
155. Müller: Foundations of the Mathematical Theory of Electromagnetic Waves
156. van der Waerden: Mathematical Statistics
157. Prohorov/Rozanov: Probability Theory
158. Constantinescu/Cornea: Potential Theory on Harmonic Spaces
159. Köthe: Topological Vector Spaces I
160. Agrest/Maksimov: Theory of Incomplete Cylindrical Functions and their Applications
161. Bhatia/Szegö: Stability Theory of Dynamical Systems
162. Nevanlinna: Analytic Functions
163. Stoer/Witzgall: Convexity and Optimization in Finite Dimensions I
164. Sario/Nakai: Classification Theory of Riemann Surfaces
165. Mitrinović/Vasić: Analytic Inequalities

166. Grothendieck/Dieudonné: Eléments de Géométrie Algébrique I
167. Chandrasekharan: Arithmetical Functions
168. Palamodov: Linear Differential Operators with Constant Coefficients
169. Rademacher: Topics in Analytic Number Theory
170. Lions: Optimal Control of Systems Governed by Partial Differential Equations
171. Singer: Best Approximation in Normed Linear Spaces by Elements of Linear Subspaces
172. Bühlmann: Mathematical Methods in Risk Theory
173. Maeda/Maeda: Theory of Symmetric Lattices
174. Stiefel/Scheifele: Linear and Regular Celestial Mechanics. Perturbed Two-body Motion—Numerical Methods—Canonical Theory
175. Larsen: An Introduction to the Theory of Multipliers
176. Grauert/Remmert: Analytische Stellenalgebren
177. Flügge: Practical Quantum Mechanics I
178. Flügge: Practical Quantum Mechanics II
179. Giraud: Cohomologie non abélienne
180. Landkof: Foundations of Modern Potential Theory
181. Lions/Magenes: Non-Homogeneous Boundary Value Problems and Applications I
182. Lions/Magenes: Non-Homogeneous Boundary Value Problems and Applications II
183. Lions/Magenes: Non-Homogeneous Boundary Value Problems and Applications III
184. Rosenblatt: Markov Processes. Structure and Asymptotic Behavior
185. Rubinowicz: Sommerfeldsche Polynommethode
186. Handbook for Automatic Computation. Vol. 2. Wilkinson/Reinsch: Linear Algebra
187. Siegel/Moser: Lectures on Celestial Mechanics
188. Warner: Harmonic Analysis on Semi-Simple Lie Groups I
189. Warner: Harmonic Analysis on Semi-Simple Lie Groups II
190. Faith: Algebra: Rings, Modules, and Categories I
192. Mal'cev: Algebraic Systems
193. Pólya/Szegö: Problems and Theorems in Analysis I
194. Igusa: Theta Functions
195. Berberian: Baer $*$-Rings
196. Athreya/Ney: Branching Processes
197. Benz: Vorlesungen über Geometrie der Algebren
198. Gaal: Linear Analysis and Representation Theory
200. Dold: Lectures on Algebraic Topology
202. Schmetterer: Introduction to Mathematical Statistics
203. Schoeneberg: Elliptic Modular Functions
204. Popov: Hyperstability of Control Systems
206. André: Homologie des algèbres commutatives
207. Donoghue: Monotone Matrix Functions and Analytic Continuation
208. Lacey: The Isometric Theory of Classical Banach Spaces
209. Ringel: Map Color Theorem
210. Gihman/Skorohod: The Theory of Stochastic Processes I
211. Comfort/Negrepontis: The Theory of Ultrafilters